Action Stations Revisited

Action Stations Revisited

The complete history of Britain's military airfields: No. 6 Northern England and the Isle of Man

Tim McLelland

Crécy Publishing Limited

First published in 2012 by Crécy Publishing Limited

A CIP record for this book is available from the British Library

ISBN 9 780859 791120

All photographs are by the author
or from his collection
unless otherwise credited.

Printed and bound in
Malta by Gutenberg Press Ltd

Crécy Publishing Limited
1a Ringway Trading Estate, Shadowmoss Road, Manchester M22 5LH
www.crecy.co.uk

CONTENTS

ACKNOWLEDGEMENTS

Creating this new *Action Stations Revisited* volume was quite a task and could not have been completed without the help and support of many enthusiastic individuals. It would be impossible to thank everyone who has helped to get this volume completed, but particular thanks must be extended to some individuals who have offered information, advice, photographs and support, all of which is naturally appreciated. My sincere gratitude goes to Peter Broom, Ken Billingham, Martyn Chorlton, Richard E. Flagg, Paul Francis, Phil Jarrett, Neil Jedrzejewski, Barry Ketley, Alex McKenzie, Chris Percy, Terry Senior and David Smith for their support and I must also thank the many individuals who kindly supplied snippets of information and some excellent photographs, all of which has helped to create an interesting and informative book.

INTRODUCTION

It is probably fair to say that in historical terms there are two distinct areas of the United Kingdom that are inevitably associated with military aviation. The most famous is the South East, where so many of the Royal Air Force's fighter stations were located, supporting the Spitfire and Hurricane fighters that defended Britain from the Luftwaffe's onslaught in 1940. The other area is one that is often referred to as 'Bomber Country' – the sprawling fields of Lincolnshire and the flat lands in the Vale of York where the RAF's bomber forces were concentrated during the Second World War. This volume of *Action Stations Revisited* is largely devoted to Bomber Country, although it is important to appreciate that this area of the United Kingdom is associated with much more than bomber aircraft. Likewise, it should be noted that the geographic coverage of this series of books requires this particular volume to reach beyond the borders of Yorkshire and Lincolnshire, far out to the west where some very different military airfields could once be found. Within these combined regions there is a fascinating and varied history to explore, much of which is often overlooked.

Naturally, many of the airfields are long gone, and more than a few have literally drifted into obscurity. This is particularly true of the many landing grounds that were adopted by the Royal Flying Corps during the First World War. They were created swiftly and often comprised of little more than cleared fields, accompanied by just a few tents and a handful of support personnel. Their existence was brief but they enabled the RFC to operate its fighter aircraft across Yorkshire and Lincolnshire, defending major cities and industrial areas from the menace of German airships. However, with the threat of attack gone, they returned to agriculture almost as quickly as they had first appeared, and 90 years later there is nothing to indicate that these sites ever had any association with military aviation. Other more substantial sites have also long since disappeared, such as the once huge RFC depot in Sheffield, which boasted many acres of hangars, support buildings and a large landing ground, all of which are long gone, swamped by the ever-expanding industrial and domestic landscape. The inexorable passage of time ultimately destroys our past.

However, some parts of our military history are much harder to eradicate. The Second World War was a seminal period that created a huge expansion of military capability. The RAF's defensive and offensive capabilities grew rapidly from the late 1930s, and as more and more fighter and bomber aircraft were created there was a corresponding need for new bases from which they could operate. In the Lincolnshire and Yorkshire region the results of this expansion policy were all too clear to see. Seemingly endless numbers of new airfields emerged, most of considerable size, sprawling across the agricultural fields and often destroying roads and property as they developed. Huge concrete runways appeared and monstrous metal and brick-built hangars grew, while support buildings emerged in numbers that sometimes equated to small villages. In a matter of just a few years the region was transformed from a tranquil area of countryside into a military establishment of truly breathtaking proportions. Airfields were everywhere, often little more than walking distance from each other, and most were soon full of the latest military hardware. The skies over Lincolnshire and Yorkshire grew busy as the pace of the RAF's activities increased until literally thousands of bombers, fighters, trainers, transports and other specialised aircraft were crammed into the region. On an almost nightly basis the skies rumbled to the sound of bombers becoming airborne, assembling into their deadly formations as they set out across the North Sea to hit the very heart of Nazi Germany. This was Bomber Country at its busiest.

Seventy years later it is difficult to imagine that so much aerial activity could be concentrated into such a relatively small region, especially when the same area is now devoid of activity and largely undisturbed by military aviation. But the massive air bases could never disappear overnight, and even though the men and machines are all long gone these huge airfields linger on. Lincolnshire is still alive with the ghosts and memories of the 1940s and it is impossible to venture far without encountering a reminder of the county's long association with military aviation. The crumbling remains of bomber airfields are scattered across the rolling fields, each former station almost encroaching upon the next. The runways can often still be seen, slowly disappearing under years of decay, while the long-abandoned hangars still stand erect, moodily highlighted against the horizon, seemingly waiting for some distant day when they will once again be called into action.

Most of these once-busy bases are doomed to slowly disappear, but some wartime sites have remained in military hands even to this day, and although the modern Royal Air Force is a mere shadow of its wartime incarnation, it is still here – but only just. For example, to the north of the region at Leeming the RAF still flies its fast and nimble Hawk jets, but with their retirement due one can only assume that even this once-busy bomber and fighter station will eventually succumb to defence expenditure cuts. The RAF's fledgling pilots learn their skills at the former bomber base at Linton-on-Ouse, but there are dark plans to relocate training to Wales, and this may well indicate that in a matter of just a few more years the Royal Air Force will have left Yorkshire completely. Who could ever have imagined that this once-vital home to the RAF's bomber force would once day be abandoned entirely?

Further south in Lincolnshire the RAF is still very much in business at Waddington and Coningsby, but to the north of Lincoln's famous Cathedral only Scampton survives, and even here there seems little hope that the RAF will stay for more than a few years. This airfield is arguably the most famous of all the RAF's sites, but much of the station has already been sold off and has become a rather dreary housing estate. The world-famous Red Arrows are still there, but even they will eventually be gone forever and it may well be that in the not so distant future the very site where the legendary Dambusters set off on their historic mission to Germany will be nothing more than an anonymous housing estate avenue.

Some miles to the west, the former RAF Binbrook lays abandoned, the dark, brooding hangars devoid of aircraft. The bombers are long gone, the Canberra jet bombers too, and even the magnificent Lightning interceptors are just a memory, as almost every inch of the base's concrete runway is progressively dug up and sold off to provide the foundations for a shiny new road somewhere or nowhere. History is literally disappearing before our eyes.

However, not everything is bad news and some historic sites have enjoyed a much happier fate. RAF Finningley was abandoned and forgotten when the RAF unexpectedly departed. It seemed likely that the airfield would be left to rot while the station's domestic site would reappear as a new prison. But fate intervened and Finningley eventually re-emerged as a new international airport, and even though it now masquerades as the ineptly named Robin Hood Airport, the former RAF airfield survives and the monstrous V-Bomber runway now supports a daily mixture of airliners and private aircraft. The last surviving flying example of the awe-inspiring Vulcan also lives here. Ironically, this very aircraft had been based here more than forty years previously, but it is now back at its former home and looks set to stay there for good.

Other airfields have also survived in civilian guise, while many others have been modified to support other activities and services, not least agricultural storage, industry and recreation. Far to the west, Warton is now the home of British aerospace expertise and remains active as the very last surviving aircraft design and manufacturing establishment. Indeed, it doesn't merely survive, it is flourishing.

But the vast majority of the almost countless airfields across the region covered by this volume have simply been forgotten by most people who encounter them, and although they may rarely be visited by anyone other than local farmers, they are still here to be seen and explored, should anyone have the will to take a look. This is, at least in historical terms, a good thing, but the more disturbing fact that cannot be ignored is that, as time progresses, fewer and fewer people even recognise or understand the history and purpose of the crumbling sites that are often on their very doorsteps. One can only imagine that if they did, many of these once-vital and hugely important military bases might survive just that little bit longer. Those of us who do recognise the importance of the history that is still laid out across Bomber Country can at least appreciate what still remains. The only certainty is that not so much as one scrap of it will last for ever.

This edition of *Action Stations Revisited* combines the information first published in three separate volumes of the 'Action Stations' series and brings the relevant entries up to date. Naturally, it is impossible to reinvent history, so the basic historical information contained in the original volumes remains unchanged. However, many years have passed since these books were first published and I have endeavoured to update each entry to take into account more recent developments, and to provide some description of the current status of each site. Space prohibits the inclusion of complete histories for every site featured (some would require a complete book), but readers who might be interested in learning more about any particular sites may well find that further publications are now available for many of them.

The appearance of countless specialised publishing companies has enabled many historians to produce surprisingly thorough and detailed histories of individual sites, and these can usually be found by trawling through the internet or by paying a visit to one's local library. Of course the internet also offers a bewildering variety of historical resources, and much more information on many of the sites featured in this book will be found there. However, the most useful (and certainly the most comprehensive) source of further information is undoubtedly the Airfield Information Exchange – an incredibly useful and interesting reference site that caters for anyone with an interest in military airfields both old and not so old. Readers who have an appetite for learning more about the sites featured in this book are certainly recommended to visit the Airfield Information Exchange forum at http://www.airfieldinformationexchange.org.

GLOSSARY

AA	Anti-Aircraft
AACU	Anti-Aircraft Co-operation Unit
AAIB	Air Accidents Investigation Branch
AAP	Aircraft Acceptance Park
AAS	Air Armament School
AASF	Advanced Air Striking Force
ABS	Air Base Squadron
(AC)	Army Co-operation
ACHU	Aircrew Holding Unit
AEF	Air Experience Flight
AES	Air Electrical School
AFC	Air Force Cross
AFDU	Air Fighting Development Unit
AFS	Advanced Flying School
AI	Airborne Interception (radar)
ALG	Advanced Landing Ground
AMWD	Air Ministry Works Department
ANS	Air Navigation School
AOC(-i-C)	Air Officer Commanding (-in-Chief)
AONS	Air Observer & Navigator School
AOP	Air Observation Post
APC	Armament Practice Camp
asl	above sea level
ASP	Aircraft Servicing Platform
ASR	Air Sea Rescue
AST	Air Service Training
ASU	Aircraft Storage Unit
ASV	Air-to-Surface Vessel (radar)
ASWDU	Air Sea Warfare Development Unit
ATA	Air Transport Auxiliary
ATC	Air Traffic Control or Air Training Corps
ATDU	Air Torpedo Development Unit
ATS	Air Training Squadron
AVM	Air Vice Marshal
BADU	Blind/Beam Approach Development Unit
BDTF	Bomber Defence Training Flight

BFTS	Basic Flying Training School
BG	Bomb/Bombardment Group (USAAF)
BW(M) or (H)	Bomb/Bombardment Wing (Medium) or (Heavy) (USAAF)
CAA	Civil Aviation Authority
CAACU	Civilian Anti-Aircraft Cooperation Unit
CACU	Coast Artillery Co-operation Unit
CAF	Canadian Air Force
C&M	Care and Maintenance
C-in-C	Commander-in-Chief
CN&CS	Central Navigation & Control School
CO	Commanding Officer
Co-op	Cooperation
CRO	Civilian Repair Organisation
DCM	Distinguished Conduct Medal
D/F	Direction Finding
DFC	Distinguished Flying Cross
DFM	Distinguished Flying Medal
DFW	Day Fighter Wing
Drem Lighting	Airfield lighting system
DSC	Distinguished Service Cross
DSO	Distinguished Service Order
DZ	Drop Zone
E&RFTS	Elementary & Reserve Flying Training School
E&WS	Electrical & Wireless School
ECM	Electronic Counter-Measures
EFTS	Elementary Flying Training School
ELG	Emergency Landing Ground
EM	Enlisted men (USAAF)
EO	Extra Over Blister hangar
ETPS	Empire Test Pilots School
E/W	East/West
FAA	Fleet Air Arm
FBG	Fighter Bomber Group (USAAF)
FEAF	Far East Air Force
FEW	Fighter Escort Wing (USAF)
FG	Fighter Group (USAAF)
FIDO	Fog Investigation/Intensive & Dispersal Operation
FIS	Flying Instructors School
FIU	Fighter Interception Unit
FPP	Ferry Pilots Pool
FRU	Forward Repair Unit (RAF)
FRU	Fleet Requirements Unit (FAA)
FS	Fighter Squadron (USAAF)
FTS	Flying Training School

FTU	Ferry Training Unit
FW	Fighter Wing (USAAF)
GED	Ground Equipment Depot
GR	General Reconnaissance
GS	Gliding School or General Service (shed/hangar)
GSU	Group Support Unit
GTS	Glider Training School/Squadron
HCU	Heavy Conversion Unit
HD	Home Defence
HE	High Explosive (bomb)
HGCU	Heavy Glider Conversion Unit
HP	Handley Page
HQ	Headquarters
ITW	Initial Training Wing
LFS	Lancaster Finishing School
LG	Landing Ground
LZ	Landing Zone
MAP	Ministry of Aircraft Production
MoD	Ministry of Defence
MoS	Ministry of Supply
MT	Motor Transport
MU	Maintenance Unit
NAAFI	Navy, Army & Air Force Institute
NCO	Non-Commissioned Officer
'Nickel'	Leaflet-dropping operations
N/S	North/South
OAPU	Overseas Aircraft Preparation Unit
OC	Officer Commanding
OCU	Operational Conversion Unit
OFU	Overseas Ferry Unit
Operation 'Market'	Airborne operation at Arnhem and Nijmegen, September 1944
Operation 'Overlord'	Invasion of Europe, June 1944
ORP	Operational Readiness Platform
OTU	Operational Training Unit
PAC	Parachute and Cable (installation), air defence equipment, 1940
(P)AFU	(Pilots) Advanced Flying Unit
PIR	Parachute Infantry Regiment
PoW	Prisoner of War
PR	Photographic Reconnaissance
PSP	Pierced Steel Planking (metal sectional runway)
Q site	Decoy site with lighting to simulate an airfield at night
RAAF	Royal Australian Air Force
RADAR	RAdio Detection And Ranging
RAE	Royal Aircraft Establishment

RAF	Royal Air Force
R&SU	Repair & Salvage Unit
RAS	Reserve Aeroplane Squadron
RAuxAF	Royal Auxiliary Air Force
RCAF	Royal Canadian Air Force
RCM	Radio Counter-Measures
RDF	Radio Direction Finding
RE	Royal Engineers
Recce	Reconnaissance
RFC	Royal Flying Corps
RFS	Reserve Flying School
RLG	Relief Landing Ground
RNAS	Royal Naval Air Service
RNVR	Royal Naval Volunteer Reserve
ROC	Royal Observer Corps
RS	Reserve Squadron
RS&RE	Royal Signals & Radar Establishment
R/T	Radio Telephony
SAC	Strategic Air Command
SAR	Search and Rescue
SBAC	Society of British Aircraft Constructors (aerospace companies)
SD	Special Duties
SFTS	Service Flying Training School
SHAEF	Supreme Headquarters, Allied Expeditionary Force
SHQ	Station Headquarters
SLG	Satellite Landing Ground
SoAG	School of Aerial Gunnery
SoTT	School of Technical Training
SP	Staging Post
TAC	Tactical Air Command (USAAF)
TBR	Torpedo Bomber Reconnaissance
TCG	Troop Carrier Group (USAAF)
TCS	Troop Carrier Squadron (USAAF)
TCW	Troop Carrier Wing (USAAF)
TDS	Training Depot Station
TDU	Torpedo Development Unit
TEU	Tactical Exercise Unit
TFW	Tactical Fighter Wing (USAF)
THUM	Temperature & Humidity Flight
TS	Training Squadron
TSCU	Transport Support Conversion Unit
TT	Target Towing
TTU	Torpedo Training Unit
UAS	University Air Squadron

USAAC	United States Army Air Corps (from 2 July 1926)
USAAF	United States Army Air Force (from 20 June 1941)
USAF	United States Air Force (from 18 September 1947)
USAFE	USAF Europe
VGS	Volunteer Gliding School
VHF	Very High Frequency
VR	Volunteer Reserve
WAAF	Women's Auxiliary Air Force
WD	War Department
'Window'	Metal foil dropped to disrupt radar systems
WRAF	Women's Royal Air Force
WS	Wireless School

Military Airfields of Northern England and the Isle of Man

The following listing of airfield sites within this volume's designated region is provided in alphabetical sequence. In addition to map and GPS references, additional information is also provided on the recorded details of each airfield's status at the time of completion. It should be noted that the runway data refers to the site's specification at the time of construction; however, many airfields were subsequently improved and modified, and the runways were often extended, sometimes quite considerably. Hangar facilities were also often changed and either expanded or revised.

001 Acaster Malbis
002 Acklington
003 Alexandra Park
004 Andreas
005 Appleton Wiske
006 Atwick
007 Barlow
008 Barrow/Walney Island
009 Barton
010 Bellasize
011 Beverley
012 Binbrook
013 Binsoe
014 Bircotes
015 Blidworth
016 Blyton
017 Boulmer
018 Brancroft
019 Brayton
020 Breighton
021 Brough
022 Brunton
023 Burn
024 Burnfoot
025 Burscough
026 Burtonwood
027 Caistor
028 Cark
029 Carlton
030 Carnaby
031 Catfoss
032 Catterick
033 Church Fenton
034 Coal Aston
035 Copmanthorpe
036 Cottam
037 Croft
038 Crosby-on-Eden
039 Cullingworth
040 Dalton
041 Dishforth
042 Doncaster
043 Donna Nook
044 Driffield
045 Dunholme Lodge
046 Dunkeswick
047 East Moor
048 Ecclesfield
049 Elsham Wolds
050 Elvington
051 Eshott
052 Faldingworth
053 Farsley
054 Finningley
055 Firbeck
056 Fulbeck
057 Full Sutton
058 Fylingdales
059 Gainsborough
060 Gamston
061 Gilling
062 Goxhill
063 Great Orton
064 Greenland Top
065 Helperby
066 Hemswell
067 Hibaldstow
068 Holme-on-Spalding-Moor
069 Hornby Hall
070 Hornsea Mere
071 Howden
072 Huggate Wold
073 Hull (Hedon)
074 Hutton in the Forest
075 Hutton Cranswick
076 Ingham
077 Inskip
078 Jurby
079 Kelstern
080 Kettleness
081 Kingstown
082 Kirkbride
083 Kirkleatham
084 Kirmington
085 Kirton-in-Lindsey
086 Knavesmire
087 Knowsley Park
088 Leconfield
089 Leeming
090 Lindholme
091 Linton-on-Ouse
092 Lissett
093 Longtown
094 Lowthorpe
095 Ludford Magna
096 Manby
097 Market Stainton
098 Marske
099 Marston Moor
100 Melbourne
101 Menthorpe Gate
102 Middleton
103 Middleton St George
104 Milfield
105 Millom
106 Morpeth
107 Murton
108 Netherthorpe
109 Newcastle-upon-Tyne: Gosforth (Dukes Moor)
110 North Coates
111 North Killingholme
112 Ouston
113 Owthorne
114 Plainville
115 Pocklington
116 Pontefract
117 Redcar
118 Riccall
119 Ringway
120 Ripon
121 Ronaldsway
122 Rufforth

123 Samlesbury
124 Sandtoft
125 Scalby Mills
126 Scampton
127 Scorton
128 Seacroft
129 Seaton Carew
130 Sherburn-in-Elmet
131 Shipton
132 Silloth
133 Skipton-on-Swale
134 Snaith
135 South Carlton
136 South Cave
137 South Otterington
138 Speke
139 Squires Gate
140 Stretton
141 Strubby
142 Sturgate
143 Tadcaster
144 Tatton Park
145 Theddlethorpe
146 Thirsk
147 Tholthorpe
148 Thornaby
149 Thorne
150 Topcliffe
151 Usworth
152 Waltham/Grimsby
153 Warton
154 Wath Head
155 West Ayton
156 West Common
157 West Hartlepool
158 Wickenby
159 Windermere
160 Wombleton
161 Wombwell
162 Woodford
163 Woodvale
164 Woolsington
165 Worksop
166 Yeadon
167 York/Clifton

The military airfields of the Isle of Man

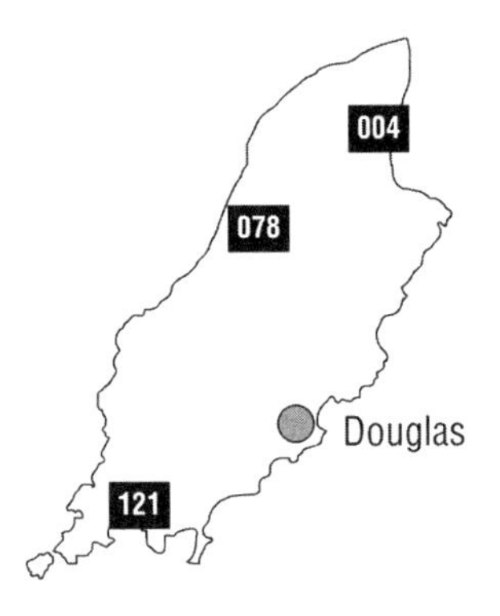

Key

000 Airfield with tarmac runway

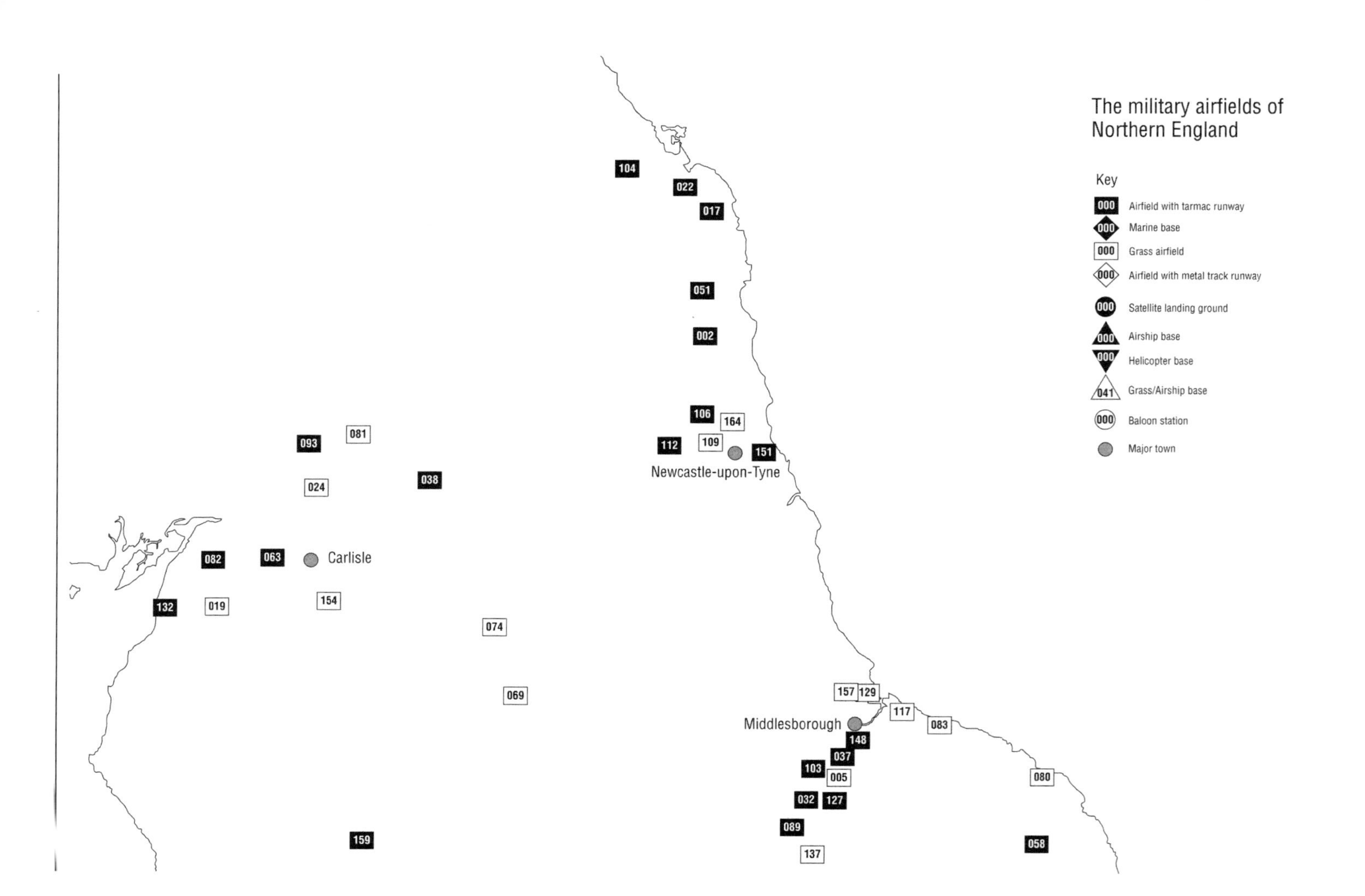
The military airfields of Northern England
Key
000 Airfield with tarmac runway
000 Marine base
000 Grass airfield
000 Airfield with metal track runway
000 Satellite landing ground
000 Airship base
000 Helicopter base
041 Grass/Airship base
000 Baloon station
Major town
Newcastle-upon-Tyne
Middlesborough
Carlisle
104
022
017
051
002
106
164
112
109
151
093
081
024
038
082
063
132
019
154
074
069
157
129
117
083
148
037
103
005
080
032
127
089
137
058
159

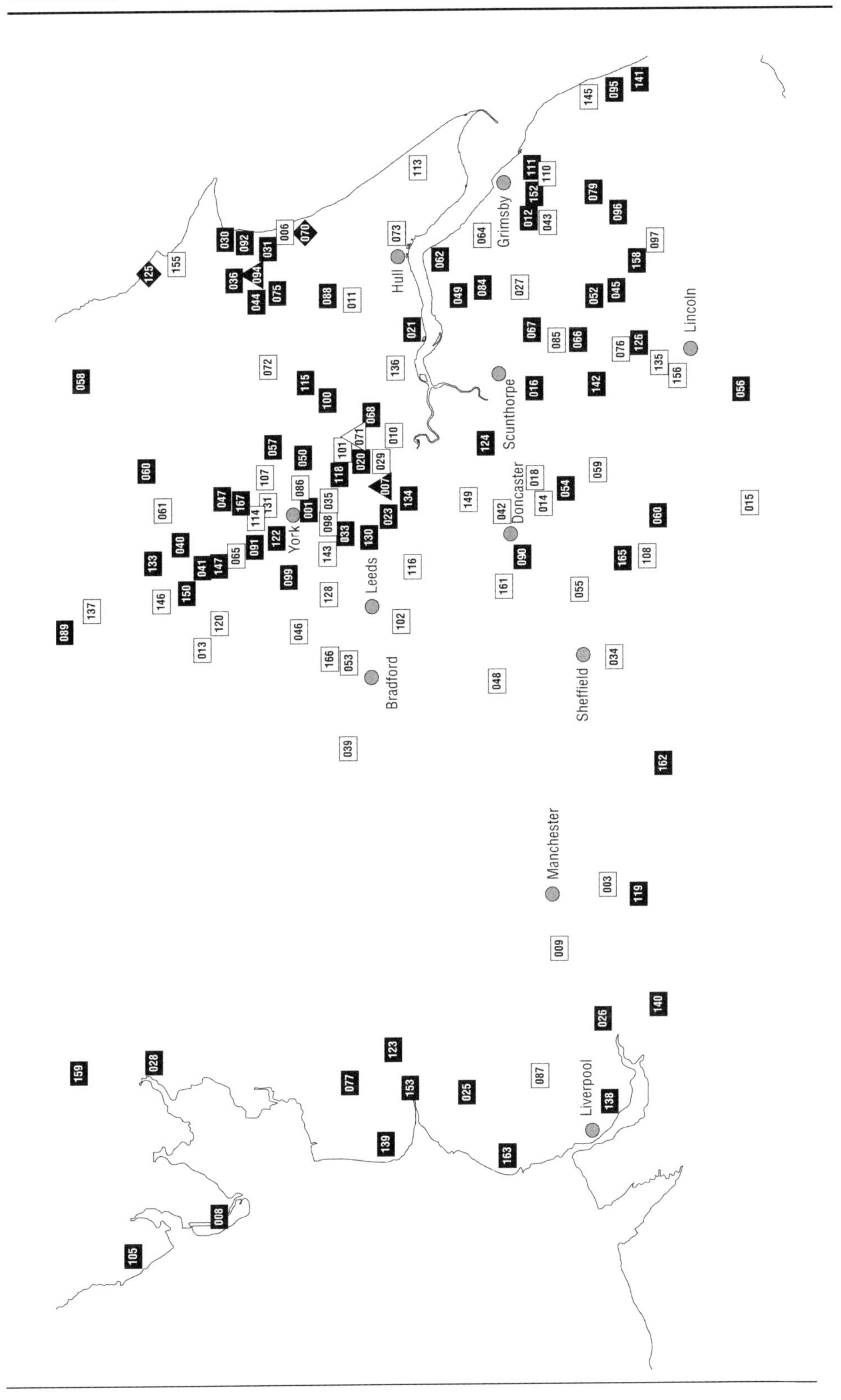
Grimsby
Hull
Lincoln
Scunthorpe
Doncaster
York
Leeds
Bradford
Sheffield
Manchester
Liverpool

THE AIRFIELDS

ACASTER MALBIS, Yorkshire

53°52'47"N/01°07'10"W; SE579429; 29.5ft asl. 6 miles S of York, W of B1122

Deep in the Vale of York, it was perhaps inevitable that this site would be plagued by low cloud, fog and mist, and it was undoubtedly the local weather conditions that shaped the history of this RAF station. Planning for an airfield here began in July 1939 after consideration of a site at Deighton had been abandoned. Intended as a fighter base, a grass field site was constructed and the station opened in January 1942 as a satellite of RAF Church Fenton. Bell Airacobras from No 610 Squadron were the first aircraft to arrive, and a short but unhappy period of operations with the American aircraft commenced here. The Airacobra was a far from perfect aeroplane and the squadron's operations were hampered by endless technical problems. These were compounded by the airfield's poor weather conditions and continual waterlogging problems, which were hardly surprising considering the close proximity of the River Ouse. It was with great relief that Spitfires arrived in March 1942 and, with a much more suitable aircraft now available, the squadron also took the opportunity to relocate to a better airfield, moving south to Digby in Lincolnshire. The last Airacobra left on 7 March.

A No 1652 HCU Halifax undergoing maintenance.

The unusual, troublesome (and short-lived) Bell Airacobras on the grass at Acaster Malbis.

Acaster Malbis was then transferred to Flying Training Command, and No 15 (Pilot) Advanced Flying Unit arrived in April with a fleet of Oxfords (its headquarters being some miles away at Leconfield near the East Coast). This unit's experience at the airfield was somewhat happier than that of 610 Squadron (possibly because of the less demanding nature of the Oxford), and activities continued until the unit moved south in January 1943.

Although in poor condition the former airfield at Acaster Malbis is still very visible.

The station was then transferred to Bomber Command. Redeveloped with three concrete bomber runways and an array of dispersals, it reopened early in 1944, although the site had been used during the previous months of closure by the Airborne Forces Experimental Establishment, based at nearby Sherburn. Although now fully equipped to accept bombers, no aircraft were assigned to the station, and a non-flying unit (No 4 Aircrew Training School) took up residence. No 91 Maintenance Unit also moved here, but flying activities were infrequent and sporadic, mostly confined to circuit training conducted by aircraft from the nearby Heavy Conversion Units. The maintenance unit was not responsible for the storage or processing of aircraft, and it mostly concentrated on the storage of weapons, with stocks of bombs usually occupying a substantial bomb dump in Stubb Wood on the eastern perimeter of the airfield.

After the end of the war the station remained as the home of No 91 MU and remained largely inactive until it finally closed in March 1957. RAF Acaster Malbis was decommissioned in 1963 and the land sold by public auction. It reverted to farmland, although many local people remember learning to drive on the disused runways.

The crumbling remains of the technical site at Acaster Malbis.

In the mid-1970s it returned to use as an airfield for light aircraft. Businessmen flew in from all over Europe for meetings with York companies, and it was particularly popular during the racing season as a fast and convenient means of transport for horses and jockeys. This usage continued until the mid-1980s, then the airfield and associated buildings eventually fell into disrepair and most of the site was simply abandoned.

Today parts of the three runways are still visible, with almost half of the main runway still largely intact. A small road (Broad Road) runs from Acaster Selby and crosses the airfield, passing adjacent to the one surviving hangar. Some of the bomber dispersals are also still visible, reminding any visitors of this airfield's potential, which was never realised. Perhaps the most significant remains are the traces of the large bomb dump, which can still be found deep inside Stubbs Wood. Ironically, it was this sub-site that was the only part of RAF Acaster Malbis to ever provide any contribution to the RAF's war effort.

Main features:
Runways: 220° 6,000 x 150 feet, 160° 4,200 x 150 feet, 290° 4,200 x 150 feet, concrete surface. *Hangars:* two T2, one B1. *Dispersals:* thirty-six spectacle. *Accommodation:* RAF: Officers 100, SNCOs 242, ORs 778; WAAF 274.

ACKLINGTON, Northumberland

55°17'59"N/01°38'21"W; NU230010. 3 miles SW of Amble, 1 mile due S of Acklington village

Aviation at Acklington can be traced back to the First World War. No 77 Squadron was formed as an HD Squadron for the protection of the Firth of Forth. Its most southerly aerodrome was called Southfields, and this was same location that Air Ministry officials visited late in 1936 with a view to constructing a more permanent and larger airfield.

The camp was officially opened in January 1938, although No 7 ATC, which was Acklington's official title, had been formed on 1 December 1937. This was later redesignated 7 ATS on 1 April 1938 and changed again to 2 AOS on 15 November 1938.

On the outbreak of the Second World War, No 2 AOS left for Warmwell and Acklington was transferred to Fighter Command and the control of 13 Group, whose HQ was at Blakelaw Estate, Ponteland Road in Newcastle-upon-Tyne. No 13 Group was one of the largest, in terms of area covered, of all other groups in Fighter Command. Under the command of AVM R. E. Saul DFC, the group's area of responsibility stretched from Catterick to southern Scotland.

Handley Page Harrow I K6941 is seen at Acklington during 1938, delivering personnel and ground equipment in support of a fighter squadron on detachment. Via Martyn Chorlton

Two Belgian pilots (Laoux and Bladt) of No 350 Squadron pose in front of a Spitfire VC at Acklington in 1943. The squadron was resident at Acklington on two occasions during that year. Via Martyn Chorlton

The action began quickly when No 607 Squadron from Usworth made the short flight in its Gladiators to Acklington on 9 October 1939. They went into battle on 17 October when three aircraft of B Flight, led by Flt Lt Sample, shot down a Do18 40 miles east of Berwick. No 607 Squadron's stay was short; by the beginning of November the unit was sent to Merville in France (via Croydon) in support of the BEF.

No 111 Squadron Hurricane Mk Is arrived from Northolt on 27 October in confident mood, but the Squadron's stay at Acklington was only a staging post in preparation for another move to Drem. Only days before departure, the CO, Sqn Ldr H. Broadhurst AFC, took off alone after receiving a report that enemy aircraft were approaching the coast. The weather was atrocious, but Broadhurst pressed on, flying only on instruments through thick cloud. On clearing an iced-up windscreen, he spotted a group of He 111s of KG26 inbound from their base in Northern Germany. A single aircraft was picked out at a range of 500 yards, but the bomber dived into cloud; however, Broadhurst remained focused and continued to close until he was just 150 yards from the Heinkel. A quick burst of fire put the ventral gunner out of action and the second burst sent the Heinkel into a flaming spiral dive into the North Sea. This was the first of many victories for No 111 Squadron and it seemed quite appropriate that this popular CO took the victory.

Another Hurricane squadron arrived at Acklington on 18 November 1939 in the shape of No 43 Squadron. A fighter squadron from birth, this unit was very keen to show how good the Hurricane was. Before its departure to Wick on 26 February 1940, the squadron managed to shoot down five enemy aircraft, all He 111s. Three of these kills were achieved in a single day, and one in particular made history as the first enemy aircraft to crash in England.

On 3 February 1940 the weather was particularly bad and, virtually without exception, Acklington was the only airfield open for operations in North East England. B Flight with three Hurricanes was 'at readiness'. Its pilots, Flt Lt Peter Townsend, Flg Off Folkes and Sgt Jim Hallowes, were trying to keep warm in a dispersal hut when the order to 'scramble' was given. There was a report that two enemy aircraft were attacking a trawler off Whitby, but on arrival only one aircraft was spotted, an He 111H of KG26. Townsend attacked first, instantly killing one member of the crew and hitting the starboard engine. Folkes followed with a concentrated attack from the rear, which caused smoke to pour from the German bomber. In a last-ditch attempt to survive, the pilot, who was the only crew member not to have been hit, made a dash for the clouds. However, the aircraft was so badly damaged that it was inevitable that the pilot would have to crash-land, as no surviving member of his crew was capable of bailing out. With the coast only 2 miles away, the Heinkel made a copybook crash-landing near Bannial Flat Farm, north of Whitby.

No 13 Group's day came on 15 August 1940 when the Luftwaffe launched a massive force of aircraft. Airfields were selected as their main targets although industrial targets were always on the agenda. Sixty-five He 111s of KG26, escorted by thirty-four Bf 110s of ZG76, were steadily crossing the North Sea, initially setting course for the Scottish coast. A group of He 115C floatplanes was also heading for the Scottish coast, the intention being that they should fly a slightly different course from the main bomber force and hopefully draw the attention of No 13 Group's fighters. This 'feint' failed to work because of poor navigation, both groups ending up together as they approached the coast.

At 1208hrs one of the many radar stations positioned along the East Coast began to plot a large group of 'bandits' approaching the Firth of Forth. The enemy force then turned south-west directly towards Tynemouth. It was now decision time for AVM Saul; he scrambled the Spitfires of 72 Squadron, sending them directly at the main force, with the plan of meeting the enemy near the Farne Islands off Bamburgh. No 79 Squadron's Hurricanes were positioned to the north of the incoming raiders while No 607 Squadron from Usworth supported No 72 Squadron. Saul had to be convinced that Tyneside was the target and that this was the only wave of bombers, because he made the daring move of bringing the Hurricanes of No 605 Squadron from Drem south to protect Tyneside.

Time was on the defenders' side, and when No 72 Squadron's eleven Spitfires had arrived off the Northumberland coast they were in the excellent position of being 3,000 feet higher than the attackers. They positioned themselves with the sun behind and, still with the height advantage, each pilot was left to choose his own target. The Bf 110s were fitted with external fuel tanks, and when it was realised that they were under attack these were quickly jettisoned. The flight had been made at the very maximum range of the Bf 110, and to save weight a fatal decision had been made to leave the rear gunners at home. In 'Custer's Last Stand' fashion, the Bf 110s formed a defensive circle in an attempt to protect themselves, although many dived for sea level and headed east for home, leaving the He 111s to fend for themselves. Enemy aircraft spiralled in flames towards the sea as chaos ensued and the Spitfires tore into as many aircraft as they could. Several He 111s dumped their bombs into the sea and headed back to Norway, while the remainder split into two groups, one heading for Tyneside with the intention of bombing Usworth while the rest turned south-east with orders to bomb Linton-on-Ouse and Dishforth.

The Hurricanes of No 79 Squadron joined the fray just as the force spilt into two, and continued to harry and attack the bombers all the way to their intended target. Once over Tyneside, the area's effective anti-aircraft guns claimed more victims while the combined force of No 79, 605 and 607 Squadrons' Hurricanes deterred the enemy from scoring a single hit on their goal, Usworth. The Luftwaffe had no idea that so many fighter squadrons were in the area and never again attempted a daylight raid on that scale against the North East.

A pivotal moment in the Battle of Britain came on 15 August 1940, when forty enemy aircraft were claimed by No 13 Group, six of them credited to No 72 Squadron.

The experience gained by both Nos 72 and 79 Squadrons would stand them in good stead because, by the end of August both units were moved to Biggin Hill, and saw more action for the remainder of the Battle of Britain. On 8 July 1941 No 72 Squadron left Acklington for the last time and went on to become one of the RAF's longest-serving squadrons.

Another new squadron was formed here on 10 May 1941. No 406 Squadron became the RCAF's first night-fighter unit, flying the Beaufighter, under the command of Wg Cdr D. G. Morris DFC. The squadron achieved its first success during a heavy air raid on Newcastle on the night of 1/2 September. Despite the fact that the Squadron had not been declared operational, Flg Off Fumerton and Sgt Bing took off from Acklington determined to test the ability of the Beaufighter. Just after 2200hrs they closed in on a Ju 88 and shot it down. By the end of the year, the Squadron had claimed five enemy aircraft destroyed, and four more were damaged before the squadron moved to Ayr on 1 February 1942.

No 1 Squadron brought the Typhoon to Acklington on 8 July 1942. These aircraft did not see action until 6 September, during one of several experimental raids by 16/KG6 operating the Me 210. During an attack on Middlesbrough, Plt Off D P. Perrin scored No 1 Squadron's first Typhoon success by shooting down an Me 210A-1 near Fell Briggs Farm, New Marske. Seven minutes later, Plt Off T. G. Bridges claimed a second kill, sending down a second Me 210A-1 near Sunnyside Farm, Fylingthorpe. No 1 Squadron saw no further action over the North East, departing for Biggin Hill on 9 February 1943.

Acklington's northern location did not lend itself to operations undertaken by the USAAF, reflected in the fact that in its entire history the airfield only housed one American unit. The 416th NF Squadron was formed at Usworth on 14 May 1943 and was hastily equipped with the Beaufighter Mk VI. The USAAF had a severe lack of any kind of serious night-fighter units and the 416th NF was one of several formed early in 1943. These new units relied heavily on the experience of the RAF and RCAF squadrons in this field, and all were trained by RAF Instructors, not only on the Beaufighter but also in night-fighting tactics. Squadron commanders protested that the American crews had not received sufficient training on the Beaufighter, but senior staff simply replied that they would gain the necessary experience in combat! By 4 August the 416th NF Squadron had left for North Africa.

The resident fighter squadron was taken over by a Polish unit on 22 September. No 316 'City of Warsaw' Squadron was not operational when it arrived with its Spitfire Mk IXs. While at Acklington, the Squadron re-equipped with the Spitfire Mk LF VB, but not long after its departure on 15 February 1944 the unit re-equipped with the Mustang Mk III, an aircraft it would use until its disbandment in December 1946.

Spitfire LF XVIe TB252 spent many years as a gate guardian, including a period at Acklington during the late 1950s and early 1960s. Via Martyn Chorlton

Work began on upgrading Acklington's runways from grass to concrete from March 1944, and construction was complete by the end of November. However, it was not until 26 February 1945 that a unit once familiar to the North East was reformed at Acklington. No 59 OTU was disbanded into the Specialised Low Attack School and the FLS at Milfield on 26 January 1944. The unit was reformed at half strength to train fighter-bomber pilots with fifty-four Typhoons, all brand-new, having only been delivered from the makers in January 1945. They were not destined to see a lot of use at Acklington; only one course was completed before No 59 OTU was disbanded for a second and final time on 6 June 1945.

While the majority of airfields closed during the post-war period, Acklington's future looked bright, and during 1945 it became a forward airfield within the new Newcastle Sector. The intention was that it would house one day-fighter unit (No 263 Squadron) and one night-fighter unit (No 219 Squadron). Their stay was brief, however; No 219's Mosquito NF XXXs had departed for Church Fenton by 2 April 1946, and No 263 Squadron's Meteors left for Horsham St Faith in September 1949.

The airfield's original role was reinstated on 1 May 1946 when it became No 2 APS, a title transferred from Spilsby. This was changed to the FATU in November 1946 and remained so until disbandment in July 1956.

Fighter squadrons of the RAF and RCAF made use of Acklington's facilities for air-to-air firing practice. When this came to an end, Acklington was elevated to a front-line fighter station again, its first occupant being No 29 Squadron with the Meteor NF.11 on 14 January 1957. Now equipped with the Meteor NF.12, the squadron left for Leuchars on 22 July 1958, bringing an end to night-fighter operations from the airfield.

Day-fighter cover at Acklington was provided by No 66 Squadron when it arrived with the Hunter F.6 on 14 February 1957. Several detachments to the Middle East were carried out during the unit's stay, but on 30 September 1960 No 66 Squadron was disbanded. Acklington's fighter days were now over, but the arrival of No 6 FTS on 4 August 1961 kept the flying alive. Initially flying the Provost T.1, these were replaced by the Jet Provost, which operated from the airfield until its disbandment on 30 June 1968.

Helicopters were destined to become the last aircraft to operate from Acklington. No 18 Squadron moved in from Gütersloh on 5 January 1968 with the Wessex HC.2, leaving for Odiham on 8 August 1969. B Flight of No 202 Squadron was the local SAR Flight and had been located at Acklington since 1957. With No 18 Squadron's departure, the flight's Whirlwind HAR.10s were the only aircraft to be seen at Acklington and, like so many other RAF airfields at that time, it would become a victim of Government cuts.

The Whirlwinds moved to Boulmer in early 1972 and Acklington was placed in a state of Care and Maintenance before its rapid run-down and closure later in the year. The old airfield was purchased immediately by the Home Office, which quickly converted the site into a Category 'C' prison, which was opened a year later. The airfield itself did not survive for much longer, becoming an open-cast mine, which has now removed all trace of a very busy station.

Remnants of the perimeter track and a few technical buildings survive within the prison area but, with the exception of a solitary pill-box, there is very little to see today. It is such a shame that one of Northumberland's, and possibly the North of England's, most important airfields has been simply erased.

Main features:
Runways: QDM 230 1,900 yards, 190 1,510 yards, 120 1,200 yards, concrete. *Hangars:* two F-Type, one J-Type Bellman, sixteen Blister. *Hardstandings:* Punched planking, thirteen TE, twelve SE. *Accommodation:* RAF: 1,324; WAAF: 393.

Meteor F.3s of Nos 66 and 92 Squadron, both from Duxford, are pictured during an APC at Acklington in 1947. Via Martyn Chorlton

ALEXANDRA PARK, Greater Manchester

53°25'13"N, 02°15'12"W; SJ832933. S of Manchester city centre, S of A5103 junction with Mauldeth Road at West Didsbury

The area of grass land that eventually became Alexandra Park Aerodrome was purchased by the War Department during 1917, having been chosen largely because of its uncluttered agricultural nature, and its location between the neighbouring districts of Fallowfield, Chorlton, Withington and West Didsbury, at the junction of Princess Road and Mauldeth Road West, some 3 miles south of Manchester's busy city centre. At the time of purchase the land was owned by the Egerton Estate.

Following the closure of the small civil airfield at Trafford Park in 1918 after only seven years of use, Alexandra Park Aerodrome was constructed and opened in May 1918 by the War Department for the assembly, test flying and delivery of aircraft built in the Manchester area by A. V. Roe & Company (Avro) at Newton Heath and the National Aircraft Factory No 2 (NAF No 2) at Heaton Chapel. The airfield took its name from the nearby Alexandra Park railway station on the Great Central Railway's branch line to Manchester. Many aircraft were brought to the airfield by rail in major sections, directly from Avro and NAF No 2 to the nearby station, although many other aircraft were transported to the airfield by road. Prior to the opening of the airfield, Avro's aircraft had been transported across the Pennines to Sheffield, where they had been tested at the Aircraft Acceptance Park at Coal Aston, so the new airfield at Alexandra Park was a very welcome asset for Avro.

DH.9 and DH.10 aircraft became very common sights at the new airfield, and acceptance flying became the airfield's primary purpose. Although a considerable amount of assembly and test flying was completed here, most of the site's activity soon became directly associated with civil flying, which resumed in the UK on 1 May 1919 at the end of the war; Lt Col Sholto Douglas arrived at Alexandra Park from Cricklewood Aerodrome, London, at the controls of a Handley Page 0/400 transporter (converted from a bomber). Carrying ten passengers in its windowless fuselage, it had taken 3hr 40min to fly from London, battling against strong headwinds.

The Avro Transport Company operated the UK's first scheduled domestic air service from Alexandra Park to Birkdale Sands (Southport) and South Shore Blackpool between 24 May and 30 September 1919, mainly using Avro 504 three-seat biplanes. Although the weather caused a few flights to be cancelled, the daily service was operated reliably, the aircraft leaving Alexandra Park at 2pm and arriving in Blackpool some 45 minutes later, having stopped over at Southport. Tickets cost 9 guineas return, or 5 guineas single. From 1922 until 1924 Daimler Airways operated daily scheduled passenger flights to Croydon Airport, followed by a regular extension to Amsterdam. The northbound flight left Croydon in the early evening, and after an overnight stop the aircraft returned south the following morning. These timings enabled Manchester passengers to connect easily with Daimler's other continental flights to and from Croydon and also with other airline services from Croydon.

On the evening of 14 September 1923 the northbound flight's de Havilland DH.34 ten-seat biplane airliner crashed near Ivinghoe Beacon in the Chilterns during an attempted forced landing in poor weather. The two pilots and three passengers were killed, making this the first fatal accident on an internal air service in the UK. The route was suspended for a period before subsequently recommencing.

In April 1924 Daimler merged with a handful of other airlines to become Imperial Airways, and the new monopoly airline promptly terminated the scheduled service from Alexandra Park; it was 1930 before Imperial again flew any schedules to any UK airport north of London.

An aerial view of Alexandra Park captured from an Avro 504 during 1923.

Aircraft competing in the King's Cup air races landed at Alexandra Park in 1922 and 1923, and there were also a number of flying displays and air shows at the aerodrome. The Lancashire Aero Club, the oldest flying club in Britain, operated from here until 1924. The terms of the lease for the land, which had been laid down by Baron Egerton of Tatton, stipulated that flying from the site would cease within five years of the war's end. Consequently, flying activity was wound down and the hangars and ancillary buildings were demolished, the aerodrome closing on 24 August 1924.

In that year Princess Road was built through the eastern part of the airfield and a council housing estate was eventually constructed on the eastern edge of the site. The remaining area of open land to the west of Princess Road is now Hough End playing fields, a council-owned facility with twenty-seven full-size pitches occupying land that was once part of the grass airfield. Some rather low-key flying still takes place on the site of the airfield, with Hough End Model Aircraft Club keeping the spirit of aviation alive.

The aerodrome hangars and ancillary buildings are now long gone and the Greater Manchester Police (GMP) Sports & Social Club at Hough End now occupies this site. GMP trains dogs and horses here and its helicopter uses the pitches at the Sports & Social Club as a landing site when needed, most often when emergency transfers to Wythenshawe Hospital are required.

The aerodrome's brief existence is commemorated on a plaque in the sports pavilion at Hough End playing fields, and a new commemorative plaque was unveiled on 7 July 2007 to mark the 90th anniversary of the airfield; it is located in the grounds of No 184 (South Manchester) Squadron, Air Cadets, in Hough End Crescent.

Main features:
Runways: none (grass field). *Hangars:* nine Belfast. *Hardstandings:* none. *Accommodation:* Avro personnel only.

ANDREAS, Isle of Man

54°22'17"N/04°25'24"W; NX426000; 101.5ft asl. 5 miles NW of Ramsey on A17, E of Andreas village

Although situated some distance from the mainland, across the Irish Sea on the Isle of Man, the creation of a fighter airfield at this location made good sense, being within easy reach of the airspace over the port of Liverpool and the surrounding industrial areas. A site of some 500 acres was purchased in 1940 and the levelling of land and the demolition of buildings and obstructions began in June. Three runways were laid in a standard triangular pattern, these being deemed adequate for aircraft of most sizes and weights, and some 24 dispersals were attached to a concrete perimeter track, these being divided into small pens for single-engine aircraft and larger pens for twins. RAF personnel arrived in 1941 and were mostly associated with airfield security, guarding the airfield, which was surrounded by a substantial barbed-wire fence, and the emerging airfield buildings on the small technical, admin and domestic sites.

By October 1941 RAF Andreas was ready to receive the first of No 457 Squadron's Spitfires from RAF Jurby as a prelude to working up to operational efficiency. This work-up would take six months; however, the transfer of 457's groundcrew and administrative staff brought welcome relief to the congestion at Jurby. It was during this period that an unfortunate accident occurred in December, when one of the aircraft was coming in to land. Crossing the end of the runway at the time was one of the builder's foremen driving a lorry. One of the Spitfire's wheels hit the cab of the lorry, causing severe injuries to its occupant, who was killed instantly. It was also during this work-up period that the Air Ministry insisted that the height of Andreas church tower be reduced, as it was a hazard, and in

A No 457 Squadron Spitfire pictured at Andreas.

A rare image of the unusual Vought Chesapeake at Andreas.

line with the southern end of the main runway. The church tower was originally 120 feet high, and the most striking feature of the Island's northern plain, being visible throughout the parish.

RAF Andreas became fully operational in March 1942, but by now No 457 Squadron was ready to move south to join 11 Group at RAF Redhill and to take part in air strikes over northern France. It was immediately replaced by its sister Australian squadron, No 452, which had been formed in April 1941.The third fighter unit to occupy RAF Andreas was No 93 Squadron, which had an entirely different background. No 93 Squadron had been involved in the development of night-fighter tactics using Havocs equipped with radar and Turbinlite searchlights. After becoming operational it was decided to split the squadron into flights attached to other night-fighter units. Following this, No 93 reformed at Andreas as an entirely new squadron equipped with Spitfires as it worked up to operational efficiency. After four months, the squadron was ready to move on to more direct action, and orders were received that would see the squadron relocate to Algiers ready for the North African landings as part of Operation 'Torch'.

By August 1944 the squadrons had gone, apart from a detachment from No 275 Squadron, which continued to operate Walrus amphibians on SAR duties. However, the arrival of No 11 Air Gunnery School early in 1943 ensured the airfield's continued activity, thanks to the unit's sizeable fleet of Ansons and Wellingtons. A rare resident at Andreas was a Vought-Sikorsky Chesapeake of the Fleet Air Arm. It belonged to No 772 Squadron (FAA), and was employed to provide simulated conditions for the Royal Navy's No 1 Radar Training School, ideally positioned on Douglas Head. The Chesapeake was replaced in October 1944, after crashing on Douglas Head, killing the pilot, Sub Lieutenant R. S. Paton. The detachment of No 772 that had been responsible for operating the Chesapeakes became No 772B Squadron in May 1945 with the arrival of Boston 111s, Corsairs and de Havilland Mosquitoes. The purpose of the Bostons was to train gunners from the naval air station at Ronaldsway, and Fairey Barracudas were added to the scene as they brought in telegraphists to have air gunnery added to their training. However, this only lasted for a short period, as No 772B was disbanded in September 1945 following the defeat of Japan.

The station wound down during 1945 and the RAF vacated the site in 1946. The airfield survived, however, and although it was used for some time as a motorbike racing circuit, it is now used sporadically for private and recreational flying. Many of the airfield buildings survive (mostly used for storage) and although the runways and taxiways are slowly crumbling they are still here.

Main features:
Runways: 060° 4,050 x 150 feet, 180° 3,600 x 150 feet, 120° 3,750 x 150 feet, tarmac surface. *Hangars:* three Bellman, eight Extra Over Blister. *Dispersals:* nine octagonal, twenty-four fighter type. *Accommodation:* RAF: Officers 67, SNCOs 144, ORs 933; WAAF 404.

APPLETON WISKE, Yorkshire

54°27'02"N/01°22'53"W; NZ402062; 197ft asl. SE of Darlington, E of A19 on Appleton to Picton road

The DH.9 was among a variety of early aircraft types that operated from Appleton Wiske.

Driving north along the small country road that links the village of Appleton Wiske with nearby Picton, a sharp right-hand turn towards the east marks the north-western boundary of what was once an airfield. It was established during the First World War and comprised just 38 acres of agricultural land mostly made up of clay, surrounded by more than a few trees. It was clearly less than ideal for flying operations, but it is believed to have been used sporadically by No 76 Squadron (flying DH.9s) while the unit was assigned to Home Defence duties. When the First World War ended the site was quickly abandoned and soon returned to agricultural use. It was reconsidered as a potential airfield prior to the beginning of the Second World War, but was found to be unsuitable for development. Today there are no traces of the former airfield, although agricultural and domestic buildings adjacent to the road include some traces of rubble that appear to owe their origins to a long-gone military presence.

Main features:
Runways: none, grass field. *Hangars:* none. *Dispersals:* none. *Accommodation:* RAF: Officers 37, SNCOs 92, ORs 420; WAAF 180.

ATWICK, Yorkshire

53°56'49"N/00°11'40"W; TA186516; 29ft asl. W of B1242, N of Atwick village, 2 miles N of Hornsea

Opened as a Class Two Landing Ground, an airfield was established at this site during the First World War. Without any permanent structures there was little evidence of the airfield's presence and the appearance of aircraft was sporadic. No 76 Squadron RFC was assigned to the local area, and its BE.9s (and other BE aircraft) undoubtedly used the field occasionally. The 50-acre site was used briefly by 504 (Special Duty) Flight of No 251 Squadron RNAS, flying DH.6s in on U-boat patrols from May 1918, and their stay extended into 1919, after which the tents and a small scattering of equipment were removed, leaving the field to return to agricultural use. Today there is no trace of the former landing field and the skies over the nearby coast are left to the seagulls.

A DH.6 on the grass at Atwick.

Main features:
Runways: none, grass field. *Hangars:* none. *Dispersals:* none. *Accommodation:* RAF: Officers 30, SNCOs 112, ORs 346; WAAF: 134

BARLOW, Yorkshire

53°44'55"N/01°00'38"W; SE653284; 20ft asl. 3 miles SE of Selby, off A1041 adjacent to Barlow village

A fairly large (880-acre) site was established adjacent to Barlow village in 1916 under the control of the RNAS, although it was leased to Armstrong-Whitworth for the construction and testing of rigid airships. The first of these (R25) was completed in 1917 and further examples were produced over subsequent years, including R33, which incorporated design modifications based on the examination of the German LZ65 that had force-landed in the UK during 1916. On the night of 23/24 September 1916 the German Zeppelin L-33 was brought down at Great Wigborough, Essex. Its commander had been participating in an air raid on London when the airship was damaged by anti-aircraft fire, then intercepted and brought down by a night-fighter whose fire failed to ignite the hydrogen. However, so much damage was done to the gasbags and fuel tanks that the ship was forced to descend. The German crew attempted to destroy the ship rather than allow it to fall into enemy hands, but so little hydrogen was left that only the doped fabric lit when they fired signal flares into the hull. L-33 was therefore virtually intact and her motors were undamaged, and the British had been handed a near perfect ship full of the latest German technology. Parts of the R33's control car is now in the care of the RAF Museum. The large airship shed (700 x 150 x 100 feet) became a dominant feature of the local skyline and survived long after the site was relinquished in August 1921. After many years as an ROAC depot, the entire site was cleared and the shed demolished. Today no traces of the site remain.

The R33 airship emerges from its hangar at Barlow in March 1919.

Main features:
Runways: none. *Hangars:* one Airship Shed. *Dispersals:* none. *Accommodation:* not known.

BARROW/WALNEY ISLAND, Cumbria

54°07'43"N/03°16'02"; SD175713; 44ft asl. 1 mile W of Barrow-in-Furness, on Red Ley Lane, N of North Scale

The origins of this relatively unknown site (known both as Barrow and Walney Island) can be traced back to 1935 when Barrow Borough Council, after raising a provision with the Air Ministry, completed proposal plans for a civic aerodrome in the Barrow-in-Furness area, although the site chosen at Walney Island was not to be the only one considered at the time. The tentative plans specified an airfield with three runways connected by a perimeter track, linked to small satellite single-aeroplane hangars. Walney Island was eventually chosen over other proposals largely because it was nearer Barrow town centre than the others. Another consideration was that the same site had also been used for aviation some years previously, having been a major centre for airship construction and operation from 1910 until 1920.

A rare photograph taken at Barrow during an air cadet parade in 1943.

It was not until 1937 that the present 600-acre site was acquired at a cost of £8,050 by compulsory purchase, and the subsequent development of the site required some clearance of local buildings, including the demolition of North End Farm. By this stage the airfield had been acquired by the Air Ministry as a potential fighter base, but building work on the site did not get under way until 1940. It was completed in late 1941 as a standard fighter-type airfield with three runways in a triangular pattern, supported by a scattering of dispersals.

On completion the station was opened to flying, and No 25 Group Flying Training Command became the first unit to take up residence, followed a number of weeks later by a smaller unit, which was soon formed into No 3 Air Gunnery School. Various fighter and trainer types became familiar sights at the airfield, although activity remained modest. A number of small improvements were made to the airfield before Christmas 1941 when No 10 Air Gunnery School arrived with more than 100 personnel, including students. In 1942 a domestic site was created for personnel, and camps were erected at sites on Cows Tarn Lane, North Scale and at a WAAF hostel on the Promenade at the foot of North Scale village. Many remnants of these auxiliary buildings remain to be seen to this day.

Over the next few years in excess of 5,000 RAF personnel received their training at Walney Island, and even after the end of the Second World War air gunners were still trained here and retained for the post-war Bomber Command. Following the closure of Cark airfield in early 1946 the RAF's Mountain Rescue Unit moved to Walney together with a Gliding School, remaining here until the airfield closed during the summer 1946, bringing with it the end of No 10 Air Gunnery School.

Walney airfield was retained by the Air Ministry for a number of years until in 1959 an offer to purchase the site came from the local shipbuilding company Vickers (the airship manufacturer that had first used this site in 1910). Vickers remains the owner to this day, albeit under the more modern name of BAE Systems. During 1964 The Lakes Gliding Club also took up residence on the airfield and, although a number of commercial ventures have been attempted over the last 20 years, no significant developments have taken place here and the site remains in use as a relatively quiet airfield, supporting business, commercial and recreational flying activities. Neatly sandwiched between two stretches of seascape, the airfield retains most of its original layout and all of the runways survive, as do the perimeter track and most of the dispersals. Two hangars are also still standing and, although little used, the airport's future seems secure.

Main features:

Runways: 120° 3,900 x 150 feet, 360° 3,300 x 150 feet, 060° 3,300 x 150 feet, tarmac surface. *Hangars:* three Callendar-Hamilton, fifteen Blister. *Hardstandings:* nineteen circular. *Accommodation:* RAF: Officers 80, ORs 144.

BARTON, Greater Manchester

53°28'9"N/02°23'13"W; SJ744973; 70ft asl. Between M62, M6 and A57, 1 mile W of Eccles

This long-established airfield has enjoyed a long association with civilian flying, but it has also had many connections with military operations. Opening in June 1930, the site was the home of Manchester's first airport and scheduled services began during the same month, with Imperial Airways operating routes to Croydon, Birmingham and Liverpool. The initial services were not particularly popular and, although both Railway Air Services and Hillman Airways established a variety of flights to and from the airport, activity at the airfield remained relatively low, and when a larger airport was opened at Ringway the services were terminated during 1938.

Fairey Aviation purchased an existing factory at Heaton Chapel in order to begin construction of Battle aircraft, and nearby Barton was selected as the flight test airfield. The first aircraft to perform test flights at Barton was in fact a Hendon, but the first Battle took to the air here in April 1937. Testing soon shifted to Ringway, but in October 1937 No 17 Elementary & Reserve Flying Training School took up residence under the control of Airwork Services, operating Tiger Moths and a handful of Avro Ansons. It remained active at Barton until the outbreak of the Second World War, when the airfield was requisitioned by National Air Communications, civil flights continuing until they were transferred to Liverpool.

Barton was then taken over by the Ministry of Aircraft Production, resulting in a variety of aircraft types appearing at the airfield for repair or temporary storage. Ansons became a regular sight as well as Hurricanes, Dominies, Battles, Fulmars and Corsairs, together with Swordfishes, which were brought in for scrapping. Percival Proctors were also manufactured at nearby Trafford Park, and more than 700 aircraft were test-flown from Barton by F. H. Hills.

Fairey Battles nearing completion prior to flight-testing at Barton.

After the war the Manchester University Air Squadron arrived in May 1947, and No 2 Reserve Flying School in October 1948. Civil flying returned with the establishment of the Lancashire Aero Club. The MUAS and No 2 RFS's Tiger Moths and Chipmunks remained at Barton until March 1953, after which the site's military associations ended, apart from a number of very popular air shows that attracted crowds from nearby Manchester. It was at one of these events that the much-loved Mosquito RR299 crashed in July 1996.

Today Barton is still an active airfield hosting many resident and visiting private aircraft. Many of the original airport buildings can still be seen, including the magnificent control tower, which remains as the oldest surviving example of its type in Europe. The airfield remains much as it did when it first opened to flying, and still suffers from the same susceptibility to waterlogging that often hampered the flight test crews and trainee flyers in their Tiger Moths.

Barton airfield in 2011, still very active but no longer associated with military aviation.

Military flying has long since gone, but Barton remains very much in business. Now known as Manchester City Airport, it has changed little since its opening, and is considered a good example of the airfields of the 1930s. There are several historical items of note, and a small museum in the visitor centre displays documents from the history of the original Manchester Airport; the Bomber Command Association also has a display. The control tower is protected by its Grade II listed building status, together with the original terminal building and hangar. The airfield is regularly used as a setting for films and TV programmes, among them *Brass* (where Barton masqueraded as Croydon Airport), *Mersey Beat*, *GBH* and *Island at War*, and the distinctive control tower often features prominently. Use of the airport by heavier aircraft is hampered by the soft peaty nature of the area – being located at the edge of Chat Moss, much heavy work would have been needed to consolidate the ground (compare the struggle building the Liverpool & Manchester Railway across Chat Moss in 1826) – and by the low-lying land and areas of nearby standing water, which encourage fog. During 2010-11 additional drainage was added to improve surface water-draining, the original clay pipes having deteriorated until they no longer functioned.

City Airport can operate as an unlicensed airfield by arrangement during the hours of darkness for commercial, military, police and air ambulance helicopters, as it can be equipped with portable runway lighting. This facility is used particularly during football matches at nearby Old Trafford (Manchester United) and City of Manchester Stadium (Manchester City).

Main features:

Runways: 09/27 2,000 x 50 feet, 02/20 1,745 x 50 feet, 14/32 1,000 x 50 feet, grass surface. *Hangars:* one T2, two Bellman, one civil flight shed. *Dispersals:* none. *Accommodation:* not known.

BELLASIZE, Yorkshire

53°44'12"N/00°45'44"W; SE817273; 10ft asl. 1 mile SW of Gilberdyke, S of B1230, adjacent to Bellasize village

As a Class One Landing Ground, Bellasize opened in April 1916 as part of the Home Defence network tasked with the protection of the region from marauding Zeppelins. The first resident aircraft were the BE.2Cs of No 33 Squadron RFC, which were operated from the airfield as required. The unit's headquarters was at Tadcaster, but detachments were maintained at sites across the region and Bellasize was an important base for the unit, being situated close to the Humber, where many Zeppelins were to be found heading inland to attack targets in England. No 76 Squadron also established a presence at the airfield some months later.

Operations from the site were often hampered by the poor condition of the grass field, which was often flooded thanks to the close proximity of the River Ouse just a few fields away. It was no surprise that the airfield was abandoned after the First World War, and without any permanent structures evidence of any military presence was soon gone.

However, shortly before the outbreak of the Second World War the site was re-examined, and in September 1939 it reopened as a Relief Landing Ground for No 4 Elementary Flying Training School, which was based at Brough (some 8 miles away) with a fleet of Blackburn B2s (which coincidentally had been built at the same site). Tiger Moths subsequently joined the unit, and by February 1942 the B2s had been withdrawn. Bellasize remained active until 1945, when the familiar sight and sound of the Tiger Moths ended and the airfield was closed in July of that year. The landing field was abandoned and gradually returned to agricultural use, and today there is no trace of the former airfield. The junction of Greenoak Lane and Bellasize Lane marks the north-western boundary of the old airfield site.

Main features:

Runways: NW/SE 2,000 x 50 feet, grass surface. *Hangars:* none. *Dispersals:* none. *Accommodation:* RAF: Officers 6, SNCOs 12, ORs 312; WAAF: 0.

BEVERLEY, Yorkshire

53°52'44"N/00°27'04"W; TA020399; 98.5ft asl. W of Beverley on A1035

Equestrian enthusiasts gathering at Beverley's well-known racecourse probably never pause to consider that the same site was once an active aerodrome. Opening in March 1916, the BE.2C fighters of No 33 Squadron RFC were the first aircraft to use the site, using Beverley together with a number of dispersal sites to mount patrols across the Humber, which was used regularly by German

A rare image of Beverley taken during its early days. Almost the entire station is visible (apart from a portion of the landing field), emphasising the modest size of these early RFC bases when compared to modern military sites. Barry Ketley

A Sopwith Camel on the grass at Beverley.

Zeppelin crews embarking on bombing missions across northern England. No 57 Squadron also established a presence here with BE.2Cs and FK.3s, flying air defence and training missions in anticipation of overseas deployment. After receiving Bristol Scouts, the Squadron departed for the Mediterranean in September 1916, while No 82 Squadron arrived from Doncaster in February of the following year. The unit's stay was brief, and a month later it left for Waddington to be replaced by No 80 Squadron from Montrose in November 1917. After working up for overseas deployment, the unit's Sopwith Camels left the following January.

Armstrong Whitworth FK8 after suffering a landing accident at Beverley. Similar catastrophic landing accidents were surprisingly common.

Beverley then became established as a training station. No 36 Training Squadron had already arrived here in July 1916 with BE.12s and Curtiss JN.3s, and over the following months Nos 78, 79, 90 and 60 Training Squadrons operated from the airfield at various times; a fairly active period of training operations continued here until November 1917, when the last of the trainer aircraft departed, leaving only the Camels of No 80 Squadron, which remained at Beverley until January 1918. After this date the airfield soon fell silent, although the station remained in use as a home for ground units, including various Army units that used it during the Second World War.

Today the airfield is long gone, with the horse-racing circuit tracing the approximate perimeter of the old airfield site. The station buildings and hangars are gone too, buried under the racecourse entrance and associated buildings. However, a few traces of the RFC base can occasionally be seen in this area, amongst the grass surface.

Main features:
Runways: none, grass surface. *Hangars:* three RFC Flight Sheds. *Dispersals:* none. *Accommodation:* not known.

BINBROOK, Lincolnshire

53°26'48"N/00°12'35"W; TF189959; 360ft asl. NW of Market Rasen, off B1203 adjacent to Brookenby village

One of the Royal Air Force's most famous stations, Binbrook owes its origins to a survey conducted in the 1930s. Construction of the airfield and facilities took some time, work being hampered by the site's remote location on a bleak hilltop, far from any major centres of population, but eventually a standard bomber airfield was created with the usual triangular runway layout and five C-Type hangars. The airfield first opened as a grass site on 27 June 1940, and the first aircraft to arrive were the Fairey Battles of Nos 12 and 142 Squadrons, which stayed only briefly before returning in September with Vickers Wellingtons. No 12 Squadron became the first established unit at the base, and by April 1941 was conducting operational missions over France. After a year at Binbrook, No 142 Squadron was transferred to the new satellite airfield at Waltham, officially named Grimsby. No 12's Wellingtons pressed on with night-bombing until September 1942, when the Squadron was transferred to Wickenby, as Binbrook was scheduled to have hard runways put down under a £200,000 contract; these were 04-22 of 2,000 yards, 09-27 of 1,415 yards and 15-33 of 1,429 yards. To obtain the required lengths it was necessary to extend the airfield boundaries in some areas, which resulted in the main runway having a slope towards the valley at its 27 end. A perimeter track was also laid at this time, and nineteen loop hardstandings for aircraft were added to the eighteen pan types that survived the runway building programme. With additional accommodation the station provided for a maximum of 2,298 male and 420 female personnel.

No 12 Squadron remained at Binbrook until September 1942, when the Wellingtons left for Wickenby (where they were replaced by Lancasters), after having completed 1,242 operational sorties. No 1481 Flight arrived during November 1941 with

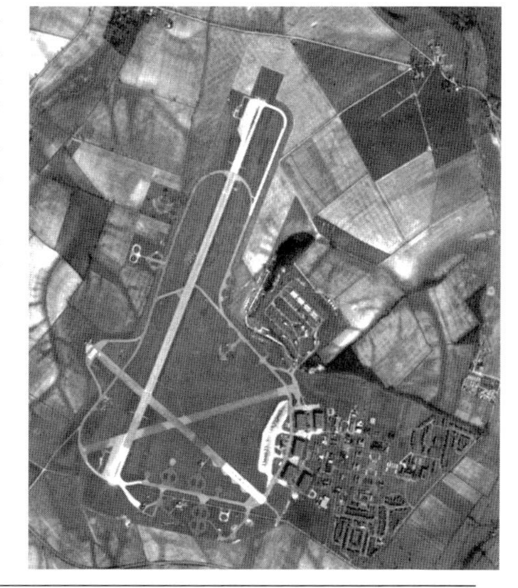

An aerial view of Binbrook taken during the 1980s. The busy flight line is clearly visible, as are the two distinct extensions of the main runway, together with the Operational Readiness Platforms at each end.

A wintry scene at Binbrook during January 1941. The airfield often fell victim to extreme weather.

target-towing Lysanders, which were supplemented by Whitleys in March the following year. Unusually, some of these target facilities Whitleys participated in bombing missions in 1942 when Arthur Harris embarked upon a policy of using every available bomber aircraft for missions against Germany. A handful of Oxfords joined the Flight too, but the unit left Binbrook when No 12 Squadron departed, enabling construction work to begin on the concrete runways.

When the airfield reopened after the building work was completed in May 1943, the first unit to arrive was No 460 Squadron, an Australian unit equipped with Lancasters, which remained at Binbrook as the only resident operational unit until the end of the Second World War. Some 5,700 Lancaster sorties were completed, the last being on 25 April 1945, against Berchtesgaden.

Former airman Henry Baskerville recalls his time at Binbrook:

'I suppose the total population of Binbrook, taking in the squadron, station and base personnel, would have numbered about two thousand. The squadron would have contributed some three hundred aircrew plus as many again ground staff, as each aircraft had a sergeant and five airmen responsible for its day-to-day maintenance. The sergeants were all Australians, part of the RAAF, but the airmen were all Britishers in the RAF. Our groundcrew were headed by Jack Jones until he was repatriated to Australia, then 'Spud' Murphy took over. His wife was named Beryl and he painted her name on the fuselage of 'O' Oboe just below the pilot's window. Each crew cherished their own plane and regarded it as much their own as the crew that flew it. They expected the aircrew to take good care of it – not to abuse the engines by running them outside their recommended specifications and not to bring it back with too many flak holes in it. For our part we appreciated them and recognised that our lives were very much dependent on their dedication and diligence. It was the captain's job to be as much an encouragement to them as it was to lead his own aircrew. At the end of our tour all of us (aircrew and groundcrew) met for an evening's celebration in the Marquis of Granby in Binbrook village.'

A No 420 Squadron Wellington at Binbrook during the summer of 1943.

No 460 Squadron left for East Kirkby in July 1945, its place being taken by No 12 Squadron, which returned from Wickenby to embark upon training duties with Lancasters; these often included the disposal of significant numbers of redundant bombs, which were dumped in the North Sea. The unit left

again in July 1946 after No 101 Squadron had arrived nine months previously, joined by No 9 Squadron in April 1946, followed by No 617 Squadron a month later. These three units eventually replaced their Lancasters with Lincolns, which in turn were eventually replaced by the Canberra.

No 85 Squadron's Canberras are lined up on their improvised flight line on one of Binbrook's disused runways.

It was at Binbrook that the world's first jet bomber entered RAF service. The Bomber Command Jet Conversion Flight was also formed at the base in December 1950 with a fleet of Meteors, followed by the Bomber Command Acceptance & Modification Unit in July 1952. The JCF's existence was fairly brief, and the BCAMU left for Hemswell during November 1953; however, No 109 Squadron formed with Canberra target-markers in January 1956, staying at the base for a year, joined by No 139 Squadron, which remained at Binbrook until December 1959.

During April 1963 No 85 Squadron arrived from West Raynham to continue Binbrook's long association with the Canberra, but by this stage the station's future was shifting towards fighter operations. The Central Fighter Establishment (CFE) arrived in October 1962, after the airfield had been reduced to Care and Maintenance status from 1 January 1960 until 1 June 1962. During this period of inactivity the main runway was extended (Operational Readiness Platforms were also

A snowy scene at Binbrook as the first of the RAF's Canberra bombers enter service. An aged Lincoln (which the Canberra replaced) is visible in the distance.

Binbrook will for ever be associated with the Lightning interceptor, having hosted Lighting squadrons from the 1960s until the base's closure decades later. The familiar sight of Binbrook's sparkling unpainted silver jets slowly gave way to drab camouflage from 1976 onwards. Mike Stroud

attached to each threshold) and facilities were improved, pending the arrival of the first fighter aircraft. The CFE brought a variety of aircraft types to the airfield, most notably the Hunter and Javelin, No 64 Squadron also moving to Binbrook when the airfield reopened, equipped with Javelins. This unit stayed until April 1965, when the Javelins departed for Singapore, after which No 5 Squadron brought its Javelins from Geilenkirchen during October.

After arriving at Binbrook the Squadron re-equipped with the supersonic Lightning. This was an aircraft that had already appeared sporadically in the skies over Binbrook as part of the CFE's activities, but with the formation of an operational Lightning squadron at Binbrook the aircraft became a regular sight, and eventually became the aircraft with which Binbrook will for ever be associated. Some Canberras remained here, however, with No 85 Squadron providing target facilities support for resident fighters and other units (also operating a few Meteors). They finally left for West Raynham in January 1972, at which stage Binbrook became an all-Lightning station. Two months later a second Lightning squadron, No 11, formed at Binbrook, and these two squadrons remained here with Lightnings over the next 20 years.

A Target Facilities Flight (TFF) had first been formed in 1966 as part of the Fighter Command Trials Unit, and this remained active with a small number of Lightnings until December 1973. The Lightning Training Flight (LTF) formed at Binbrook on 1 October 1975 and remained active as the sole Lightning training unit until disbandment in April 1987. The Lightning Augmentation Flight (which provided aircraft and pilots from reserve resources) formed out of the Instant Readiness Reserve Unit in 1981, and was short-lived, although a few aircraft did appear wearing the unit's markings. Operations continued without interruption, the short-ranged Lightnings rarely venturing too far from home unless accompanied by a tanker. Aircraft were regularly assigned to QRA (Quick Reaction Alert) duties, and Lightnings could often be seen scrambling from the QRA shed at the southern end of the main runway, roaring skywards to intercept Soviet bombers far out over the North Sea.

However, with Phantoms and the new Tornado F3 interceptors fulfilling the air defence role, the aged Lightning had reached the end of its useful life by the end of the 1980s. Following the disbandment of No 5 Squadron in 1987, No 11 Squadron disbanded in May 1988 and the mighty Lightning was finally retired. This also spelled the end for Binbrook, and over the next few months the station was vacated and finally abandoned.

After a long period of inactivity some limited civilian flying returned when Global Aviation set up a storage and repair facility for former RAF Jet Provost aircraft, although its operations eventually shifted to Humberside Airport. In July 1992 Binbrook's runway was briefly back in use when

A hapless Lightning F6 on the runway at Binbrook after suffering a nose-wheel malfunction.

Lightning XR724 was delivered by air from Shawbury for preservation at its former base. This aircraft remains at Binbrook to this day in open storage, battling the ravages of the brutal weather conditions, having earned the distinction of being the very last Lightning to use Binbrook's runway. Prior to this unique event the airfield had been extremely busy for a brief period in the summer of 1989 when it was used as the filming location for the movie *Memphis Belle*. B-17 Fortresses rumbled around the local vicinity for some time, and even a fake control tower was constructed in front of the hangars – the original RAF tower had been demolished on safety grounds.

However, with the arrival of the last Lightning in 1992 Binbrook's flying days ended for good, and the airfield slowly fell into decay. Unlike many former bomber airfields, which lay untouched for many years, there was a significant and almost hasty effort to remove Binbrook's runways and taxiways, and today almost all of these have been obliterated, their traces now being visible only from the air. The thresholds of the main runway are still largely intact, however, and the QRA

A last look at Binbrook's main runway, looking south towards the old Lightning QRA shed.

(Quick Reaction Alert) shelter still stands, albeit in poor condition. Although agriculture has now sprawled across the site of the former airfield, the massive hangars still dominate the skyline, but the long-term future of one of them is in doubt following a fire early in 2011. The former admin area is now the home to various private companies, and although the RAF is long gone many of the original station buildings still remain and some sense of the RAF's presence is still discernable among the buildings and hangars.

The lonely Lightning, draped in tarpaulins and car tyres, has now been joined by a Sea Harrier, but there appears to be no prospect of either aircraft being placed on proper display, and their presence seems only to reinforce the rather gloomy atmosphere that pervades what was once such a busy and famous station. The former domestic site was sold off and eventually became a village in its own right, but with the airfield having been destroyed with such haste, the renaming of the old RAF site as Brookenby was perhaps the final insult to what was once one of the RAF's most iconic homes.

Main features:
Runways: 217° 7,500 x 150 feet, 269° 4,200 x 150 feet, 327° 4,200 x 150 feet, concrete and tarmac surface. *Hangars:* five C-Type. *Dispersals:* thirty-six Heavy Bomber. *Accommodation:* RAF: Officers 156, SNCOs 416, ORs 1,437; WAAF: 388.

BINSOE, Yorkshire

54°13'5"N/01°36'58"W; SE251803; 423ft asl. 6 miles NW of Ripon between A6108 and B6267 on Binsoe Lane

As a Class Three Landing Ground, this site was in effect a simple field cleared of obstructions and reserved for flying operations. It comprised some 35 acres with a rectangular layout just short of 12,000 feet in length and some 1,200 feet wide. Few aircraft ever operated here, the only likely visitors being the Avro 504 and BE aircraft operated by No 76 Squadron RFC, which would probably have used the field for occasional deployments. When the First World War ended the site was abandoned and it quickly reverted to agricultural use. Today there is no trace of this field's links with aviation and the area has long since become a peaceful part of the Yorkshire countryside.

Main features:
Runways: none, grass surface. *Hangars:* none. *Dispersals:* none. *Accommodation:* not known.

BIRCOTES, Yorkshire

53°26'7"N/01°02'41"W; SK635935; 105ft asl. 1 mile W of Bawtry on A631

Situated just a few miles from the sprawling acres of RAF Finningley, the airfield at Bircotes is often overlooked by historians, especially when one considers that the famous headquarters of No 1 Group Bomber Command was just a mile away in Bawtry. However, Bircotes was an unsophisticated but important airfield in the RAF's inventory, first opening in November 1941. Originally called RAF Bawtry, the station's name was changed in June 1942 in order to avoid confusion with the Bomber Command HQ.

Communications aircraft were the first to arrive here and kept the field active, serving the HQ, which was just a short drive away. No 1 Group Communications Flight remained at Bircotes until November 1945. More substantial aircraft arrived in the shape of Hampdens, Manchesters and Wellingtons operated by No 25 Operational Training Unit (OTU) at RAF Finningley. This unit made use of Bircotes as a Relief Landing Ground (RLG), joined by Masters and Oxfords from Hucknall's No 16 Flying Training School, which used the airfield for a similar purpose in 1943. No 82 OTU brought its Wellingtons to Bircotes in 1943 while work was completed at its home base at Gamston. This deployment was followed by Ansons and Wellingtons from No 18 OTU, which remained here until November 1944. No 28 OTU operated from Bircotes during the summer of 1944, after which it was used less frequently although it remained in use by No 25 Maintenance Unit and by No 61 MU until 1948 (two other Maintenance Units also used it as a sub-site at various times).

Although a relatively inactive station, Bircotes had its fair share of incidents, including the forced landing of a Manchester (L7478) between the airfield and RAF Finningley. Wellington N1375 landed short of the field in June 1942 and eventually came to rest on the grass minus its undercarriage. Another Wellington (DJ976) overshot the field in September and crashed a mile to the west of the station.

By the end of 1944 all flying had ended at Bircotes (communications flights were shifted to nearby Finningley) and the station closed down. Although only a simple grass field, Bircotes did have a concrete perimeter track and most of this can still be seen today, together with one surviving hangar in the south-west corner of the site. Further along the perimeter track to the north there are traces of admin buildings still lurking in the undergrowth.

Less than a mile to the east, RAF Bawtry was a Royal Air Force station located at Bawtry Hall in Bawtry village and was No 1 Group RAF Bomber Command headquarters and administration unit during and following the Second World War. Bawtry Hall itself was erected around 1785 by a prosperous wool merchant from Wakefield, Yorkshire. During the Second World War the RAF took it over and it became an RAF command centre. RAF Bawtry did not have its own airfield, but instead took advantage of RAF Bircotes, which was only a 5-minute drive away. Bawtry Hall served the Royal Air Force from 1941 until 1984, first as HQ for No 1 Group, Bomber Command during and after the Second World War, then as Strike Command HQ up to and including the later stages of the Cold War. The famous bombing mission over the airfield at Port Stanley by Vulcan bombers during the Falklands conflict was coordinated from the operations room at Bawtry Hall. RAF Bawtry became the centre of the RAF Meteorological Service for many years, but ceased military operations in 1986. It is now a hotel complex, although much of the RAF's underground bunker has (perhaps not surprisingly) survived.

Main features:
Runways: N/S 4,600 x 50 feet, E/W 4,500 x 50 feet, NW/SE 4,000 x 50 feet, grass surface. *Hangars:* one T1, one B1, one Bessonneau. *Dispersals:* thirty circular. *Accommodation:* RAF: Officers 84, SNCOs 193, ORs 487; WAAF: 80.

BLIDWORTH, Nottinghamshire

53°04'51"N/01°07'11"W; SK590540; 325ft asl. SE of Mansfield between A614 and A60 on Blidworth Lane

No 35 Satellite Landing Ground was established here in August 1941. Used initially by No 27 Maintenance Unit (based at Shawbury), the site was parented by No 51 MU at Lichfield. The airfield was only small and comprised little more than a cleared landing strip of some 3,000 feet. Used mostly for the temporary storage of aircraft, the types brought here were mostly fighters, including Westland Whirlwinds and Lockheed P-38 Lightnings. Some were stored adjacent to the landing strip, although most were positioned in the nearby woodland to provide a degree of camouflage. No 25 (Polish) Elementary Flying Training School also used the site as a Relief Landing Ground during 1941-42, its Tiger Moths being the only obvious signs of activity here, apart from the occasional ferry flights.

At the end of the Second World War the site was redundant and was vacated in 1945. The airfield quickly returned to agricultural use and today the only remaining traces of the RAF's presence is the former Guardroom, which is now a privately owned bungalow.

Main features:
Runways: none, grass surface. *Hangars:* none. *Dispersals:* none. *Accommodation:* not known.

BLYTON, Lincolnshire

53°27'5"N/00°41'38"W; SK868957; 72ft asl. 4 miles NE of Gainsborough on B1205 Kirton Road

Construction of an airfield here began early in 1942, and a standard bomber layout was adopted with three runways in a triangular arrangement. The dispersals were largely confined to the eastern side of the airfield, while the admin site was on the opposite side of the site, close to Blyton village. The usual thirty-six pan hardstandings were provided off the concrete perimeter track, although one was lost on the east side to a T2 hangar erected north of Cold Harbour Farm. A B1 hangar was positioned south of Cold Harbour and a second T2 on the technical site, situated south-west of the runway heads 03 and 11. Bomb stores were located in fields between runway heads 14 and 21. Six domestic, two WAAF, two communal and sick quarters sites were dispersed among fields north of Blyton village on either side of the A159. Total accommodation provided for 1,966 males and 389 females.

No 199 Squadron reformed here on 7 November 1942 with Wellingtons, but the unit's stay was only brief and, after flying some operational missions, the Wellingtons left for Ingham just three months later. No 1 Group Air Bomber Training Flight had arrived in September 1942 with a handful of Oxfords, and these were joined by No 1481 Flight's Defiants a few days later, supplemented by Whitleys and Lysanders. After a stay of only two months the unit moved to Lindholme.

With all of these units gone, Blyton was assigned to training, and in January 1943 No 1662 Heavy Conversion Unit (HCU) formed here with sixteen Lancasters and sixteen Halifaxes, although the former were replaced by more Halifaxes by the end of the year. The unit remained busy training bomber crews until 6 April 1945, when it disbanded. Some flying was exported to Sturgate when this airfield became available as a Relief Landing Ground, and Elsham Wolds was used for the same purpose. The sound of Halifaxes pounding Blyton's airfield circuit by day and night became a familiar one in the local area, but by the end of 1944 they had been replaced by Lancasters.

When the HCU disbanded Blyton fell silent, and it was not until October 1951 that any significant activity returned, when the airfield was adopted as an RAL for No 101 Refresher Flying School at Finningley. The unit's Wellingtons, Oxfords and Harvards (and eventually Meteors) were

A No 1481 Flight Wellington at Blyton.

seen at Blyton regularly until January 1954. The station was also a sub-site for No 61 Maintenance Unit at Handforth, but with the RLG status gone the airfield was no longer used for flying. It was short-listed as a potential USAF base, but by 1960 it and the station had been abandoned.

It then lay unused, serving only as a navigational visual reporting point for crews returning to Finningley, and although even this vague association with flying is also now long gone, the basic airfield layout can still be seen. Small stretches of the secondary runways have survived, but substantial portions of the old main runway are still intact (mostly used for car-racing), and along the eastern edges of the airfield the perimeter track can still be found, together with some traces of dispersals, a hangar base and other remains. Now known as Blyton Park, the site's owners do at least express some interest in aviation:

> 'Although aircraft rarely land at Blyton now and they are usually helicopters, we still see some wonderful sights in the skies above the field. In 2010 we welcomed the Hurricane from the Battle of Britain Flight, which did a fly-past at the BAS Grasstrack round in August. In addition, the sole surviving airworthy Vulcan

A 1988 aerial view of Blyton showing the surviving portions of the three runways.

Although a great deal of the former airfield has gone, Blyton's layout is still visible from the air.

Few buildings remain at Blyton, and most are little more than rubble.

Blyton's former main runway is now used for motor-racing.

bomber passed over at low altitude on 11 September as we were preparing for the Rallycross Championship. Finally the Red Arrows can often be seen in the distance as they practise aerobatics from their home at RAF Scampton, which was of course home to that most famous squadron of all – The Dambusters of 617 Squadron in their Lancasters.'

Main features:
Runways: 212° 6,000 x 150 feet, 272° 4,200 x 150 feet, 322° 4,200 x 150 feet, concrete surface. *Hangars:* two T2, one B1. *Dispersals:* thirty-six Heavy Bomber. *Accommodation:* RAF: Officers 148, SNCOs 422, ORs 1,106; WAAF: 364.

BOULMER, Northumberland

55°25'13"N/01°36'01"W; NU259133. 4 miles W of Alnwick, S of minor road between Longhoughton and Boulmer

Boulmer had a rather inauspicious career during the war, but despite this it is still an active RAF station. Reopened in 1953 as an Air Defence Centre radar station, it was later redesignated as a Group Control Centre, becoming responsible for the radars at Buchan (Scotland) and Killard Point (Northern Ireland). It continued to be developed into a Sector Operations Centre and served at the front line of detection of Soviet aircraft during the Cold War.

Back in 1940, a decoy airfield was needed and a location was chosen just inland and south-west of the village of Boulmer, 10 miles north of Acklington. By the time this site was completed, Acklington would have become very familiar to the enemy and consequently there were very few deliberate attacks on the Boulmer decoy. On the night of 16/17 September 1940 the decoy was hit by a pair of HE bombs, which damaged a nearby cottage. The airfield also came close to being bombed on 3 June 1941 when an enemy plane apparently passed low over it but chose to bomb and machine-gun the village instead.

The neighbouring coast near Boulmer lent itself perfectly for the location of an aerial gunnery range, which was first used by the Lysanders and Bothas of No 4 AGS in early 1942.

Twelve miles south of Boulmer, Eshott became the home of No 57 OTU. This was a very large unit, which quickly overwhelmed that airfield with its Spitfires, Masters and Battle target tugs, all of which made full use of Boulmer's facilities. The latter was therefore a prime candidate to become a satellite for Eshott and was quickly pressed into use as a Relief Landing Ground. With great pressure to finish Boulmer because of the amount of aircraft operating from Eshott, the airfield was officially taken over by No 57 OTU on 1 March 1943. The unit's Advanced Training Squadron was the first to use Boulmer's new tarmac runways, followed by a flight of Masters.

Modern-day equipment at Boulmer is the Westland Sea King HAR.3 operated by No 202 Squadron. Crown Copyright via Martyn Chorlton

Milfield, to the north, suffered the same congestion problems as Eshott, and on many occasions Hurricanes and Masters from No 59 OTU also used Boulmer throughout 1943 and 1944.

No 57 OTU remained at Boulmer until its disbandment on 6 June 1945. Eshott was also being closed down, so Boulmer was placed under Care and Maintenance, becoming a satellite for Acklington.

Still retained by the Air Ministry, Boulmer was reactivated in 1953, the site having been chosen as an Air Defence Control Centre. A new site for this complex collection of buildings was chosen next to the B1339 north of Lesbury. Within a year the building was completed and the station officially reopened as No 500 Signals Unit.

The airfield's runways were in sufficiently good condition for use by No 6 FTS, which moved to Acklington in July 1961. Boulmer became an RLG for the Jet Provosts of the FTS until the unit's disbandment in 1968.

B Flight of No 202 Squadron, with the Whirlwind HAR.4, had originally moved to Boulmer in 1972 after Acklington's closure, but never made it its permanent home. Boulmer's first permanent flying unit, A Flight of No 202 Squadron with the Whirlwind HAR.10, arrived in 1978. The Flight re-equipped with the more capable Sea King HAR.3 in 1979 and continues to operate this helicopter today, providing Search and Rescue for a vast area that stretches throughout northern England, southern Scotland and the Lake District and as far as the Norwegian coast.

An aerial view of the eastern side of the airfield, now incorporated into a public road. The remnants of all three runways can still be seen; note how close to the coast the airfield actually is. Via Martyn Chorlton

Boulmer is now the home of the School of Aerospace Battle Management and is the HQ of the UK ASACS (Air Surveillance and Control System). The airfield's layout is very unusual because the station is divided into two modern camps, neither of which encroaches or utilises the flying area of the wartime airfield. Large sections of the runways and perimeter track remain, all best viewed from the minor road that joins Lesbury and Boulmer. This road follows the line of the eastern perimeter track, and another minor road that leads to Seaton Point and Marmouth Scars was also part of the wartime airfield.

The original airfield was accessed from the Long Houghton-Boulmer road, and this is the location of No 202 Squadron and its Sea King helicopters. The helicopter station is a self-contained RAF camp, complete with its own hangar, tower, technical and domestic buildings and, like ASACS, is surrounded by a very high secure fence. Behind this fence, a connection with the Cold War role of the airfield is represented by a No 56 Squadron Phantom FGR.2.

Main features:
Runways: QDM 014 1,800 yards, 062 1,400 yards, 152 1,300 yards, concrete. *Hangars:* four Dorman Long Blister. *Hardstandings:* twenty-five SE. *Accommodation:* RAF: 617; WAAF: 106.

BRANCROFT, Nottinghamshire

53°28'13"N/00°59'59"W; SK664975; 46ft asl. 5 miles SE of Doncaster off A614 adjacent to Doncaster-Sheffield Airport

This minor site was one of many fields in the region that were adopted as Royal Flying Corps landing grounds for fighter aircraft assigned to patrols over the Humber region, where marauding Germans regularly made their way to targets over England. No 33 Squadron's A Flight operated here in 1918 with Avro 504s on a sporadic basis, and the field was used as required for air defence of the nearby cities of Doncaster and Sheffield. No permanent structures were established here (in fact, the site was retained by a farmer), and after the First World War it was returned to agriculture. Not surprisingly, there are no surviving traces of the RFC's presence, and the field is now simply an area of cultivated grassland adjacent to the old Bawtry Road, and just a few feet from the perimeter of the former RAF Finningley, which now serves as a regional airport.

Main features:
Runways: none, grass surface. *Hangars:* none. *Dispersals:* none. *Accommodation:* not known.

BRAYTON, Cumbria

54°46'16"N/03°17'14"W; NY172425; 256ft asl. 10 miles NE of Maryport, NW of Baggrow village on B5299

Constructed on the remains of Brayton Hall, this small airfield was acquired in 1941 but remained unused due to administration and planning problems. It was not until April 1942 that it was made available for use, and a month later a trial landing was performed by an Avro Anson; this led to the arrival of the first Wellington bombers, which were to be temporarily stored at the site. No 12 Maintenance Unit was the sole user of the airfield, and the unit remained here until January 1946. Wellingtons were the main type to he stored here although some B-17 Fortresses, B-25 Mitchells and Halifaxes were also seen here at various times. Aircraft were sometimes placed in camouflaged hides, although the size and the numbers of stored aircraft (the site had a capacity of around 100) ultimately made this a pointless exercise. The last aircraft to leave the airfield were three Vengeances and a Wellington, which departed in December 1945.

With all traces of the airfield long gone, a golf course now sits at what was once the junction of the two grass landing strips. However, Brayton survives in a surprisingly intact state today. The combined bungalow-style watch office/guardroom has unfortunately badly deteriorated in more recent decades, losing its distinctive twin roofs, but other nearby support buildings survive in better condition, as does a Robin hangar across the generations-owned grass fields as a sawmill. As ever there are various odd tracks leading here and there, but the average eye would never realise that this used to be an airfield that played a most important – if unpublicised – part in winning the Second World War.

Main features:
Runways: N/S 4,000 x 50 feet, NE/SW 4,000 x 50 feet grass surface. *Hangars:* none. *Dispersals:* none. *Accommodation:* not known.

BREIGHTON, Yorkshire

53°48'23"N/00°54'25"W; SE720349; 19.5ft asl. 5 miles NW of Howden, S of Highfield village, off B1228

RAF Breighton first opened in January 1942 as part of No 1 Group Bomber Command, acting as a satellite field to Holme-on-Spalding-Moor. Oddly, the station's name was taken from the smaller of two adjacent villages. A standard bomber airfield was developed with three standard-sized runways and a pair of T2 hangars, together with the usual array of thirty-six dispersals. The only slight variation from standard bomber design was the common intersection of all three runways – something more common to FAA airfields. They were subsequently lengthened, resulting in 'add-on' perimeter tracks to each new threshold.

A Whitley from No 78 Squadron, during the time it was based at Breighton in 1940.

The first unit to arrive was No 460 Squadron RAAF with Wellingtons. After a period of operations, the unit partially re-equipped with Halifaxes (a Conversion Flight being established at Holme), but by October the unit was equipped with Lancasters and Manchesters. Operations continued (during which time the Manchesters were withdrawn), and in May 1943 the Squadron moved to Binbrook (the Conversion Flight having gone to Lindholme). Breighton was transferred to No 4 Group and in June No 78 Squadron arrived from Linton-on-Ouse with Halifaxes. This squadron embarked upon many combat sorties (including participation in D-Day) and by the end of the war the unit had flown some 5,120 operational missions, most of them from Breighton. The last sortie was completed on 25 April 1945 and, after a brief period of relative inactivity, No 4 Group shifted its activities to transport operations, and No 78 Squadron began flying transport sorties with its fleet of war-weary Halifaxes. During its use by Bomber Command 169 aircraft dispatched on operations from Breighton failed to return or crashed in the UK. In addition to twenty-nine Wellingtons, No 460 Squadron lost fifteen Lancasters, while No 78 Squadron lost 125 Halifaxes. On 7 May 1945 the Group became part of Transport Command and Dakotas were slowly delivered to replace the bombers. On 1 September the Squadron finally left Yorkshire and headed for Almaza in Egypt.

An overhead view of Breighton airfield taken during 1944.

An oblique reconnaissance photograph of Breighton and the surrounding area, taken in 1945.

The end of the Second World War inevitably left many bomber airfields redundant, and Breighton was no exception. It closed in 1946 and remained abandoned for more than ten years until it was once again brought back into use as part of Britain's nuclear deterrent force. Three huge concrete launch pads were constructed out on the airfield adjacent to the main and secondary runways. These were designed to accommodate the Thor missile, together with the necessary support equipment and launch facilities. In August 1959 RAF Breighton re-emerged as the home to Thors, operated by No 240 Squadron. With the mighty missiles established at the base, more arrived a year later in the shape of Bloodhound SAMs from No 112 Squadron, and launch positions for thirty-two of them were constructed in an area to the east of the third runway. A former member of No 112 Squadron recalls his time at Breighton:

> 'The local people were very interested in what we were up to and it didn't take them long to supposedly work out which of our Bloodhounds had nuclear warheads ... they reckoned that they were the ones with the scarlet nose cones! We didn't disillusion them – we weren't allowed to talk about it. In fact, the scarlet nose cones (which we called 'Noddy Caps') were simply electromagnetic radiation covers that we used to protect the electronics when the missiles were disconnected from the electronic protection circuits during removal, transportation and reloading! We were billeted some at RAF Church Fenton (mostly singles) and at the RAF Married Quarters at Acomb, in York. The only time we ever stayed overnight at Breighton

was when there was a flap on, then a crew slept there ready to launch if necessary – but it never happened. We never had any contact with the Thor Squadron next door, although our task was to defend them. They were Bomber Command and we were Fighter, and anyway their missiles were American while ours were British!'

A 1990 view of Breighton airfield, most of which has survived, although industrial development is slowly creeping across the site.

The rapidly deteriorating remains of the Bloodhound missile site at Breighton.

The runways were never used by planes, only as a road from one end of the Squadron to the other. Although flying had ended at Breighton, there was no doubt that the RAF was back, but the Thors did not stay in the UK for long and by 1963 had been returned to the US. The Bloodhound missiles were also withdrawn shortly afterwards, and in March 1964 the RAF left Breighton for a second (and final) time.

The airfield was once again abandoned and remained silent for many years, apart from the occasional visit from agricultural aircraft. But Breighton returned to flying with the arrival of the Real Aeroplane Company, which now occupies a small collection of buildings in the south-west corner of the old airfield. Part of the adjacent taxiway has been resurfaced to create a small runway (11/29 of some 2,000 feet), which is used for private and recreational flying and for the operation of the various preserved warbirds now based here. One of the old T2 hangars still survives, and by driving through the various clusters of industrial businesses that now occupy parts of the airfield it is possible to find the fascinating rows of concrete pads that once accommodated the Bloodhound missiles. Further investigation will also reveal the remains of the three huge concrete Thor pads that are now occupied by buildings, cars and assorted junk.

Main features:
Runways: 087° 6,000 x 150 feet, 172° 4,200 x 150 feet, 040° 4,200 x 150 feet, tarmac surface. *Hangars:* two T2, one B1. *Dispersals:* thirty-six Heavy Bomber. *Accommodation:* RAF: Officers 153, SNCOs 338, ORs 732; WAAF: 191.

BROUGH, Yorkshire

53°43'16"N/00°33'53"W; SE948259; 20ft asl. 1 mile SE of Brough village, south of A63

Nestled on the banks of the Humber Estuary, the airfield at Brough was constructed as a direct result of a survey made by a Mr Mark Swann on behalf of the Blackburn company. He was asked to investigate locations in the region that could be developed into a manufacturing and testing base for new landplanes and seaplanes. A large steel hangar was erected in 1916 together with a slipway, which enabled aircraft to be deposited on the Humber for flight trials.

The first design to use the facilities was the GP Seaplane (which later became the Kangaroo), but when the First World War began the airfield was placed under military control and became No 2 (Northern) Marine Acceptance Depot, responsible for the handling of most RNAS seaplanes throughout the war.

Above: *A Blackburn TB at Brough in 1915.*

Left and below: *Phoenix Cork Mk II N87 after completion inside Brough's flight sheds. The same buildings later accommodated a wide variety of combat aircraft, culminating with the Buccaneer and Hawk jets.* Phil Jarrett

The magnificent Phoenix Cork flying boat pictured outside the flight sheds at Brough during August 1918, ahead of the slipway leading into the Humber Estuary. Phil Jarrett

In the post-war era Blackburn set up a small freight and passenger service, but it was largely unsuccessful and in 1924 the company began training RAF Reserve officers. A flying school was set up and quickly became a very active RAF Reserve school. Meanwhile Blackburn resumed aircraft design and construction, absorbing the company's Leeds works by 1932. Many well-known and significant aircraft types were subsequently developed, all of which were constructed, tested and delivered directly from Brough. Among the long list of types was the B2 trainer, Seagrave, Shark, Botha, Firebrand, Firecrest and Skua. The pace of activity increased as the Second World War approached, and throughout the conflict Brough continued to produce countless combat aircraft including a significant number that were manufactured on behalf of other companies; Swordfishes and Barracudas were among the non-Blackburn types to be manufactured here. Blackburn also assumed responsibility for US-built aircraft that were adopted for Fleet Air Arm service, overseeing a Modification Centre established at Roosevelt Field in the USA. Training also continued at Brough and the airfield became the home of No 4 EFTS with Tiger Moths, the nearby airfield at Bellasize being used as a Relief Landing Ground throughout the wartime years.

A Blackburn Shark pictured during its short trip from the flight sheds at Brough to the slipway into the Humber Estuary from where its first test flights were conducted. Phil Jarrett

A rare 1920 aerial view of Brough showing the first Blackburn sheds, which eventually developed into the huge British Aerospace factory that now occupies the same site. Phil Jarrett

After the war Blackburn's future looked rather bleak, and the company soon merged with General Aircraft to become Blackburn & General Aircraft Ltd. The first direct result of this merger was the appearance of the huge Universal Freighter, which was gradually developed into the famous Beverley transport. It was this aircraft that also prompted the construction of a concrete landing strip, which was literally squeezed in between the factory and the perimeter hedgerows. But despite the availability of the small runway, there was still insufficient space for the Beverley's heavyweight operations, so some of the flight testing was transferred to the much larger airfield at Holme-on-Spalding-Moor. When the company was subsequently awarded a contract to produce what became the mighty Buccaneer strike aircraft, all the flight testing had to be conducted from Holme, and the completed aircraft were partially dismantled and transported to Holme by road before being reassembled for flight.

Sadly, Brough's airfield was only suitable for the smallest and lightest of aircraft, and expanding it proved to be impossible. Even civilian operations were hampered by the airfield's size (together with the continuing flight test operations), and from the mid-1960s a small number of civil charter

A pair of Blackburn Kangaroos pictured outside their flight test shed at Brough during 1920. Phil Jarrett

A rarc view inside the Blackburn factory showing the NA.39 undergoing stress and fatigue tests. The NA.39 became the Buccaneer, which served with great distinction with both the RAF and Fleet Air Arm, together with the South African Air Force. Phil Jarrett

Right: *In April 1950 the mighty Blackburn Universal Freighter rests between test flights at Brough.*

Below: *This picture demonstrates the capacity of the Universal Freighter to accommodate a large and heavy cargo load. The Freighter was eventually developed into the mighty Beverley transport aircraft for the RAF.* Phil Jarrett

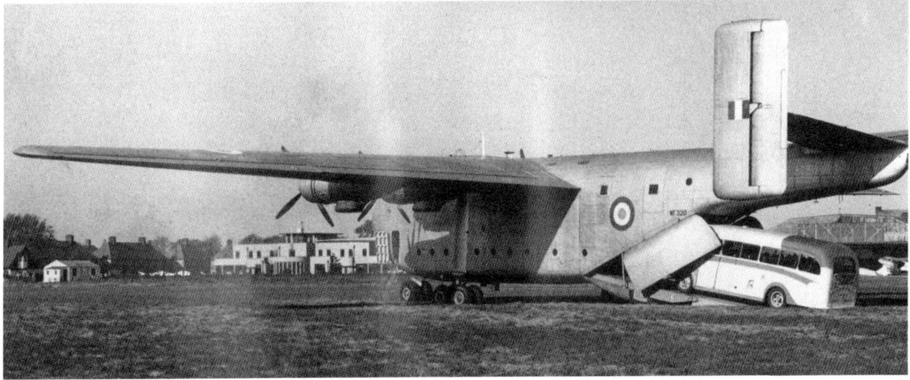

A 1997 aerial view of Brough showing the factory site and the short concrete runway used most recently for Hawk delivery flights.

operations were established here, together with a regular service to Luton. But the use of the airfield for civil flying was short-lived and, with most test-flying transferred to Holme, it was soon virtually disused. The EFTS left Brough at the end of the Second World War, test-flying shifted to Holme, and civilian operations proved to be impractical. The only signs of activity were the occasional communications flights bringing staff to and from other factories (Blackburn became part of the Hawker Siddeley Group), and the appearance of an occasional private flyer.

Despite the airfield's virtual demise the Brough factory remained firmly in business, producing components for many aircraft types (Nimrod, BAe 146, Buccaneer, Phantom, Harrier and Airbus), and it eventually became responsible for the manufacture of Hawk trainers, a task that continues today as part of BAE Systems. Remarkably, the airfield's tiny runway has been brought back into use, and as Hawks are completed at the factory they are towed onto the airfield and flown out to Warton for testing prior to delivery. The flights are sporadic and few in number, but the tiny runway is still in use for the time being. It can easily be seen from the adjacent lanes, and the cluster of finger dispersals is still visible in an adjacent field next to the railway line.

Sadly, the longer-term future of both factory and airfield are in doubt. Hawk production will end and plans have been drawn up for housing developments, which will probably encroach upon the airfield. On 27 September 2011 BAe announced that it intended to bring the manufacturing of aircraft at Brough to an end, effectively closing the world's oldest aircraft factory. A 'Battle for Brough' campaign was immediately launched to save the historic site. However, this probably means that in a few years from now the historic factory and airfield will be gone, the magnificent Buccaneer that has stood on guard next to the runway will be gone too, and nothing will remain to remind visitors of a site that played such an important part in Britain's military history.

Main features:
Runway: 305° 3,400 x 100 feet, concrete and tarmac surface. *Hangars:* six Over Blister, one Double Over Blister, two civil flight sheds. *Dispersals:* eight finger type. *Accommodation:* not known.

BRUNTON, Northumberland

55°31'28"N/01°40'35"W; NU205257. 9 miles N of Alnwick, between villages of Brunton and Tughall, next to main East Coast railway line

Construction of a new airfield began in mid-1941, with 600 acres of land being requisitioned for the site north of Brunton stretching across to the small village of Tughall. Simultaneously a Q decoy site was constructed 6 miles to the north, near the village of Elford.

On 2 August 1942 No 59 OTU moved from Crosby-on-Eden to Milfield, and simultaneously the latter became the parent of Brunton. Two days later, seventeen Hurricanes arrived from Longtown (Crosby's satellite), beginning their first training sorties the following day.

The threat of invasion was always in the minds of the military planners. One contingency plan was called the Saracen Scheme, a method of quickly converting OTUs into fighter squadrons. In No 59 OTU's case, the unit would become No 559 Squadron, and in the event of an invasion could be reallocated to a different airfield, in this case Woolsington. The title of No 559 Squadron was actually used at Brunton in March 1943, briefly transferring to Milfield and later returning as an independent squadron flying the night-fighter variant of the Hurricane.

No 59 OTU had been operating the Hurricane since it began flying training in March 1941; however, the unit was earmarked to become the main OTU for the Typhoon. The transition between the two fighters was very slow because of early teething problems with the Typhoon. By May 1943 the big Hawker fighter began to arrive at Milfield and in turn also started to use Brunton.

This aerial view of Brunton shows how complete this disused airfield is. All three runways, perimeter track and the majority of the 'panhandle' dispersals are intact. The airfield is best viewed from the railway bridge that crosses the main line in the lower left-hand corner of the photograph. Crown Copyright via Martyn Chorlton

No 59 OTU was disbanded on 26 January 1944 and was absorbed into No 1 Specialised Attack Instructors School, which in turn was absorbed into the FLS. Brunton remained a satellite, but the majority of FLS operations were flown from Milfield.

The airfield was quiet for much of 1944, although the FLS did use it for refuelling and re-arming exercises as well as using the airfield as a miniature bombing range. Later in the year Milfield received another unit. No 56 OTU, which had been disbanded in October 1943, was reformed on 15 December 1944 with more than 130 Typhoons, Tempests, Masters and support aircraft. Brunton was first used by No 56 OTU in January 1945 for its Typhoon and Tempest operations, as it would have been impossible to operate effectively from one airfield despite the fact that the FLS had departed from Milfield at the end of 1944.

All aircraft operated by No 56 OTU at Brunton were training pilots in their last stage of the course syllabus, which may explain why only one incident involving a Tempest occurred on the airfield. On 9 April 1945 Tempest V EJ845 swung on take-off and left the runway, destroying the windsock in the process. Sadly, Flt Lt I. W. Smith RCAF, an experienced pilot, was killed. A few weeks later all OTU operations came to an end at Brunton, and a very swift closure of the airfield came at the end of May 1945.

Aviation has since returned to Brunton, which remains virtually intact with most of the main runway still in use. Light aircraft operate occasionally from here, and the Border Parachute Centre has made the old wartime airfield its home.

The best way to view Brunton is probably from a railway bridge that is accessible via a wartime road on the western side of the airfield. From this viewpoint it is easy to imagine wartime aircraft operating, and the line of air raid shelters that can be seen are in excellent condition. In the distance several period buildings remain, the majority being utilised by the local farm, which owns the bulk of the site.

In 2002 a Coastal Emitter Site was built at Brunton, encased in a large camouflaged dome. The site is under the control of RAF Spadeadam, located on the edge of Wark and Kielder Forests. The radar provides aircrews with an electronically controlled area off the Northumbrian coast where they can practise manoeuvres and tactics, specifically involving the use of chaff and flare. A typical military sign with a red border tentatively indicates the presence of the military after an absence of 57 years.

A wet day helps to show up the perimeter track and the end of the east-west 1,100-yard runway, which is still complete. Martyn Chorlton

Main features:
Runways: QDM 030 1,600 yards, 150 1,100 yards, 079 1,100 yards, concrete. *Hangars:* four Blister. *Hardstandings:* twenty-five. *Accommodation:* RAF: 654; WAAF: 108.

BURN, Yorkshire

53°44'53"N/01°05'09"W; SE603283; 16.5ft asl. 2 miles S of Selby, E of A19

Opening as a Royal Air Force station during November 1942, the airfield at Burn was constructed in accordance with standard policy for new bomber airfields. The three runways were laid in a standard triangular pattern although the airfield layout was to some extent influenced by the close proximity of the surrounding roads and the busy LNER railway line, which clipped the airfield boundary (and was identified as a flight hazard throughout the airfield's existence). The total of thirty-six proposed hardstandings were pan-types, but two were lost to hangar construction and two

While workers attend to the airfield shrubbery, a Halifax bomber gets airborne on a training sortie.

loop standings were added to replace them. Hangars consisted of a T2 and a B1 on the south-west side of the technical site, which lay between runway heads 02 and 08. A second T2 stood on the western side, between runway heads 08 and 16 near Burn village. The bomb stores were positioned towards the Selby Canal on the north side. Nine domestic sites for 1,805 males and 276 females were dispersed north-west beyond Burn Lane and Brick Kiln Lane, as were the two communal and sick quarters sites. However, while it was common at most new stations for the planned camp not to be completed until many weeks after the airfield came into use, delays and shortages saw this work continue at Burn into the summer of 1943.

The first aircraft arrived while construction was still under way, No 1653 Heavy Conversion Unit arriving with Liberators (together with examples of the Anson, Lysander, Ventura, Blenheim and Oxford) during June 1942. The unit stayed at Burn until October, after which the Wellingtons of No 431 Squadron RCAF arrived. The unit quickly assumed operational duties and earned the distinction of flying Bomber Command's 100,000th sortie of the second World War. During July 1943 the squadron moved to Tholthorpe (re-equipping with Halifaxes in the process) and the brief presence of AOP Austers (Nos 658 and 659 Squadrons) came to an end shortly before the Canadians left.

A period of relative inactivity followed without any resident units, and Burn was assigned to the role of Relief Landing Ground. However, the bombers returned in 1944, and No 578 Squadron arrived with Halifaxes during February of that year. Burn's Halifaxes contributed greatly to the war effort, and when the last operational sortie was completed on 15 March 1944 the Squadron had totalled some 2,721 Halifax sorties since re-equipping with the type. Two of the squadron's Halifaxes had completed more than 100 operational sorties. A claim to fame, apart from No 578's VC, 143 Distinguished Flying Crosses and 82 Distinguished Flying Medals over its fourteen-month life, was its consistent bombing accuracy, resulting in the granting of a squadron crest by His Majesty King George VI in February 1945 with the motto 'Accuracy'. Yet another accolade was earned by the Squadron's groundcrews, whose outstanding servicing of the Bristol Hercules XVI engines resulted in the award by the Bristol Aeroplane Company of a shield, now on display at the Yorkshire Air Museum at Elvington near York.

Fredrick Blackmore MBE recalls joining No 578 Squadron:

> 'I arrived at Burn as a Flight Engineer. After nine operations I was taken ill and my crew were detailed to take part in a operation to bomb Dusseldorf on 22 April 1944. Unfortunately they were shot down, but all survived. I spent the rest of my operational flying career as a spare engineer to various other crews.
>
> Aircraft flown was the Halifax Mk III with four Hercules engines and a crew of seven – Pilot, Navigator, Wireless Operator, Bomb Aimer, Flight Engineer and two Gunners – mid-upper and rear.
>
> The following extract is taken from official records: "Squadron Target No 7: Stuttgart – 20th/21st March 1944. Ten aircraft were detailed but LK D and LK R had to abandon the operation, 'R' with an unserviceable rear turret and with a mid upper turret and compass unserviceable. Wing Commander Wilkerson flew LK Z on this raid and on return reported a big explosion which he thought to be a large gasometer blowing up. This was also reported by Flight Sergeant Sparkes in LK P, who noted a pillar of smoke up to 24,000 feet, visible 30 miles away. Seven of the attacking aircraft returned to base. Flying Officer Hope-Robinson in LK Y landed at Husbands Bosworth due to fuel shortage. Flying Officer Starkoff in LK D, having abandoned the operation, unloaded his bombs onto a flak position and noted that the flak ceased after the bombs exploded. The weather was cloudy, some wintry showers, but visibility good. 578 Squadron were carrying some incendiaries and most reports agreed that the attack was well concentrated. There was little opposition, though fighters were active on the return trip. The Squadron stood down for the rest of the 21st!"'

When the unit disbanded on 15 April, Burn's active status finished, and by the end of the year the airfield was closed to flying. It was used briefly as a storage site for military vehicles (Royal Army

Service Corps), but was soon abandoned and left to slowly deteriorate for another forty years. More recently it was feared that a particle accelerator facility would be built on the site, probably resulting in the destruction of most of the surviving airfield structure. However, this plan was later dropped and a more recent attempt to create a refuse incinerator site is still being considered. But for the time being the remains of this airfield are still very much in existence. The three huge runways are still intact and used infrequently by the Burn Gliding Club. The taxiways are also still in evidence, but the dispersals are mostly gone, together with the hangars and other buildings.

Main features:
Runways: 200° 6,000 x 150 feet, 264° 4,200 x 150 feet, 335° 4,200 x 150 feet, concrete surface. *Hangars:* two T2, one B1. *Dispersals:* thirty-six Heavy Bomber. *Accommodation:* RAF: Officers 129, SNCOs 279, ORs 1,397; WAAF: 276.

BURNFOOT, Cumbria

54°59'14"N/02°59'05"W; NY370662; 23ft asl. 5 miles N of Carlisle, between M6 and A7, W of Sandysyke

Designed as a Relief Landing Ground for nearby Kingstown (some 6 miles away), the airfield at Burnfoot was an unsophisticated facility comprising little more than a cleared grass field straddling a small lane. A handful of Blister hangars was assembled here but building work was kept to a minimum. The site opened in July 1940 as part of No 15 Elementary Flying Training School, and the small airfield was soon occupied with Tiger Moths and Magisters that plodded around the local area during daylight hours before returning to the parent airfield at Kirkbride before nightfall. Burnfoot became a surprisingly busy site as training requirements increased, and by the summer of 1942 the Magisters had all been withdrawn, leaving more space for the ubiquitous Tiger Moth.

As the Second World War ended the training activities slowly diminished and in 1947 the EFTS was disbanded. Also the need for a satellite field had long since disappeared and Burnfoot was abandoned towards the end of 1945. The Blister hangars were demolished and the scattering of other small buildings soon disappeared. Today there are no traces of the former airfield, although the boundaries of the landing strip are still intact and the small lane that links Sandysyke with a small farm still crosses the old airfield.

Main features:
Runways: none, grass surface. *Hangars:* none. *Dispersals:* none. *Accommodation:* not known.

BURSCOUGH, Lancashire

53°35'41"N/02°52'02"W; SD427112; 85ft asl. 2 miles N of Ormskirk, N of B5242

Although a somewhat unusual location for a naval airfield, Burscough was in fact well positioned just a few miles from Southport, affording easy access to the West Coast and the many ports in this area. Construction began late in 1942, and the 650-acre site comprised four runways, laid out in typical naval fashion, and all quite short in comparison to the much larger bomber runways being laid around the UK at that time. Although large hangars were confined to just a pair of standard T2s, a large number of small S-Sheds appeared within the clusters of dispersals scattered around the airfield perimeter.

The station opened on 1 September 1943 as HMS *Ringtail*. The first units to be established here were Nos 886, 808 and 897 Naval Air Squadrons, forming No 3 Naval Fighter Wing. The unit moved to Lee-on-Solent in February 1944, replaced by No 4 Naval Fighter Wing, comprising Nos 809, 879 and 807 Naval Air Squadrons. Other units also appeared at Burscough for brief periods, including No 1837 NAS and No 1820 NAS, which operated the rarely seen Helldiver from the station during the summer of 1944. A wide variety of naval aircraft types frequented the airfield not only as part of the many short-lived unit assignments, but also during visits and exercise deployments.

Sadly, not all of the flying activity took place without incident. No 1771 Squadron moved to Burscough in March 1944, presumably for further training prior to embarking on HMS *Trumpeter* in June. On Sunday 28 May 1944 one of the unit's Fireflies (Z1906) was carrying out camera attack exercises with other aircraft from the Squadron at 3,000 feet. The exercise was conducted some 2 miles east of Burscough, between the villages of Rufford and Holmeswood. The crew on this fateful day were Sub Lt Maurice Walton Williams RNZNVR from Napier, Hawks Bay, New Zealand, and Sub Lt Peter Graham Sunderland RNVR from Harrogate, both aged just 22. During a tight turn Z1906 was observed to go into a spin from which it failed to recover and struck the ground, bursting into flames on farmland and killing both crew members instantly.

An aerial image of Burscough, probably taken in 1945.

Most of the inhabitants of the nearest village were attending the morning service at the local church and, although accustomed to the frequent aerial activity over their homes, they had been alerted by the unusual engine note of an aircraft out of control. Those in the local Auxiliary Fire Service rushed to collect their equipment and raced to the scene, while to most observers the column of black smoke, also visible from the airfield at Burscough, told its own grim story. Those first on the scene found the aircraft completely in flames and were forced to take cover as ordnance carried by the Firefly began to explode. Through the flames the bodies of both crew members were discernible to some witnesses, but clearly beyond help, not that the would-be rescuers could get near the conflagration. By the time heavier fire-fighting equipment could get to the scene from the airfield the aircraft was completely burned out and all that was left to do was the grim task of recovering the bodies of the unfortunate airmen, who are both now buried at nearby Burscough.

Training activities here included No 735 NAS, which arrived in March 1944, and B Flight of No 735 NAS, which became No 707 NAS at Burscough, flying Avengers, Barracudas, Ansons and Swordfishes on AI and ASV radar training duties. This unit left for Gosport in August 1945 and No 735 NAS disbanded in April 1946.

With the Second World War over, the Fleet Air Arm's activities rapidly diminished, and although Burscough had effectively expanded during 1945 (acquiring nearby Woodvale as a satellite station) it was soon redundant, and HMS *Ringtail* closed on 15 June 1946. It remained in military hands as a storage facility until 1957, when it was finally sold off. Today most of the runway structure has gone, with only one of the four runways still fairly intact. Parts of the perimeter track remain but the rest of the former FAA station is little more than a memory. The adjacent industrial estate contains a number of roads that remind visitors of the area's previous incarnation as HMS *Ringtail*, and within the estate the two T2 hangars have survived and remain in use as storage sheds. Some of the old dispersals are still to be seen and a surprisingly large number of S-Sheds are still standing. Sadly, the magnificent three-storey control tower was demolished some years ago.

Main features:

Runways: 080° 3,720 x 150 feet, 120° 3,075 x 150 feet, 210° 3,015 x 150 feet, 350° 3,000 x 150 feet, concrete and tarmac surface. *Hangars:* two T2. *Dispersals:* fifty loop. *Accommodation:* FAA: Officers 189, ORs 1,204; WRNS 370.

BURTONWOOD, Lancashire

53°24'41"N/02°39'10"W; SJ567907; 59ft asl. 3 miles NW of Warrington, W of B1122, adjacent to J8 of M62

Drivers on the busy M62 motorway are unlikely to notice some odd patches of concrete just beyond Junction 8 as they head towards Liverpool. Indeed, there is very little to indicate that there was ever a connection with aviation in this area, and few drivers would ever be aware that they are driving along the foundations of what was once a huge runway from where mighty B-36 strategic bombers roared into the sky.

The airfield on this site first appeared in 1940, having been designed as a home for No 47 Maintenance Unit, although by the time the airfield opened the MU had become part of the Civil Repair Organisation. Aircraft began to arrive at Burtonwood from May of that year and the site quickly filled with aircraft. Most were housed here for relatively short periods, but almost a hundred were delivered to the site on a monthly basis.

No 37 MU assumed control of the base from April 1940 and the airfield was gradually extended in order to accommodate the unit's requirements with additional hangars and dispersals being constructed; the MU also adopted a number of satellite airfields for even more storage. By 1942 Burtonwood was concentrating on modification work, particularly that required by American-supplied aircraft that were destined for RAF and FAA service. It was somewhat ironic that, even at this stage, Burtonwood had developed an association with the USA, for when the MU closed on 15 July 1942 the RAF transferred the base to USAAF control.

Left: *A rare aerial image of Burtonwood taken during 1945.*

Below: *A Boeing B-50 roars onto final approach at Burtonwood.*

The Americans immediately began further development of the site until it eventually became one of the USAAF's most important assets (it was in fact the largest airfield in Europe), preparing and repairing thousands of aircraft and items of equipment for wartime service. However, conditions for the American personnel were hardly luxurious, as one former airman recalls:

B-17 Fortresses in temporary storage at Burtonwood.

> 'Living conditions in the Nissen huts on Burtonwood were not very good. In fact, in my thirty-three years of service with the Air Force, this was the worst living conditions that I experienced. They were worse than those I experienced in Turkey or Vietnam. One must realise, though, this is only a few short years after the Second World War and we were rapidly building up our forces to counter the Russians in a Cold War that sometimes got hot. When I arrived at Burtonwood I was the thirteenth man in our hut, but many members were moved on to man other bases and we got down to six men to a hut. Our Nissen huts were heated by two coke stoves and we were issued a bed with the blankets, sheets, etc, a foot locker and a stand to hang our clothing. Our clothing was covered with a piece of tarpaulin. We later built with plywood a covered place to hang our clothing. We two-tone painted our hut and bought curtains for our four windows. Because of our efforts to improve our living conditions in our hut, we were awarded two morale flights to Rome, Italy, and Copenhagen, Denmark.
>
> There was one large toilet and shower facility for two squadrons (approximately 500 men). The shower facility consisted of approximately twenty shower heads in one open stall. Approximately fifty wash basins were provided in one open area. The water was seldom hot and a cold shower was very invigorating, especially in the winter time. In Site 4 was our only dining facility, which fed the entire base. Even with these austere conditions, I look back at my first assignment in the Air Force with nothing but good feelings. I enjoyed England and the English people and greatly admired their outlook on life.'

When the Second World War ended the RAF resumed control of the base and No 37 MU returned together with No 276 MU, which had a responsibility for handling American aircraft. However, the

A nostalgic image of the MATS terminal at Burtonwood, with a C-47 in front of the base's unique control tower.

A B-66 Destroyer pictured on a rainy day at Burtonwood.

In October 1956 a mighty RB-36 roars over Cheshire en route to Burtonwood's runway.

RAF's presence slowly diminished and in 1948 the Americans returned, this time in support of post-war bomber and transport operations. With an extended main runway (stretching out to the east to a total length of 9,000 feet) and a huge concrete apron to the south, Burtonwood provided a home for B-29 Superfortresses deployed to Europe. It also provided a staging point for countless USAF transports, particularly during the Berlin Airlift.

In November 1953 the 53rd Weather Reconnaissance Squadron was established here with WB-29 and WB-50D aircraft, remaining at the base until April 1959. The USAF's MATS (Military Air Transport Service) continued to use Burtonwood as a 'hub' until 1958, when operations shifted to Mildenhall. By this stage the C-47, C-54 and the mighty C-124 were familiar shapes in the region, and eventually even the more modern C-133 and C-130 Hercules frequented the base. Fighter types were also seen (F-84 and F-86 aircraft were overhauled here), but the most impressive sight (and sound) was undoubtedly the sporadic appearance of the huge B-36 bomber, which visited Burtonwood as part of training exercises.

Regular flight operations ended early in 1959 with the USAF's presence having largely transferred to Mildenhall, although the Americans maintained a presence here until June 1994, the US Army adopting the base from 1966 as an equipment and supplies depot. Helicopters were often seen on the airfield, but the only other aviation assets here in the 1960s were the gliders of No 635 VGS, which remained here until September 1983.

With the departure of the US Army in 1994 the airfield was abandoned, the sprawling acres of runway and hardstandings being left to ruin. The construction of the M62 motorway eventually caused the destruction of the old main runway, which formed the foundations of the motorway as it crossed the airfield site. Urban and industrial expansion has slowly encroached upon the site and it is now completely obliterated, a shopping mall now standing where the bombers and transports once touched down. The only vague reminders of the once huge airfield are the traces of the enormous MATS ramp, which lay in the grass adjacent to Burtonwood Road roundabout off the A574. It is particularly sad that what was once a truly huge and very significant military airfield has been completely destroyed.

Main features:
Runways: 270° 5,250 x 150 feet, 220° 4,350 x 150 feet, 150° 4,350 x 150 feet, concrete and tarmac surface. *Hangars:* more than twelve various. *Dispersals:* fifty finger type. *Accommodation:* Officers 708, ORs 19,280.

CAISTOR, Lincolnshire

53°30'18"N/00°21'46"W; TA086021; 75.5ft asl. 1 mile W of Caistor, E of B1434 on Caistor Road

Originally developed as a Satellite Landing Ground for Kirton-in-Lindsey, the airfield at Caistor was first authorised in June 1939 and flight operations began here in 1940. Various aircraft types used the airfield on a sporadic basis, including Hurricanes from No 85 Squadron. Caistor also served as a Relief Landing Ground for No 15 Pilot Advanced Flying Unit based at nearby Kirmington, and Oxfords became a familiar sight on the airfield during 1942. Later in that year the airfield was allocated to RAF Manby, but it was not used and for some time was placed under Care and Maintenance. The site was used again occasionally towards the end of the Second World War (mostly by Cranwell), but the airfield was never particularly busy and it was never developed beyond its original status as an unsophisticated grass landing field.

However, long after it had been abandoned by the RAF, it was reacquired in the late 1950s as part of Britain's plan to deploy the Thor ICBM. Three concrete launch pads were built in the centre of the old airfield and support facilities were developed, leading to the reopening of RAF Caistor in 1959. The Thors remained here until May 1963, when No 269 Squadron (the resident Thor unit) disbanded. The airfield was then abandoned once more and the area returned to agricultural use.

Almost fifty years later the old Blister hangar still stands, together with a scattering of admin huts. The Thor pads have survived and remain in remarkably good condition, hidden within the sprawling agriculture. The site is on private land and cannot be accessed, but the pads are certainly still there, as are the post-war RAF entrance and admin buildings, all of which are now abandoned and looking eerily forlorn.

An aerial view of Caistor in 1994, the Thor launch pads still clearly visible.

Caistor's Thor complex was similar to other such bases, with three launch pads linked to support facilities and ringed by security fencing.

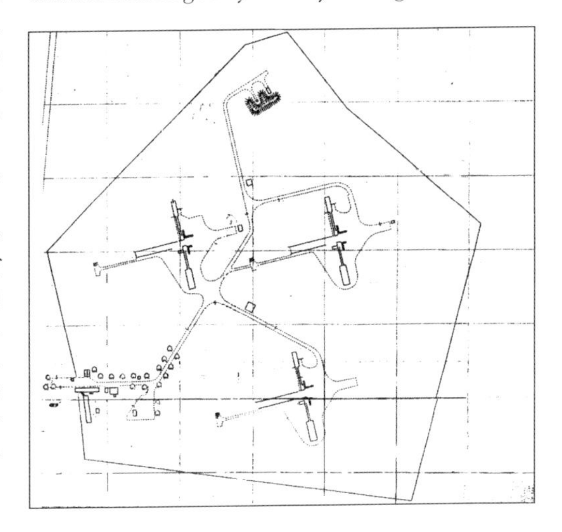

Main features:

Runways: NE/SW 4,200 x 100 feet, SE/NW 4,350 x 100 feet, E/W 3,600 x 100 feet, N/S 3,300 x 100 feet, grass surface. *Hangar:* one Double Blister. *Dispersals:* six fighter type. *Accommodation:* RAF: Officers 76, SNCOs 41, ORs 461; WAAF: 0.

CARK, Cumbria

54°09'48"N/02°57'31"W; SD375745; 20ft asl. S of Flookburgh off B5277, on Moor Lane

Land was acquired here in October 1940 and construction of an airfield soon began, intended for use as a satellite landing ground for another airfield being constructed nearby at Barrow. The location for this satellite field on a peninsula on the edge of Morecambe Bay was less than ideal, being just 20 feet above sea level and surrounded by high ground. It also suffered from fog and mist

Oxfords and Ansons at rest on the airfield at Cark.

and had a generally poor weather record. Although Barrow's fighter aircraft were to have used the airfield at Cark, by the time it was completed plans had changed and the station was assigned to Flying Training Command. In March 1942 the Staff Pilot Training Unit began operations here, training students for Air Observer Schools and Observer Advanced Flying Units. Equipped with a fleet of Ansons, a small number of Miles Masters were also used by the unit. Apart from a brief disbandment in November 1942, the SPTU remained at Cark throughout the Second World War.

Anti-Aircraft Cooperation Flights also operated from Cark at various times, the first being F Flight of No 1 AACU with a mixed fleet of aircraft including Hawker Henleys. Arriving in January 1942, the unit became No 1614 Flight and eventually No 650 Squadron. Operations were normally conducted in support of No 9 Light Anti-Aircraft Practice Camp at nearby Flookburgh, and No 14 LAAPC at Nethertown. Provisional plans were drawn up to expand the airfield (including extensions to two of the runways) but as the war drew to a conclusion these plans were abandoned and the RAF vacated the site in 1947. No 188 Gliding School was the last unit to use the airfield (in May of that year), having first arrived in February 1944.

Following closure the airfield was left unused for many years, but some recreational flying has now returned to the site, with a section of one of the old runways restored to support various light aircraft. Although intact, the original runways are overgrown, but the perimeter track remains visible and traces of the old dispersals can still be found, reaching out into the adjacent fields.

Main features:

Runways: 250° 3,900 x 150 feet, 170° 3,600 x 150 feet, 310° 3,600 x 150 feet, tarmac surface. *Hangars:* nineteen Extra Over Blister, one Bellman. *Dispersals:* six twin fighter type. *Accommodation:* RAF: Officers 69, SNCOs 271, ORs 1,019; WAAF: 204.

CARLTON, Yorkshire

53°42'59"N/01°01'48"W; SE641248; 29.5ft asl. 1 mile NW of Carlton village, W of A1041

This small and undeveloped site was in effect little more than a grass field reserved for flying, available to Home Defence units during the First World War. It came into use early in 1916, primarily as a landing ground for B Flight of No 33 Squadron, but it is not known whether the site was used much, if at all. It was abandoned in 1917 and quickly returned to agricultural use. There is some doubt as to the precise location of the field, and without any surviving traces of the brief military presence the landing field appears to have slipped into permanent obscurity.

Main features:
Runways: none, grass surface. *Hangars:* none. *Dispersals:* none. *Accommodation:* not known.

CARNABY, Yorkshire

54°03'40"N/00°15'26"W; TA141642; 39.5ft asl. S of A614, 1 mile SW of Bridlington

One of the most unusual airfields developed during the Second World War, Carnaby was opened for flying in March 1944, construction work having commenced a year previously. As part of No 4 Group RAF, the airfield was designed as an emergency landing site, capable of accepting aircraft of any type or size, particularly lost crews or damaged aircraft returning from operations over Europe. With an extremely long and wide runway, clear of all obstructions and adjacent to the coast, the site was ideal for the task, enabling crews to divert here instead of being obliged to attempt landings at smaller airfields where the chances of a successful recovery might not be so good. In addition to the massive runway, a linking perimeter track was created to the south of the runway and an unusual loop of dispersals was attached to the western end. No permanent structures were created other than a control tower, the remaining admin and domestic accommodation being of a distinctly temporary nature (mostly wooden huts). Equipped with FIDO (Fog Investigation Dispersal Operation) equipment, Carnaby was a valued asset and more than 1,500 emergency landings had been completed here by March 1946 when the station closed.

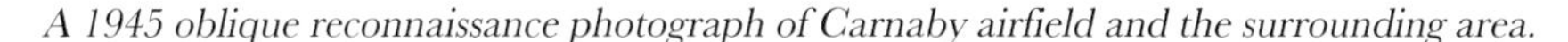

A 1945 oblique reconnaissance photograph of Carnaby airfield and the surrounding area.

A reconnaissance image of Carnaby, illustrating the unique dispersal loop and the Thor launch pads that were subsequently constructed inside it.

Delia Smith recalled her experience of Carnaby:

'The aerodrome was 3 miles long and constructed early in the war. The workmen would have no idea what a great part the airfield was to play in the defeat of Germany. On either side of the emergency runway, for 1,800 yards in double lines, ran the pipelines of FIDO filled with petrol; they were to operate in 3 seconds and be fully operational in less than 5 minutes. It was first used in April 1944. The official description of the system was of a great glow in the sky accompanied by dense volumes of thick black oily smoke; when the smoke disappeared the fog was tinged red, growing brighter and brighter until the people in Bridlington could read newspapers by it. The blaze could be seen for more than 50 miles. Some fire brigades from many miles away actually clanged themselves to Bridlington to find that there was no fire. What they were witnessing for the first time was the faithful watchdog of distressed aircraft, the Carnaby FIDO.

From April 1944 to the end of hostilities 1,600 aircraft made emergency landings at Carnaby alone, and this figure did not include normal or routine landings. Once, seventy-five four-engined aircraft, returning from an attack on Germany, landed in just under an hour, and thirty-seven were once landed in 25 minutes. A low-flying German bomber dropped a bomb on the runway, which landed 25 yards from the control tower. Fortunately, it failed to explode and several aircraft were landed safely and their crews saved.

Another incident concerned a Halifax. One night on its way back from Norway it was attacked and a hole blown in its side. There was no sign of the air gunner. The crew knew they'd lost him; imagine their surprise when, on landing safely at Carnaby, they found the air-gunner suspended only by his parachute straps, alive, in spite of the nightmare journey.'

Carnaby remained silent for some time after the Second World War, but the RAF returned in September 1949 when the airfield was reopened as a Relief Landing Ground for No 203 Advanced Flying School, based at nearby Driffield. The unit's Vampire trainers became the first jet aircraft to frequent the area, and during 1950 the entire unit was based here while Driffield's runways were resurfaced. The unit continued to use Carnaby (and was eventually renamed as No 8 Flying Training School) until 1954, when Full Sutton assumed the RLG role and Carnaby was once again vacated by the RAF. One of the last uses for Carnaby as an airfield was in 1951 when Blackburn used the site for testing the HP.88, a small crescent-winged jet that was manufactured for Handley Page to test the company's new wing design (which was eventually developed into the Victor bomber). The HP.88 made its first flight here (the home airfield at Brough being too small), and a number of test flights were performed until the aircraft was destroyed during a subsequent test flight at Stansted.

The unique Handley Page HP.88 conducted its series of test flights from Carnaby.

The RAF returned once more in 1959 when No 150 Squadron formed here with Thor ICBMs. Three concrete pads were constructed in the middle of the old dispersal loop and over to the east a compound was created for the Bloodhound missiles from No 247 Squadron, which arrived in July 1960. The Thors left early in 1963 and the Bloodhounds departed towards the end of the year, then the RAF abandoned Carnaby once more in 1964. Subsequently the runway was briefly used for car-racing, but the site was finally sold off for development in 1972. The massive (and historically significant) runway became the base for a huge industrial estate and the concrete Thor launch pads were dug up, together with the unique dispersal loop that surrounded them. From the air the proportions of the monstrous runway are still very clear, but from the ground it is difficult to imagine that any airfield was once here. All traces of the Thor and Bloodhound site have been obliterated – a sad outcome for a fascinating site.

Main features:
Runway: 259° 9,000 x 700 feet (plus over-run), concrete and bitumen surface. *Hangars:* none. *Dispersals:* twenty-four finger type. *Accommodation:* RAF: Officers 18, SNCOs 36, ORs 494; WAAF: 0.

CATFOSS, Yorkshire

53°55'10"N/00°16'20"W; TA135484; 36ft asl. 1 mile NE of Brandesburton, E of A165

Although almost every British military airfield was named after the nearest town or village, RAF Catfoss was one of a few exceptions, taking the name of nearby Catfoss Grange, the estate's land being requisitioned for the new aerodrome, designed to enable training operations to take place over the nearby air gunnery ranges on the coast.

No 1 Armament Training Camp was established here in January 1942 with a mixed fleet of Audax, Hart, Gordon, Gauntlet and Wallace aircraft. Flights to and from the ranges took place all day as students were taught the necessary skills of air-to-ground gunnery and bombing. Significantly larger aircraft arrived in September 1935 when the Heyfords of No 97 Squadron appeared. During April 1938 the Armament Training Camp became No 1 Armament Training Station, and by now the aircraft fleet included Tiger Moths in addition to the longer-term residents. Training continued until the outbreak of the Second World War, at which stage the unit was transferred to Northern Ireland where training could continue without any undue risk of attack.

A rare picture of Catfoss during its early days, illustrating the distinctly temporary nature of the accommodation and hangars.

After a major reconstruction programme the airfield resumed activity in 1940, although a detachment of No 616 Squadron's Spitfires had already been based here by then. No 16 OTU established a detachment at Catfoss for a few months, but the station's future was to be within Coastal Command. No 2 (Coastal) Operational Training Unit formed here in October 1940, equipped with

Handley Page Heyford J9130 lumbers into the air. Heyfords were a common sight at Catfoss.

Part of a line-up of early biplane fighters on show at Catfoss during what appears to be a VIP visit. One of the station's hangars is visible in the distance. Barry Ketley

An Avro Tutor is pushed back into a hangar at Catfoss with what appears to be a gaggle of Hawker Furies in the background. Barry Ketley

An unusual and fascinating aerial view of Catfoss in 1932 illustrating the scattered bombers, the Bessonneau hangars and the tents that were occupied by the groundcrews. Also worthy of note is the surprisingly small overall size of the technical and admin site when compared to more modern RAF stations. Phil Jarrett

large fleet of aircraft that eventually comprised Blenheims, Ansons, Beaufighters and Oxfords. Some flying activity was exported to Lissett and almost all of the station's operations shifted there from late 1942 for a few months while concrete runways were laid. Most of the unit's activities eventually centred on the training of Beaufighter crews, most of which were destined for overseas theatres, and training continued until February 1944, when No 2 (C)OTU disbanded. The Central Gunnery School then moved to Catfoss with a large and varied fleet of aircraft, which included both bombers and fighters. This unit eventually moved to Leconfield in December 1945, after which the airfield was abandoned.

It was not until 1947 that new plans for the airfield were revealed, and it was expected that the site would become an airport for nearby Hull. The plan never progressed and Catfoss remained derelict until 1958, when the RAF returned and three Thor launch pads were constructed adjacent to the old runways. No 226 Squadron was here until March 1963, when the Thor ICBMs were withdrawn and the RAF abandoned the station once more. Since then the site has been partially redeveloped. Substantial parts of two of the old runways have gone and industrial sites sprawl across part of the airfield and the old admin sites. Two of the Thor launch

This Wellington was delivered to the Central Gunnery School at Catfoss after completion at the Vickers factory.

Catfoss as seen from the air during 1991. Two of the Thor missile launch pads are still clearly visible.

Although a great deal of the former site at Catfoss has survived, the most prominent reminder of the RAF's presence here is a beautifully maintained memorial wall. Barry Ketley

pads have survived in surprisingly good condition and the remains of the control tower are still here. One of the A-Type hangars still stands, and other buildings can still be seen among the industrial developments. Parts of the perimeter track are still here too, together with a few of the dispersals.

Main features:
Runways: 193° 5,250 x 150 feet, 295° 5,130 x 150 feet, 290° 4,680 x 150 feet, concrete and wood chippings surface. *Hangars:* one A-Type, one A1 Type, one T1, fifteen Extra Over Blister, two Bellman, one B1. *Dispersals:* twenty-five circular, fifteen oval. *Accommodation:* RAF: Officers 165, SNCOs 281, ORs 1,054; WAAF: 0.

CATTERICK, Yorkshire

54°21'58"N/01°37'11"W; SE248967; 170ft asl. Adjacent to A1(M), half a mile S of Catterick

Although Catterick is inevitably associated with the Army, the same location has also been an active site for military flying since 1914, when the Royal Flying Corps first established a base here. No 14 Training Squadron formed at Catterick on 1 February 1915 as part of No 8 Wing with a mix of Farman types assigned to basic flying training duties. The 19th Wing assumed control of the site in May 1916, by which time No 6 Training Squadron had arrived for a short stay.

As the First World War developed, a variety of operational units were formed and built up at Catterick, each staying only briefly before relocating once fully established. The resident training squadrons amalgamated during July 1918 to become No 49 Training Depot Station, and the combined unit operated a variety of aircraft types until disbandment during March 1919. By this stage the station and adjacent airfield had been developed quite significantly, but after the end of the war it was placed in long-term storage and it was not until the early 1920s that some of the station's facilities were improved in anticipation of a return to a more active status.

In October 1926 aircraft returned to the site when No 26 Squadron's Armstrong-Whitworth

Hawker Demon J9933 pictured on a sortie from Catterick.

Right: *An unusual pre-war image of Catterick's main site.*

Sopwith Pup 'Ickle Poop' at Catterick in 1918.

Atlas biplanes arrived. Working in cooperation with Catterick's Army units, the Squadron operated here in isolation until No 41 Squadron's Hawker Demons arrived in September 1936. Further development of the airfield continued, and a pair of C-Type hangars were erected, although the overall size of the site remained unchanged, being constricted by the River Swale to the east and the Great North Road (the A1) to the west. The resident Atlases were eventually replaced by Audaxes, Hectors and finally by Lysanders, while the Demons were replaced by Hawker Furies until Spitfires arrived in January 1939. The two squadrons departed towards the end of 1939 and Catterick became a Sector Station as part of No 13 Group.

A large number of RAF squadrons were then assigned to the station for short periods, tasked with the defence of North East England. Among them were Nos 17, 54, 63, 64, 68, 130, 134, 145, 222, 313, 401, 403, 600 and 609 Squadrons, operating Spitfires, Hurricanes, Mustangs, Blenheims and Defiants. Nos 1472 and 1490 Flights were also resident at Catterick in 1943, with Hurricanes and Tomahawks, assigned to specialist training duties (including target-towing). Although most of the time spent here by these squadrons was uneventful, there were some incidents that served to illustrate the very real risks encountered by units even when they were far from the South East.

On Tuesday 18 July 1939 the pilot of Spitfire K9888, undertaking a cross-country navigation exercise, left Catterick at 10.56hrs to fly over the Pennines to an airfield on the western side – Dumfries was quoted in the pilot's inquest, whereas Kingstown airfield is also quoted in a number of more modern documentations of the incident. The pilot was instructed to fly a direct route across the Pennines, probably through the Stainmore gap. On leaving Catterick he passed Bowes, suggesting that he was taking a route across Stainmore towards Brough, Appleby and Carlisle, which would take him clear of the Cross Fell area. The aircraft was flying over the Appleby area when it encountered a bank of very thick fog. The pilot had been briefed to return to base if he came into cloud, and it was thought that he had cleared the hills but decided to turn back towards Catterick. While he was making the turn the aircraft flew back into the hills and struck the ground on Great Dun Fell. The pilot was killed in the accident just 30 minutes after leaving Catterick. Workers at the nearby Silver Band Mine were the first on the scene, having heard the aircraft fly overhead, then the crash higher up the hill. They set out from the mine to try and locate the site, but in the thick fog this took some time, and sadly nothing could be done for the pilot when they found him. The flight commander at Catterick was later blamed for this accident, as he had not ensured that the weather forecast was suitable before authorising the flight, although no action was taken as the pilot had probably left it too late in turning back towards Catterick.

The last operational squadron to fly from Catterick was No 222, and when its Spitfires departed in February 1944 the airfield became almost inactive, seeing only occasional visitors connected with either the RAF station itself or the nearby Army units. Catterick became the home of the RAF Regiment and, despite being a very significant RAF station, there was very little flying action other than the glider activity provided by No 645 Volunteer Gliding School, which arrived in March 1960. However, the airfield did occasionally host some visiting communications aircraft and helicopters, and after the RAF's School of Firefighting arrived a number of retired RAF aircraft made their final flights to Catterick, ready to end their days as training airframes, dumped around the airfield perimeter where most were eventually burned to destruction. Among the very last aircraft to arrive for the School was Vulcan XH561.

The RAF Regiment moved to Honington in May 1994 and the RAF station closed on 1 July of that year. The gliders left for Topcliffe in 2003 and the Army assumed control of the station and airfield. Today the C-Type hangars can still be seen from the A1(M) and the small concrete runway is still visible, although it is essentially abandoned. The wrecked hulks of the fire training aircraft are long gone, and although the airfield remains virtually intact it has remained silent for many years and looks set to remain so.

Main features:
Runways: 285° 3,600 x 150 feet, 220° 4,200 x 150 feet, 270° 6,000 x 150 feet, grass/tarmac surface. *Hangars:* six Extra Over Blister, two Over Blister, two C-Type. *Dispersals:* five fighter type. *Accommodation:* RAF: Officers 110, SNCOs 110, ORs 1,210; WAAF: 426.

CHURCH FENTON, Yorkshire

53°49'56"N/01°11'57"W; SE528376; 30ft asl. 3 miles SE of Tadcaster, south of B1223

The Royal Air Force's association with Church Fenton began during the late 1930s when construction of a grass airfield began on a site adjacent to the village railway station. In addition to a large and well-drained landing field, a pair of short-bay-pattern C-Type hangars appeared, although a planned third hangar was abandoned and replaced by a less-ambitious T-Type structure. The station (partially completed) opened in the summer of 1937, and Gloster Gladiators from No 72 Squadron arrived in June of that year followed by No 213 Squadron's Gauntlets a month later. These aircraft stayed until May of the following year, when they were replaced by No 64 Squadron with Demons.

When the Second World War began the station quickly became part of No 13 Group, and No 72 Squadron (now equipped with Spitfires) moved to Leconfield, being replaced by a detachment from No 245 Squadron, together with No 242 Squadron equipped with Hurricanes. No 12 Group resumed control of the station in August 1940 and a variety of operational squadrons were then assigned here for relatively short periods, each tasked with the air defence of the local region. Church Fenton hosted a large number of squadrons and units resident during the wartime years, including Nos 25, 26, 46, 71, 72, 85, 87, 96, 124, 183, 242, 249, 288, 308, 456, 488, 600 and 604 Squadrons, operating types such as the Spitfire, Beaufighter, Typhoon, Defiant, Blenheim and Mosquito, although the Hurricane was by far the most common sight on the airfield during the early months of the conflict. Perhaps the most unusual aircraft type seen here was the unsuccessful Brewster Buffalo, operated by No 71 (Eagle) Squadron. Each squadron's stay was usually fairly short (moving to and from the more active regions of the UK), and units continued to be rotated through the base during the early days of the Second World War until November 1940, when the station was assigned to the increasingly important training of night-fighter crews.

A one-off formation of three aircraft types operated by No 72 Squadron, flying over the control tower at Church Fenton.

An interesting view of Church Fenton's flight line, the modern Tutors and much-modernised control tower contrasting with the Second World War-vintage hangar.

Another view of the flight line at Church Fenton, with No 7 FTS Jet Provosts and Tucanos occupying the apron where Hornets and Javelins once stood.

No 54 OTU formed on 25 November 1940 with a mixed fleet of Oxfords, Blenheims, Defiants, Havocs and Masters, and the station transferred to No 81 Group. Training continued until May 1942, when the station returned to operational flying, largely in response to the growing number of German raids in the Midlands and northern England. No 25 Squadron's Beaufighters (later replaced by Mosquitoes) became the main residents until December 1943, and a variety of day- and night-fighter units continued to rotate through the station until the end of hostilities.

While many stations wound down in 1945, night-fighters remained active at Church Fenton after the war, and in April 1946 the first jets arrived when No 263 Squadron's Meteors assumed the station's day-fighter commitment, with Mosquitoes still assigned to night operations, courtesy of No 264 Squadron. In April 1947 the sleek de Havilland Hornet arrived when Nos 19 and 41 Squadrons moved from Wittering, joining the resident Mosquitoes to add to the busy pace of activities at the station. When the Mosquitoes left they were replaced by RAuxAF Meteors (No 609 Squadron), and yet more Meteors arrived in September 1957 when No 85 Squadron moved from West Malling. After the departure of the Hornets in 1951 Church Fenton became an all-jet station, and the Meteors were joined by Hunters and Javelins, which re-equipped Nos 19 and 72 Squadrons.

The station remained very active as a base for night-fighter operations until June 1959, when the noisy Javelins and Hunters left for Leconfield and the airfield was transferred to Flying Training Command. Despite being a long-established base for fighter aircraft, Church Fenton was far from ideal for jet operations, with two fairly modestly sized runways, few dispersals and only limited hangar facilities, compared to the larger airfield at nearby Leconfield. Although various training units eventually reformed here (including Leeds University Air Squadron), it was not until April 1962 that significant flying activity returned when No 7 FTS began Jet Provost operations here, joined by an Advanced Training Flight (flying Vampires) in 1966. The Jet Provosts, Vampires and UAS Chipmunks ensured that the airfield was again very busy, and nearby Elvington was regularly employed as a Relief Landing Ground.

The FTS disbanded in 1966 but training activities continued with the Primary Flying Squadron, which arrived (with its Chipmunks) from South Cerney in 1967. This became No 2 FTS in 1970, and Bulldogs joined in 1973 as part of the Royal Navy element of the unit. The FTS remained here until 1974, when it disbanded, the Royal Navy's element moving to Leeming. Church Fenton was then placed under Care and Maintenance, although the airfield facilities were still used as an RLG for Linton-on-Ouse, where most of the RAF's basic jet training was now concentrated.

Four years later an increased requirement for basic training saw the station reopen when No 7 FTS was formed here and intensive training resumed with Jet Provosts. These faithful aircraft remained in business until 1990, when the new Shorts Tucano replaced the increasingly aged Jet Provosts. Despite a significant modification programme being initiated, Church Fenton was listed for closure in 1992, and the FTS disbanded on 31 March. However, the station did not close and remained under RAF control as an RLG for Linton, eventually becoming home to the Joint Elementary Flying Training School, equipped with Slingsby Fireflies. These basic trainers remained here until 2003, at which time the airfield again became a rather less active RLG site again, although No 9 AEF and the Yorkshire UAS continued to operate here with Bulldogs and (eventually) Tutors.

The days of intensive operational flying are now little more than memories, the roar of the Javelins and Hunters having faded into history. The annual air displays that attracted huge crowds from around the region are also long gone, and Church Fenton is almost forgotten, but it survives in good condition with both of the wartime runways available for use. Tucanos still buzz around the airfield circuit and the hangars still house a handful of Tutors and various private aircraft, but the tranquil skies around the base contrast sharply with the days when Church Fenton was one of the RAF's busiest airfields.

Main features:
Runways: 168° 4,800 x 150 feet, 245° 3,600 x 150 feet, concrete and tarmac surface. *Hangars:* ten Extra Over Blister, one Bellman, two C-Type. *Dispersals:* four twin fighter type. *Accommodation:* RAF: Officers 119, SNCOs 150, ORs 976; WAAF: 330.

COAL ASTON (Sheffield), Yorkshire

53°19'42"N/01°27'54"W; SK357814; 633ft asl. S of Sheffield on A6102, off A61 (Norton College)

A 1937 image of Coal Aston showing part of the main site and the grass landing field.

This relatively unknown site first appeared in 1915 when the Royal Flying Corps established a small landing field here, which was intended to become a training site. However, it was first used as a base from which the RFC's fighter aircraft could operate while engaged on air defence duties over the area, particularly the city of Sheffield, which was then a mile or so away.

No 33 Squadron's BE.2C aircraft were the first aircraft to be based here in 1916. Towards the end of the year the threat of attacks from German airships had diminished and No 33 withdrew, leaving Coal Aston under the control of No 17 Training Squadron. The station slowly expanded to take on second-line duties, particularly the repair and overhaul of RFC aircraft, and after a very substantial building programme the site became No 2 (Northern) Aircraft Repair Depot in 1917. A large number of hangars appeared (at least ten) and the technical site was extended to the south, eventually resulting in a narrow-gauge railway being laid to transport materials to and from the airfield itself. The domestic area also grew considerably. Aircraft types handled by the depot included the RE.8, Sopwith 1½ Strutter, BE.12, Sopwith Scouts, FE.2B and various others. The repair of engines was also a function of the depot, and a salvage section was also formed here, which became responsible for the recovery of all RFC aircraft across northern England.

Coal Aston was an important and busy station but, like many others, it became far less active when the First World War ended, and by 1920 the RFC had vacated the site. It was sold in 1922 and Sheffield Corporation considered retaining it as a municipal airport, but (unlike many other cities across the UK) the idea was eventually abandoned. The landing field was left unused and the hangars and other buildings slowly crumbled, before being demolished in order to make way for two new schools. Most of the domestic site was also bulldozed.

A rare image of Coal Aston showing just part of the huge depot. The grass landing field was to the upper right of the picture.

The deserted main runway at Sheffield City Airport, where only a handful of military aircraft ever touched down.

Today a college is situated on the site where the hangars once stood, but the old landing field is still here, now used as a sports pitch. The surrounding area is now part of Sheffield's urban sprawl and it is almost impossible to accept that so much of the local area was once part of a surprisingly large RFC station.

Just a mile further along the A6102 to the east, another site is still visible, which (because of its proximity) has often been confused with Coal Aston. This was once RAF Norton, designated as No 16 Balloon Centre. Initially known as Lightwood, it opened in 1939 and eventually accommodated three balloon squadrons. It became RAF Norton in July 1943 as the home of No 3 Ground Radio Servicing Squadron. Although not an airfield, the site did host a number of popular air displays, which included visits from Vulcans, Hunters, Shackletons and many other types, and the station's 'gate guard' Spitfire now resides in the RAF Museum at Hendon. RAF Norton closed in January 1965, and although part of the site is now an NHS facility the mortal remains of the balloon sheds can still be seen.

Sheffield itself boasted a more substantial airfield, which opened in 1997. Having been the largest European city to be without its own airport, the new facility was long overdue. Limited scheduled services began and some military aircraft also used the site, not least Finningley's Jetstreams, which occasionally performed approaches. At least one Hercules used the airport, as did other types including the RAF's BAe 146 and Air Atlantique's former ETPS Twin Pioneer, while a former RAF Chipmunk was even based here for some time. Sadly, the airport was poorly designed with a meagre runway of less than 4,000 feet, no radar, and a poisonous (almost self-fulfilling) business clause that provided for the site's re-sale for just £1 if it proved not to be financial viable after ten years. Not surprisingly, the airport was closed in April 2008 and sold for development. The runway is still intact but no longer used, and only a police helicopter frequents the site.

Mention should also be made of a small private landing strip that is sometimes referred to as Coal Aston. This is in fact at Apperknowle, and should not be confused with the much larger RFC site that was more than a couple of miles away.

Main features:
Runways: none, grass field. *Hangars:* eight RFC shed type. *Dispersals:* none. *Accommodation:* not known.

COPMANTHORPE, Yorkshire

53°54'43"N/01°07'39"W; SE574465; 36ft asl. 3 miles SW of York off Temple Lane, S of Copmanthorpe village

This relatively unknown site emerged during 1915 as a replacement for a similar facility that was already established on York racecourse. Designed as an operating base for RFC Home Defence units, the station hosted a variety of aircraft types, beginning with BE.2C aircraft operated by No 33 Squadron. These arrived early in 1916 and were joined by No 57 Squadron in June of that year. The resident BE.2Cs and Avro 504s were tasked with the air defence of the local area, most notably the airspace around York, which had already been subject to the attention of German Zeppelin raids. Two Flights were also established at nearby Tadcaster, but during August 1916 both squadrons moved en masse to this base, enabling No 76 Squadron to reform at Copmanthorpe. This unit remained active here until 1919, when it too moved to Tadcaster and the airfield was then abandoned.

The scattering of admin buildings and hangars was eventually demolished and the landing field was returned to agricultural use. Today there is virtually no evidence of any military presence at this site, although the general outline of the airfield can still be determined by the layout of the farmland just to the north of a small enclave of housing off Temple Lane. A small stretch of crumbling concrete among the farm buildings here may well be the only surviving trace of this long-abandoned site.

Main features:
Runways: none, grass field. *Hangars:* various, temporary. *Dispersals:* none. *Accommodation:* not known.

COTTAM, Yorkshire

54°03'58"N/0°28'45"W; SE996644; 485ft asl. 5 miles N of Driffield, W of B1249

Heading north along the B1249, a left turn onto York Road leads to a seemingly insignificant junction with a small farm track that heads off into the surrounding fields. There is almost no evidence to confirm that this was once the intersection of two bomber runways, part of Cottam airfield, which came into being during September 1939. The site was created as a satellite field for nearby Driffield, but by 1940 it had become established as a fully fledged bomber station.

However, it was not an ideal location for an RAF station, being perched on a hill with no prospect of good road access. The North Yorkshire Moors created harsh wind conditions that would inevitably cause problems for aircraft in the circuit, and by the time the airfield was completed it had been accepted that it could not be used for bomber operations. Towards the end of 1940 No 4 Group's Target Towing Flight arrived with Lysanders and Battles, but the airfield remained largely unused. Some visiting aircraft were seen here, most notably a USAAF B-24, which landed here while attempting to locate nearby Carnaby, but the airfield was left virtually abandoned until No 244 Maintenance Unit arrived in September 1944, at which stage it became an open storage area for bombs and ammunition. No 91 MU also used Cottam as a sub-site and moved its HQ here in 1947 as an Explosives Holding Unit, but by 1949 the HQ had moved again and Cottam returned to sub-site status until the early 1950s, at which stage the RAF presence gradually wound down and the site was abandoned for good.

An aerial view of the little-used airfield at Cottam.

Cottam was a classic example of poor planning, and represented what was in effect a very expensive mistake. The runways and perimeter tracks were never used for their intended purpose and today have almost been obliterated. Only part of the main runway remains visible close to

One of only a handful of surviving buildings at Cottam, pictured in 2011.

The seldom-used former main runway at Cottam, still visible in 2011.

Cottam House, although a farm track runs along the entire length of the main runway's position, while another track follows the path of a secondary runway. Just one circular dispersal can still be seen, as a reminder of the many concrete pans that were pointlessly constructed here.

Main features:
Runways: 084° 5,400 x 150 feet, 126° 4,200 x 150 feet, 013° 3,800 x 150 feet, concrete and wood chippings surface. *Hangars:* one T2. *Dispersals:* twenty-four 150-foot circular concrete, three 150-foot circular tarmac. *Accommodation:* RAF: Officers 60, SNCOs 214, ORs 783; WAAF: 118.

CROFT, County Durham

54°27'23"N/01°33'29"W; NZ287068; 175ft asl. SW of Dalton-on-Tees, off A167

Designed as a satellite airfield for RAF Middleton St George, the airfield site at Croft was constructed during 1941. Unusually, the village from which the station's name was derived was some distance away, the nearest village (Dalton) having already been associated with another airfield that was being developed there. To confuse matters still further, the local residents often referred to the airfield as Neasham. The standard three-runway bomber airfield was completed during 1941 and it opened as part of No 4 Group in October of that year.

No 78 Squadron became the first resident unit, equipped with Whitleys but converting to Halifaxes in March 1942. Like many other bomber units, No 78 was a major participant in the legendary Thousand Bomber raids that took place in the summer of 1942. Sir Arthur Harris knew that the future of Bomber Command was still in doubt and he approached both Winston Churchill and Sir Charles Portal with the bold idea of assembling a force of 1,000 bombers and sending them out in one massive raid on a German city. Churchill and Portal were both impressed, and they agreed. Although Harris had only a little over 400 aircraft with trained crews that were regularly used for front-line operational work, he did have a considerable number of further aircraft in the conversion units attached to groups with four-engined aircraft and in Bomber Command's own operational training units, 91 and 92 Groups. This secondary Bomber Command strength could be crewed by a combination of instructors, many of them ex-operational, and by men in the later stages of their training. To complete the 1,000 aircraft required, Harris asked for the help of his fellow commanders-in-chief in Coastal Command and Flying Training Command, and both officers were willing to help. Sir Philip Joubert of Coastal Command immediately offered to provide

250 bombers, many of them from squadrons that had once served in Bomber Command. Flying Training Command offered fifty aircraft, but many of these were later found to be insufficiently equipped for night-bombing and only four Wellingtons were eventually provided from this source.

All now looked well. The target figure of 1,000 bombers was easily covered and detailed planning for the operation commenced. The tactics to be employed were of major concern, not only for the success of this unprecedented raid but as an experiment upon which future operations could be based. The tactics eventually adopted would form the basis for standard Bomber Command operations for the next two years and some elements would remain in use until the end of the war. The major innovation was the introduction of a bomber stream in which all aircraft would fly by a common route and at the same speed to and from the target, each aircraft being allotted a height band and a time slot in the stream to minimise the risk of collision. The recent introduction of Gee made it much easier for crews to navigate within the precise limits required for such flying, although there would always be wayward crews who would drift away from the stream. The hoped-for advantage from the bomber stream was that the force could pass through the minimum number of German radar night-fighter boxes. The controller in each box could only direct a maximum of six potential interceptions per hour. The passage of the stream through the smallest number of boxes would, therefore, reduce the number of possible interceptions, particularly if the bomber stream could be kept as short as possible and pass through the belt of boxes quickly.

An aerial view of Croft taken during 1944.

This led on to the next decision, to reduce still further the time allowed for the actual bombing at the target. Where four hours had been allowed earlier in the war for a raid by 100 aircraft, and two hours had been deemed a revolutionary concentration for 234 aircraft at Lubeck, only 90 minutes were allowed for 1,000 aircraft in this coming operation. The big fear in these matters was always that of collisions, but on this occasion this was accepted in return for the opportunity to allow the bomber stream to pass through the night-fighter boxes quickly, to swamp the Flak defences at the target and, above all, to put down such a concentration of incendiary bombs in a short period that the fire services would be overwhelmed and large areas of the city would be consumed by conflagrations.

As in previous raids, the coming operation would be led by experienced crews whose aircraft were equipped with Gee. 1 and 3 Groups were selected to provide these raid leaders in the Thousand Bomber plan. But as the planning period came to an end, potential disaster struck. The Admiralty refused to allow the Coastal Command aircraft to take part in the raid. This was obviously a further step in the long-running battle between the RAF and the Royal Navy over the control of maritime air power, and the Admiralty realised that a success for this grandiose Bomber Command plan was not likely to help its prospects for building up a force of long-range aircraft for the war against the U-boat – and it was quite correct in that belief.

Harris now appeared to be falling well short of the dramatic figure of 1,000 aircraft with which he intended to carry out what was evidently a massive public relations exercise. Bomber Command therefore redoubled its efforts. Every spare aircrew member and aircraft was gathered in by the operational squadrons, but the decisive reinforcement came from Bomber Command's own training units, which committed more crews from the bottom half of their training courses. Every effort was made to provide the training crews with at least an experienced pilot, but forty-nine aircraft of the 208 provided by 91 Group would take off with pupil pilots.

When the operation was eventually mounted, 1,047 bombers would be able to take off, all but the four from Training Command being provided by Bomber Command's own resources, in spite of the fearful risk of sending so many untrained crews. When Churchill and Harris discussed the possible casualty figures, Churchill said that he would be prepared for the loss of 100 aircraft. The force about to be dispatched was more than two and a half times greater than any previous single night's effort by Bomber Command. In addition to the bombers, forty-nine Blenheims of 2 Group reinforced by thirty-nine aircraft of Fighter Command and fifteen from Army Co-operation Command would carry out Intruder raids on German night-fighter airfields near the route of the bomber stream.

Final orders were ready on 26 May, with the full moon approaching. The force stood ready, waiting for the weather. Harris hoped to use the 1,000-bomber force more than once if conditions permitted, before the extra aircraft gathered together were dispersed to their normal locations. His first choice of target was Hamburg, the second largest city in Germany, a great port and, an attraction for the Admiralty, builder of about 100 U-boats each year. But the weather over Germany was unfavourable for three days running, and on 30 May Harris had to decide to send the bombers to his second target choice, Cologne, the third largest city in Germany. Soon after noon on that day, the order to attack Cologne went out to the groups and squadrons and the raid took place that night.

The first Thousand Bomber raid was a great success, but a follow-up to Essen two nights later was not. The moon phase then passed and the training aircraft returned to their normal work, but they were recalled once more for a further massive raid on Bremen during the end of the June moon period, although the figure of 1,000 aircraft was not quite reached on that raid. Harris had originally hoped to assemble 1,000 aircraft for one or two raids in every moon period, but he abandoned this idea and the full 1,000 operation using so many training aircraft was not carried out again after the Bremen raid, although smaller numbers of training aircraft were called upon from time to time later in the year.

The Thousand Bomber raids certainly made their mark on history and were another great turning point in Bomber Command's war. The new tactics were mainly successful; there were never any serious casualties through collision, and the time over target would progressively be shortened until 700 or 800 aircraft regularly passed over the city they were bombing in less than 20 minutes! The morale of Bomber Command was certainly uplifted by this great demonstration of air power and by the wide publicity that followed. That same publicity also confirmed Bomber Command's future as a major force, and it can be said that, although there were bad as well as good times to come, Bomber Command never looked back after the Thousand Bomber raids. These events also placed Sir Arthur Harris firmly in the public eye, where, as 'Bomber' Harris, he would remain for the rest of his life.

The rest of the midsummer weeks passed with the front-line squadrons being pressed hard when the weather and moon conditions were favourable – and sometimes when they were not so favourable. The shorter nights again restricted raids to the coastal targets, the Ruhr and the Rhineland. There was another concentration of sustained effort against Essen in June, but this important target remained elusive of Bomber Command success. A similar campaign against Duisburg fared little better. There were minor operational changes. Harris started to restrict the practice whereby freshmen crews were introduced to operations gradually by being sent to lightly defended, close-range targets on the French coast. New crews were still allowed their one leaflet flight to France or Belgium, but after that they were expected to go to any target in Europe.

Harris was forced to agree to the temporary detachment of six more squadrons – even one of Lancasters – and one of his operational training units to Coastal Command to help with the U-boat war.

Air and ground crews from No 431 Squadron pose with their Halifax bomber at Croft during 1944.

There was a further draining away of operational effort when that unsatisfactory new aircraft, the Manchester, disappeared from 5 Group's order of battle at the end of June – although the Lancasters being sent to this Group would soon more than replace the loss. There were only minor changes in 2 Group, although one feature was to be the portent of a brilliant future for a new type of aircraft.

In the early morning after the first Thousand Bomber raid, five small twin-engined bombers of wooden construction flew to the smoking city of Cologne to take photographs and throw a few more bombs into that unhappy place. The de Havilland Mosquito had arrived. By the time the war ended, this aircraft would perform an undreamed-of range of tasks for Bomber Command.

The exact number of aircraft claiming to have bombed Cologne is in doubt; the Official History says 898, but Bomber Command's Night Bombing Sheets indicate that 868 aircraft bombed the main target, with fifteen aircraft bombing other targets. The total tonnage of bombs was 1,455, two-thirds of which was incendiaries. German records show that 2,500 separate fires were started, of which the local fire brigade classed 1,700 as large, but there was no sea of fire as had been experienced at Lubeck and Rostock because Cologne was mainly a modern city with wide streets. The local records contained an impressive list of property damaged: 3,330 buildings destroyed, 2,090 seriously damaged and 7,420 lightly damaged. More than 90 per cent of this damage was caused by fire rather than high-explosive bombs. From that total of 12,840 buildings were 2,560 industrial and commercial buildings, though many of these were small ones. However, thirty-six large firms suffered complete loss of production, seventy suffered 50-80 per cent loss, and 222 up to 50 per cent. Among the buildings classed as totally destroyed were seven official administration buildings, fourteen public buildings, seven banks, nine hospitals, seventeen churches, sixteen schools, four university buildings, ten postal and railway buildings, ten buildings of historic interest, two newspaper offices, four hotels, two cinemas and six department stores. Damage was also caused to twelve water mains, five gas mains, thirty-two main electricity cables and twelve main telephone routes. The only military installation mentioned is a Flak barracks. In domestic housing, the following dwelling units (mainly flats/apartments) are listed: 13,010 destroyed, 6,360 seriously damaged and 22,270 lightly damaged. These details of physical damage in Cologne are a good example of the results of area bombing. Similar results can be expected from those Bomber Command raids that were successful during subsequent years.

The estimates of casualties in Cologne are, unusually, quite precise. Figures quoted for deaths vary only between 469 and 486. The lower figure comprises 411 civilians and 58 military casualties, mostly members of Flak units; 5,027 people were listed as injured and 45,132 as bombed out. It was estimated that between 135,000 and 150,000 of Cologne's population of nearly 700,000 fled the city after the raid. The RAF casualties were forty-one aircraft lost, including one Wellington that is known to have crashed into the sea. The lost aircraft comprised twenty-nine Wellingtons, four Manchesters, three Halifaxes, two Stirlings, one Hampden, one Lancaster and one Whitley. The total loss of aircraft exceeded the previous highest loss of thirty-seven aircraft on the night of 7/8 November 1941, when a large force was sent out in bad weather conditions, but the proportion of the force lost in the Cologne raid – 3.9 per cent – though high, was deemed acceptable in view of the perfect weather conditions, which not only led to the bombing success but also helped the German defences.

Bomber Command later estimated that twenty-two aircraft were lost over or near Cologne: sixteen shot down by Flak, four by night-fighters and two in a collision. Most of the other losses were due to night-fighter action in the radar boxes between the coast and Cologne. Bomber Command also calculated the losses suffered by each of the three waves of the attack – 4.8, 4.1 and 1.9 per cent – and assumed that the German defences were progressively overwhelmed by bombing and affected by smoke as the raid went on. Further calculations showed that the losses suffered by the operational training unit crews – 3.3 per cent – were lower than the 4.1 per cent casualties of the regular bomber groups, and also that those training aircraft with pupil pilots suffered lower casualties than those with instructor pilots!

Another Victoria Cross was awarded for an action on that night. A Manchester of No 50 Squadron, piloted by Flying Officer L. T. Manser, was caught in a searchlight cone and seriously damaged by Flak on the approaches to Cologne. Manser held the plane steady until his bomb load was released and, despite further damage, set course for England, although he and his crew could have safely bailed out after leaving the target area. The Manchester steadily lost height, and when it became obvious that there was no hope of reaching England Manser ordered his crew to bail out, which they all did safely. In holding the plane steady for the last man to leave, Manser lost the opportunity to save himself and was killed. He is buried at Heverlee War Cemetery in Belgium.

After participating in these infamous missions over Germany, the squadron returned to Middleton in June 1942, Croft having been assigned to No 6 (RCAF) Group. In October of that year No 419 (Canadian) Squadron arrived with Wellingtons, although the unit soon moved to Middleton, its aircraft being reassigned to No 427 Squadron, which remained at Croft until May 1943, by which time it had flown some 261 operational sorties and had lost ten Wellingtons in the process. Croft now saw its first Halifaxes, with No 1664 HCU being formed here on 10 May for a period of intensive multi-engine training that continued until December.

The station then returned to operational duties and Nos 431 and 434 Squadrons arrived, remaining at Croft until the end of the war, by which time they had re-equipped with Lancasters. No 434 Squadron flew the majority of the station's operational missions and had achieved some 1,926 Halifax sorties by the end of the war (and lost twenty-eight aircraft), followed by 390 Lancaster sorties with three losses. Both of the resident squadrons were reassigned to the Tiger Force in June 1945, and both left Croft in order to retrain in Canada (although, with hostilities ending, they were soon disbanded).

The airfield was mostly disused from this time and was used only infrequently as a satellite field for No 13 OTU, which was operating Mosquitoes from Middleton. By the summer of 1946 the airfield was abandoned and the station closed towards the end of that year, after which the site remained disused. Today the airfield is still mostly intact, although most of the land has long since returned to agricultural use. The main runway is still largely intact, as is one of the secondary runways, which now forms part of a racetrack that runs across much of the former airfield. Almost all of the station's buildings have gone, but from the threshold of the old main runway it is all too easy to picture the countless Halifaxes and Lancasters that lumbered into the air here, laden with their deadly bomb loads.

Main features:
Runways: 085° 6,000 x 150 feet, 026° 4,200 x 150 feet, 145° 3,800 x 150 feet, tarmac surface. *Hangars:* two T2, one B1. *Dispersals:* thirty-six Heavy Bomber. *Accommodation:* RAF: Officers 199, SNCOs 425, ORs 1,018; WAAF: 250.

CROSBY-ON-EDEN, Cumbria

54°56'14"N/02°88'40"W; NY481605; 160ft asl. 5 miles E of Carlisle, on A689

Created as a base for Operational Training Units, the airfield at Crosby-on-Eden first opened on 20 February 1941 as part of No 81 Group. Flying commenced in the same month when No 59 OTU arrived from Turnhouse with a large fleet of Hurricane fighters (more than seventy) in addition to a handful of Masters and a couple of Battle target tugs. Training activities quickly built up and activity soon required a satellite airfield to be established at nearby Longtown. The OTU remained at Crosby until August 1942, when it was replaced by No 9 (Coastal) Operational Training Unit with a fleet of Beauforts and Ansons, which arrived from Aldergrove.

As part of No 17 Group the airfield quickly became busy again after a modest extension of the main runway was completed (which necessitated the closure of a local road). Beaufighters began to arrive towards the end of 1942, and by the spring of 1943 the OTU was busy training crews on a mixed fleet of Beaufighters and Beauforts; this training activity continued until August 1944, when the unit finally disbanded. However, the unit was renamed on the same day as No 109 Operational Training Unit and it quickly re-equipped with Dakota transports as part of No 44 Group, tasked with the multi-engine training of transport crews. As part of this task a small number of Horsa gliders were also used by the unit for glider towing training, although the bulk of the unit's aircraft

A large number of Czech child evacuees were ferried to Crosby-on-Eden by No 570 Squadron, courtesy of its mighty Stirling aircraft.

comprised a large fleet of more than thirty Dakotas. The unit disbanded on 10 August 1945 to become No 1383 (Transport) Conversion Unit, and remained in business until 6 August 1946, when it was finally disbanded for good. The station then closed and the airfield was abandoned.

However, only a decade later the airfield was reopened for flying as a fledgling airport serving Carlisle and the surrounding area. The airport struggled to survive but private and business flying continued through successive years and the site remains active to this day as Carlisle Airport. The wartime airfield remains largely unchanged and even the old T2 hangars have survived, with both the main and a secondary runway still actively used for flight operations. The Solway Aviation Museum was established here and a small collection of aircraft (including a rare former RAF Phantom) is on display adjacent to the airport terminal. More recently the site was also used to stage a huge pop concert, and while television viewers watched the very modern Lady Gaga perform, glimpses of Crosby's military history could still clearly be seen in the background.

Main features:
Runways: 252° 6,000 x 150 feet, 192° 4,200 x 150 feet, 315° 4,200 x 150 feet, tarmac surface. *Hangars:* five Blister, three Bellman, three T2. *Dispersals:* thirty-three fighter type, nineteen loop, two 150-foot circular, four 100-foot circular. *Accommodation:* RAF: Officers 156, SNCOs 339, ORs 1,588; WAAF: 254.

CULLINGWORTH, Yorkshire

53°48'59"N/01°54'08"W; SE065355; 875ft asl. 3 miles S of Keighley at junction of A629 and B6429

This relatively unknown site was one of many locations selected to act as a temporary base for RFC fighters assigned to the air defence of the local area. It was declared fit for flight operations in 1916 and No 33 Squadron deployed a small number of aircraft here on a regular basis. No 76 Squadron took over from No 33 in October 1916 and fighter detachments continued from Cullingworth as required. The site was usually inactive, however, and when hostilities ended in 1918 it was quickly abandoned and all traces of the RFC's presence quickly disappeared. Today the former airfield site is still clearly visible beyond a cluster of farm buildings, but there is nothing here to remind visitors of the site's connections with Britain's military history.

Main features:
Runways: none, grass field. *Hangars:* none. *Dispersals:* none. *Accommodation:* not known.

DALTON, Yorkshire

54°10'38"N/01°21'39"W; SE418759; 68ft asl. 6 miles NE of Ripon, at junction of A167 and A168

This bomber airfield was constructed in an unusual location, bounded by surrounding roads and a river. It was also extremely close to its parent airfield at Topcliffe and another airfield at Skipton-on-Swale, which inevitably led to difficulties with conflicting traffic in such a small area.

This wartime image of Dalton shows what appears to be a FIDO fog dispersal system laid out on either side of the main runway. Also visible is the way in which the east-west runway has been significantly extended to form what became the main runway.

The station opened in November 1941 as a grass field, acting as a satellite to Topcliffe, the first aircraft arriving that month in the shape of Whitleys from No 102 Squadron. Conversion to Halifaxes took place at Dalton (a Conversion Flight forming here), and the unit moved to Topcliffe in June 1942, after which Nos 76 and 78 Squadron's Conversion Flights arrived with more Halifaxes. No 1652 HCU was established at Dalton in July, but the airfield closed on 31 August for reconstruction. It reopened in November 1942 with three concrete runways (which were shorter than standard, thanks to the airfield's confined location), and No 428 Squadron was established here with Wellingtons. In January 1943 Dalton was transferred to No 6 RCAF Bomber Group, and No 424 Squadron arrived from Leeming during May for a short two-week stay. When it left, No 428 Squadron remained active here until June, when it moved to Middleton St George.

This Wellington from No 438 Squadron suffered severe damage from anti-aircraft fire on a raid to Duisburg in April 1943.

Jack Powley was a wireless operator with the unit and his flight log recalls his vivid account of the squadron's raid on Duisburg:

> 'Airborne at 2125 and so we go once again for what is know as "Happy Valley". Surrounded by the same belt of searchlights as Essen and so we must expect a warm reception. Weather reports are once again not very reliable. We fly above cloud all the way across the sea, having been told at briefing that we should expect to find the cloud broken over the target. However, as we approach the target area we find we are still in 10/10 cloud and can't see a damned thing – not even the Flak bursting. We see a glow on the clouds below us and knowing we are near the target we figure the glow must be from fires on the deck. We decide to make our run in and all at once we are boxed in by heavy Flak. Willie [Sgt Williamson] takes violent evasive action but it gets hotter. Then all at once we sure stopped something. The old kite shakes from end to end and we begin to lose height. We get rid of our bombs but still we lose height and Willie is fighting like hell to control her. At this time he shouts, "OK fellows, grab your chutes!" We were hit at 2320 at a height of about 14,000 feet and lost height rapidly to 10,000. The kite starts to behave a bit better and Willie seems to have her more or less under control but she won't climb at all. But as long as the damned thing will fly that's all we worry about. If anyone can get us back to England then Willie can – so we hope for the best.

Crumbling remains are all that survive at Dalton, as illustrated by this 2006 image.

> There is no reply from Bertie [Sgt Bertrand], our rear gunner, so Watty [Sgt Watkins, navigator] goes back to see what is wrong. He almost walks out of the end because the turret had disappeared completely. Poor Bertie, he didn't even know what hit us. We tried to follow the route out as ordered, that is across Belgium and Northern France. Watty hadn't much to navigate on apart from a few "loop bearings" I was able to get him. Searchlights try to pick us up and there wasn't much we could do about it, not being able to take evasive action. As Willie said at the time, "They don't give a guy a bloody chance."
>
> For over two hours we flew over enemy territory in this state and gradually losing height most of the way. We crossed the French coast at about 0120 (although we didn't see it) at 2,000 feet. Here the searchlights coned us and Willie put her nose down and made for the drink. The question was now, would she pull out? At 900 feet Willie pulls her up and the searchlights have lost us. We didn't see the English coast when we crossed it. We shot off a couple of distress cartridges and eventually searchlights directed us to some place. Gas was getting pretty low and we had to get down. Before coming in to land we took up crash positions because the hydraulic system was u/s. Willie put the old bus down as though she were an injured baby, and boy, was she injured! To Willie I say, "A bloody good show." Not forgetting Watty for his DR navigation. Some of his equipment was blown up together with our serial connections. Not forgetting Park [Sgt Parker] either for his great efforts. As for Bertie, only a miracle could have saved him and I say may he rest in peace. Duration of trip: 5hrs 20mins.'

After No 428 Squadron's departure the station then shifted to training activities, and No 1666 Heavy Conversion Unit formed here on 15 May 1943 with Halifaxes. After only a few months the unit left, and in July No 1691 (Bomber) Gunnery Flight arrived with Spitfires and Hurricanes (and a handful of Martinets). No 420 Squadron moved to Dalton in October 1943 for a short stay, but when it departed in December Dalton's flying activities were restricted only to those performed by the Gunnery Flights. An Aircrew School (a non-flying unit) was established here, but by the summer of 1945 even this remaining unit disbanded and flying at Dalton came to an end. No 91 Maintenance Unit used the airfield for bomb storage during the early 1950s, but flying activities effectively ended in 1945, the station closing in 1955.

A significant amount of the airfield structure has survived and most of the former main runway can still be seen, together with portions of the secondary runways, dispersals and perimeter track. The former technical and admin site is now lost under a large industrial area, but two of the hangars have survived. A gaggle of rusting trucks lurks within a tree-lined yard where the main runway's threshold once stood, and where the mighty Halifaxes once roared by.

Main features:
Runways: 297° 5,400 x 150 feet, 236° 4,680 x 150 feet, 176° 4,200 x 150 feet, tarmac surface. *Hangars:* two T2, one B1. *Dispersals:* thirty-six Heavy Bomber. *Accommodation:* RAF: Officers 60, SNCOs 288, ORs 787; WAAF: 216.

DISHFORTH, Yorkshire

54°08'26"N/01°25'02"W; SE381717; 108ft asl. 2 miles N of Boroughbridge, on A1(M)

Dishforth was one of a group of airfields constructed in this area of Yorkshire, scattered around the Vale of York. Work on creating an airfield close to the village of Dishforth began in the mid-1930s, and by 1936 a large bomber airfield had been completed with a grass surface of some 200 acres with a crescent of Type-C hangars fronting a conventional bombing circle, five in number with the fifth tucked in behind the fourth and backed by the technical workshops and administrative offices. Behind these were the communal buildings and barracks, all in brick, steel and concrete and centrally heated. The station opened in September 1936, and by January 1937 work was far enough advanced for No 10 Squadron and its Heyfords to be brought up from Boscombe Down; the following month No 78 Squadron arrived from the same station with the same aircraft type.

An aerial reconnaissance image of Dishforth showing construction of the airfield still under way. Only a small portion of the main runway has been laid, although the perimeter track and dispersals are already clearly evident. Also of interest is the camouflage paint applied to the hangars.

In March No 10 Squadron became the first to convert to the Whitley, then considered a heavy bomber. No 78 also received Whitleys, but the supply was slow and it was not fully equipped until the summer. An the outbreak of hostilities No 78 was

A gaggle of RAF fighters take a short break at Dishforth during a cross-country flight in 1937. One of Dishforth's resident Heyfords is visible in the foreground. Phil Jarrett

designated as a reserve squadron in No 4 Group, so it did not go to war from this station. However, No 10 Squadron dispatched the first sorties from Dishforth on 8 September when eight Whitleys were sent to distribute leaflets over Germany. No 78 was moved to Linton-on-Ouse in December 1939 in an exchange with No 51 Squadron. In July 1940 No 10 Squadron moved to Leeming and No 78 returned from Linton to embark upon operations from Dishforth. No 78 operated here until the following April, when it left for Middleton St George. No 51 Squadron remained at Dishforth until May 1942, when it was assigned to Coastal Command and shifted to Chivenor.

During the early war years a number of pan-type hardstandings were laid down on the airfield, some of which were built in an adjacent field on the western side of the Great North Road. Bomber Command's plans to create a Royal Canadian Air Force Group resulted in No 425 Squadron forming at Dishforth in June 1942, flying Wellingtons. It undertook its first raid on the night of 5 October, and ten days later a second RCAF squadron (No 426) was formed here, also flying Wellingtons and becoming operational on the night of 14/15 January 1943. At this time Dishforth came under No 6 Group, the formal transfer having occurred on New Year's Day. No 425 was one of the Wellington squadrons picked to increase bombing capability in the Mediterranean, and it left Dishforth in May for North Africa. The following month its sister squadron was transferred to Linton-on-Ouse so that hard runways could be laid at Dishforth.

This work (carried out by F. Haslam Ltd) took nearly six months and brought the airfield up to Class A standard. The main runway was 1,976 yards long with the shorter secondary runways at 1,500 and 1,488 yards. The existing perimeter taxiway was enlarged, and the pan hardstandings built during the preceding year were linked to it, increasing the total to thirty-six. Additional domestic buildings brought the maximum personnel that could be accommodated to 1,782 males and 332 females, making this one of the RAF's largest operational stations.

At the beginning of November 1943 No 1664 Heavy Conversion Unit and its Halifaxes moved in from Croft, soon to be joined by No 425 Squadron, which had returned from North Africa. Re-equipped with Halifaxes, No 425 departed for Tholthorpe a month later. For the rest of the war in Europe, Dishforth was to maintain a training role, No 1664 HCU eventually being disbanded in early April 1945. Combined with the break-up of No 6 Group in June, this meant that the station was soon abandoned by the Canadians. The war had seen 128 of Dishforth's bombers go missing or crash in the UK, comprising ninety Whitleys, thirty-seven Wellingtons and a single Halifax.

Later in the summer of 1945 Transport Command took over the station and Nos 1659 and 1665 Heavy Transport Conversion Units arrived, later merged into No 1332 Transport Conversion Unit. These units were engaged in training crews to operate Liberators and Yorks. In 1948 another change of designation for the resident unit came with the formation No 241 Operational Conversion Unit. The ubiquitous Hastings transport entered RAF service at Dishforth, and both Nos 47 and 297 Squadrons trained on the type here before moving to Germany to participate in the Berlin Airlift. No 240 OCU arrived in March 1951 with Valettas, merging with the resident unit to form No 242 OCU, which operated a large fleet of various transport types, often requiring the use of Relief Landing Grounds at Rufforth and Topcliffe.

The mighty Beverley transport arrived at Dishforth during April 1957, and transport operations continued here until January 1962, when the OCU departed for Thorney Island (No 30 Squadron having left in November 1959). The airfield then became relatively inactive with only the Chipmunks of the Leeds UAS and 9 AEF based here until 1966, when flying ended completely. However, Dishforth soon became active again, albeit in a much reduced capacity, as a Relief Landing Ground for Linton-on-Ouse, where Jet Provost training was consolidated during the 1960s and 1970s. Jet Provosts were seen in Dishforth's airfield circuit almost every day, and although the RAF's training activities have dwindled considerably in recent years Linton's trainers (now turboprop-powered Tucanos) still use Dishforth's runways when required. However, the station is now under Army control and is devoted largely to helicopter operations with Lynx and Apache aircraft now occupying the huge apron where the monstrous Beverleys once stood.

The Great North Road has been rebuilt as the A1(M), but thanks to Dishforth's slightly elevated position a great deal of the airfield can still be seen from the motorway. The hangars are still very much in use and the post-war control tower (placed rather unusually on the opposite side of the airfield from the hangars) is still active.

Main features:
Runways: 340° 6,000 x 150 feet, 220° 4,200 x 150 feet, 280° 3,800 x 150 feet, concrete and tarmac surface. *Hangars:* Five C-Type. *Dispersals:* thirty-six Heavy Bomber. *Accommodation:* RAF: Officers 260, SNCOs 360, ORs 1,479; WAAF: 478.

DONCASTER, Yorkshire

53°30'49"N/01°06'32"W; SE591022; 30ft asl. Off A368 Bawtry Road, S of Doncaster Racecourse

The very first 'Aviation Gathering' took place at Doncaster Racecourse in October 1909. The original caption explains that M Delagrange's Bleriot monoplane is being prepared for a demonstration flight in front of the grandstand. Phil Jarrett

Doncaster enjoys a significant (if little-known) place in Britain's aviation history, as the site where the very first British 'air show' was staged. Doncaster Racecourse hosted an 'Aviation Gathering' during October 1909 and both fixed-wing aircraft and balloons were seen over the area for a few days. However, it was not until 1916 that a more significant aviation presence came to Doncaster when a Royal Flying Corps site was established on the flat fields embraced by the racecourse.

No 15 Reserve Squadron was assigned to the site, tasked with training and the air defence of the local area, with the threat of Zeppelin raids on the nearby town of Doncaster being evident. Some technical and admin buildings (including hangars) were constructed, and in July 1918 the site became No 47 Training Depot Station. Activities continued until the middle of 1919, by which time the airfield was no longer required and was duly abandoned, returning to the more peaceful pursuit of horse-racing.

A decade later, Doncaster Council decided that a municipal airport should be built for the town and, as the RFC site was long abandoned, it was decided that a larger site could be created adjacent to the Racecourse on the opposite side of Bawtry Road. The grass field opened for civil flying in 1939, although some military activity was already in place by this stage with No 616 Squadron RAuxAF having made its headquarters here.

When war was declared later that year the airfield was assigned to the RAF and quickly became a Relief Landing Ground for RAF Finningley, which was situated just a few miles to the east. No 271 Squadron took up residence with a substantial fleet of aircraft tasked with transport duties, and

Above: *A Cody Mk 1 aircraft performing for the thrilled crowds at Doncaster Racecourse.*

Above right: *Doncaster's unusual airport hangar, long since demolished.*

Right: *The long-gone control tower at Doncaster.*

although the Squadron operated detachments at various sites Doncaster was also designated as a detachment when the unit moved its HQ to Down Ampney in February 1944. The diverse range of aircraft types operated by the Squadron reflected its role, which encompassed the operation of civilian aircraft that had been requisitioned for military wartime use; Doncaster frequently hosted a large and varied selection of types, which included HP.42s, Hudsons and Dakotas. The Squadron also operated a fleet of Harrow air ambulances, known locally as 'Sparrows'.

Brooklands Aviation (later joined by Scottish Aviation) created a maintenance base at the airfield as part of the Civilian Repair Organisation. Various aircraft types were repaired and modified at the facility, most notably Wellington bombers. While these aircraft came and went, the resident transport squadron kept the airfield busy, and Finningley continued to use the site as a satellite when required. The RAF's Dakota Modification Centre was established at Doncaster in March 1944, and for some months the airfield was littered with numerous examples of the aircraft, most being new-build aircraft delivered from America and modified at Doncaster for RAF service. However, by the end of 1944 the Dakota activities had transferred to Kemble and flying activity slowly wound down until only No 9 Refresher Flying School's Tiger Moths remained as residents. These were later replaced by Chipmunks and Prentices, but by the 1960s these too had gone and, with the RAF station here having closed in 1945, the last connections with military flying were severed and the site was fully returned to civilian hands

Private flying developed here for many years and Doncaster became a popular and active site. But the expansion of the adjacent town continued eastwards, and by the 1980s it was clear that the airport's future was far from secure. With no will for investment or improvements, the site slowly fell into decay and by the early 1990s most private flying had ended. In 1994 the site was formally closed and the few airport buildings were soon demolished. The grass landing field was drastically redeveloped and has now disappeared under a vast business and commercial trading estate, punctuated by housing and landscaped grass and wet land. All that remains of the former RAF base is a small area of concrete, some wooden huts and a hangar that is now the home of Aeroventure, a modest museum collection embracing a large number of aircraft components and pieces of wreckage, but also numerous helicopters, a Hunter, Lightning, Sea Prince, Gannet, Harrier and Jet

Provost. A Dakota is also slowly being reassembled here, reminding visitors of the days when the area was once inhabited by countless examples of this famous aircraft type.

Main features:
Runways: N/S 4,000 feet, SE/NW 4,000 feet, NE/SW 3,600 feet, grass surface. *Hangars:* three Bellman. *Dispersals:* thirteen circular. *Accommodation:* RAF: Officers 58, SNCOs 247, ORs 664; WAAF: 154.

DONNA NOOK, Lincolnshire

53°28'30"N/0°09'14"W; TF430997; 10ft asl. Off Marsh Lane, NE of North Somercotes

Donna Nook is a name that is inevitably linked to the Royal Air Force's scattering of weapons ranges, which can be found at various points along the British coastline. However, few people are aware of this site's original use as an active airfield. The site was first used by the RAF in 1926 when a bombing and gunnery range was set up on the beach, and it remained in use throughout the Second World War with a wide variety of aircraft types making use of its target facilities.

In 1940 an area of land just to the south of the range buildings was cleared for use as an airfield, primarily for communications aircraft associated with the range and as a suitable emergency landing ground for aircraft in the region. The main user was No 1 Air Armament School, based at nearby Manby, although the airfield was used only sporadically; indeed, it was designated as a wartime decoy site, even though it had its own decoy field at Marsh Chapel. North Coates eventually adopted the airfield as a Relief Landing Ground, and that station's Mobile Torpedo Unit relocated to Donna Nook for a short period. It was also used by No 61 Maintenance Unit when the war ended, but by 1948 the airfield was effectively abandoned.

A USAF F-111 delivers a stick of retarded bombs over the Donna Nook range during the 1980s.

However, the RAF maintained its presence at the nearby range even though most of the range traffic shifted a few miles south to Theddlethorpe where better Gee and Gee-H facilities were available. By 1973 the growing popularity of Mablethorpe as a holiday resort resulted in an expansion of housing and camp sites, which rendered Theddlethorpe too dangerous for further use and the range was closed, enabling Donna Nook to assume the role. Through the 1980s and 1990s the range was very busy with a steady supply of customers using the beach and raft targets, all monitored and calibrated from a watch tower on the coast. Most RAF and NATO types were seen here, perhaps the most spectacular being the USAF's A-10 Thunderbolts, which inevitably caused a stir when their huge Gatling guns were fired.

Today the range is much less active, as military operations across the UK have wound down. However, it is still used on a daily basis, both for weapons drops and for flare launches. Spectators are welcome to watch the occasional periods of activity from the coast, and many visitors come to the range when the local seal population takes over at various points in the year. The nearby airfield is, however, long gone, and only a few crumbling concrete structures remain among the coastal scrubland.

Main features:
Runways: NE/SW 4,200 x 50 feet, NW/SE 4,200 x 50 feet, grass surface. *Hangars:* Two Blister, one T2. *Dispersals:* none. *Accommodation:* RAF: Officers 67, SNCOs 115, ORs 1,303; WAAF: 104.

DRIFFIELD, Yorkshire

53°59'42"N/0°28'56"W; SE996564; 75.5ft asl. 2 miles SW of Driffield on A614

An early image of Eastburn's No 21 Training Depot, which eventually became RAF Driffield. The cause of this aircraft's unfortunate fate is unknown, but landing and take-off accidents were common during the site's early years. Barry Ketley

RAF Driffield dates back to 1917 when an airfield was established on a site at Eastburn that had been used for some time as a landing ground for use by locally based Home Defence units operated by the Royal Flying Corps, particularly No 33 Squadron. After the site was developed into a more permanent airfield the first unit to be based here was No 2 School of Aerial Fighting, which arrived in October 1917; it soon became No 16 (Training) Group, and moved to Marske. Meanwhile part of the unit had been developed into No 3 School of Aerial Fighting & Gunnery, but this too quickly left for Eastburn, and it was not until July 1918 that the airfield became active again as No 21 Training Depot Station with a variety of almost 100 trainer aircraft on strength, such as Pups, 504s and SE.5a fighters.

After the First World War a couple of operational units arrived here in 1919 (Nos 202 and 217 Squadrons), but by early 1920 operations had wound down and the station closed during the summer. The airfield, with its many buildings (including substantial flight sheds), was held in reserve for some time, but was eventually abandoned. However, little more than a decade later the site was identified as suitable for development into a new bomber airfield, and in 1935 construction began of a larger airfield, new admin, domestic and technical buildings, and five huge C-Type hangars.

No 77 Squadron Whitleys shiver in the snow at Driffield.

No 426 Squadron's Liberator transports at Driffield.

A No 104 Squadron Wellington pictured at Driffield in 1941.

The first aircraft to appear at the new RAF Driffield were Vickers Virginias from Nos 58 and 215 Squadrons, which came here early in September 1936 from Worthy Down to form the new Driffield Bomber Wing. The airfield had in fact been under construction for some time and had officially opened some two months previously; although the local area remained quiet for some weeks, bomber operations soon built up and the station was very active by 1937 when B Flight of No 58 Squadron was expanded to become No 51 Squadron, after which it moved to Boscombe Down. Meanwhile B Flight of No 215 Squadron became No 75 Squadron on 15 March, but stayed at Driffield as part of the Bomber Wing. During the following year the Virginias were gradually replaced by Whitleys.

When the Second World War began, two Whitley units (Nos 77 and 102 Squadrons) were at Driffield, and both squadrons embarked on 'Nickel' (leaflet) raids and bombing missions over Germany and (eventually) Italy. Not surprisingly, the station's importance was recognised by the Luftwaffe and it suffered from a heavy raid in August 1940 that temporarily closed it for some months. Hangars were damaged and more than a dozen Whitleys written off.

Whitleys from No 102 Squadron at rest on the grass at Driffield. Of interest is the variation in the style of national insignia applied to the two aircraft. Barry Ketley

A panoramic view of the local area around Driffield, taken in 1945.

The station reopened for a third time in January 1941 and for a brief period a succession of fighter units were based here, including No 485 Squadron with Spitfires and 213 Squadron with Hurricanes. Later in the year bombers returned, and in May 1941 the station's aircraft resumed operational missions over Germany. Wellingtons became the most common sight on the airfield, although when No 1502 BATF and 1484 TTF arrived aircraft such as the Oxford, Martinet and Lysander were also commonly seen. The airfield remained very busy until December 1942, when it was closed to flying so that concrete runways could be laid together with taxiways and dispersals to accommodate heavy bomber aircraft.

A 1960s image of Driffield showing the Thor launch pads and a visiting MAC C-124 Globemaster.

The mighty C-Type hangars at Driffield, still in use but no longer accommodating aircraft.

Driffield reopened yet again on 6 June 1943 and a variety of training units moved in, although it was not until August 1944 that the first heavies arrived in the shape of Halifaxes belonging to No 466 Squadron, followed by No 462 Squadron. Operations continued from here until the end of the war, when the Halifax units were reassigned to transport operations; these occupied Driffield until the summer of 1945, when the station was placed under Care and Maintenance.

Once again the station survived, and in September 1946 No 10 Air Navigation School arrived from Swanton Morley with a mix of Wellingtons and Ansons. As part of Flying Training Command, Driffield hosted other units including No 204 Advanced Flying Training School, but it was No 203 AFS that brought the first jet aircraft to Driffield when its Meteors and Vampires arrived in September 1949. The unit temporarily moved to its Relief Landing Ground at Carnaby while Driffield's runways were improved for jet aircraft, after which it returned as No 8 FTS in June 1954. Driffield then transferred to Fighter Command and No 219 Squadron (followed by No 33 Squadron) arrived with Venom night-fighters. The unit's short stay was followed by the arrival of the Fighter Weapons School from Leconfield with Meteors, Hunters, Vampires, Venoms and Sabres, but by March 1958 the unit had gone and Driffield was silent once more.

The end of flying activities here was not the end of Driffield's history, however, as it became a Thor nuclear missile base with three launch pads being built adjacent to the runways. No 98 Squadron arrived in August 1959 and the station maintained a missile alert facility until 1963. The airfield was mostly inactive, although occasional aircraft were still to be seen, including occasional visits by USAF transport aircraft, delivering or removing missiles for deployment here or at other missile sites. With the withdrawal of the Thors the station was again placed under Care and Maintenance, but it became active again for a few months from September 1967 when Hawker Siddeley deployed its test fleet (mostly Buccaneers) here while the company's airfield at Holme-on-Spalding-Moor was improved.

When the Buccaneers departed in 1968 the airfield fell silent for many years and it was not until 1977 that the Army assumed control of the base. Most of the station's admin and technical buildings were retained for use by the Army, but sadly the airfield was eventually destroyed, the runways, taxiways and Thor pads being removed in order to create a driving facility for military vehicles. Today the huge hangars still dominate the site of the former airfield and the main concrete apron is still to be seen. Sadly, only small tracks trace the position of the old runways and perimeter track.

Main features:
Runways: 236° 6,000 x 150 feet, 281° 4,200 x 150 feet, 336° 4,200 x 150 feet, concrete surface. *Hangars:* five C-Type. *Dispersals:* thirty-six Heavy Bomber. *Accommodation:* RAF: Officers 189, SNCOs 373, ORs 1,322, WAAF: 422.

DUNHOLME LODGE, Lincolnshire

53°17'29"N/0°30'12"W; SK998781; 118ft asl. 5 miles N of Lincoln between A15 and A46, NW of Nettleham

Fans of the world-famous Red Arrows who gather at the eastern perimeter of RAF Scampton's airfield to marvel at the Royal Air Force Aerobatic Team's daily rehearsal flights are probably unaware that just to the east of the adjacent A15, across the rolling fields, lies the site of a famous wartime RAF airfield.

Situated just 2 miles from RAF Scampton, the airfield at Dunholme Lodge was constructed in 1942 as part of wartime expansion plans with a standard layout comprising three runways and a scattering of bomber dispersals. Unusually, one of the secondary runways was afforded a slightly greater length than standard, but in all other respects the airfield was completed as a typical bomber aerodrome with a bomb bump, three hangars and an extensive technical and admin site. Scheduled for development to Class A standard, work began on runway construction in September 1942 with Wimpey as a major contractor. The three runways were 04-22 of 2,000 yards, 16-34 at 1,400 yards, and 10-28 at 1,500 yards. However, this last was extended to 1,700 yards by No 5002 Airfield Construction Unit in February-April 1944. Twenty-five pan hardstandings and fourteen loop type were provided along the perimeter track, another three of the original pans being isolated and not linked to taxiways. The technical site was north-east between runway heads 22 and 34 near the Lodge. One T2 hangar stood on the technical site, a second further south between heads 04 and 34, near the bomb stores, and a B1 between heads 04 and 10. Dispersed in the countryside around Welton were seven domestic and two communal sites and the sick quarters, providing for a maximum 1,637 males and 468 females.

A No 44 Squadron Lancaster preparing to depart on a mission to Berlin.

Opening in May 1943, the first aircraft had arrived during the previous summer when No 1485 Flight and No 5 Group's Air Bomber Training Flight operated a handful of target facilities and trainer aircraft here. However, the first bomber squadrons to arrive were Nos 44 and 619 Squadrons, which transferred from Waddington and Coningsby in May 1943 and April 1944 respectively. Dunholme Lodge was reopened after the airfield was reconstructed in May 1943 as part of No 5 Group, and became the home of the veteran No 44 Squadron – removed from its long association with Waddington. No 44's Lancasters were not joined by another squadron until April the following year, when No 619 moved in from Coningsby.

In September 1944 the station was reallocated to No 1 Group, and both Lancaster squadrons moved out, to Spilsby and Strubby respectively. The following month No 170 Squadron, recently reformed as a Lancaster squadron at Kelstern, took up station, but its stay was little more than a month before it was moved to Hemswell. Unconfirmed reports state that the airfield was vacated to reduce local night congestion, there being several other stations close by with overlapping circuits. This may be true, as the airfield was then used mostly for daylight traffic, namely the reception, storage and delivery of Hamilcar gliders.

During Bomber Command's offensive operations from Dunholme Lodge 120 Lancasters either failed to return or were destroyed in crashes. Operations with Lancasters had commenced immediately when the airfield reopened in 1943, and Dunholme was also designated as a satellite airfield for Scampton – hardly surprising when the two airfields were so close to each other and the busy air traffic pattern had to be coordinated with great care. When No 619 Squadron departed to Spilsby in

September 1944 the Lancasters of No 170 took their place for a couple of months, but by the end of the year all the Lancasters had gone and Dunholme's flying activities diminished significantly, with only aircraft from Scampton using the airfield as a satellite.

It seems likely that it was Scampton's close proximity that led to the seemingly premature wind-down of operations at Dunholme, but the airfield remained in use as required by Scampton, and General Aircraft also used the site for glider modification work until the end of the Second World War. Flying activity ended completely in 1945, although the RAF retained the site and, after many years of inactivity, a portion of the airfield was rebuilt to accommodate the Bloodhound missiles of No 141 Squadron, which arrived here in 1959. The missile battery (together with support facilities) operated from Dunholme until March 1964, when the missiles were removed and the station was finally closed down. It remained abandoned for some time and eventually the site reverted to agricultural use.

A Bristol Bloodhound surface-to air missile at Dunholme Lodge.

Today almost all of the RAF's presence has been obliterated, with only a small portion of the secondary runway still intact, although Horncastle Lane runs along most of what was once the shortest of the three runways. The northernmost curve of this road (as it turns west towards the A15) marks the point where this runway once intersected with the main runway. With all traces of the airfield gone, it is almost impossible to imagine that at this very point, nearly seventy years ago, Lancasters roared by and lifted their heavy loads skywards towards Germany.

Main features:
Runways: 220° 6,000 x 150 feet, 164° 5,100 x 150 feet, 278° 4,200 x 150ft, concrete and tarmac surface. *Hangars:* two T2, one C-Type. *Dispersals:* thirty-six Heavy Bomber. *Accommodation:* RAF: Officers 144, SNCOs 369, ORs 1,124; WAAF: 468.

DUNKESWICK, Yorkshire

53°55'13"N/01°32'19"W; SE303472; 216.5ft asl. 4 miles S of Harrogate, W of A61 between Harewood and Weeton

This site was one of many in the area assigned to Home Defence activities during the First World War. Although never technically designated as an airfield or landing ground, the site was cleared of obstacles and an area of some 40 acres was made available to RFC aircraft for use as required. These sites could be used either by day or night, although as Dunkeswick was a particularly rudimentary facility it seems unlikely that any aircraft were ever accommodated here beyond daylight hours.

The first unit to make use of the site was No 33 Squadron, which deployed some of its aircraft here in 1916 while defending the area against German Zeppelin raids. In October 1916 the defence role was transferred to No 76 Squadron, and aircraft from this unit were also seen at Dunkeswick until the end of the war, at which stage the site was quickly abandoned and swiftly returned to agriculture. With no permanent structures having been built here, there is nothing visible today to indicate its use for flying so many years ago. Off the A61, Weeton Lane and Green Lane trace the perimeter of the field where the RFC fighters could once be seen.

Main features:
Runways: none, grass field. *Hangars:* none. *Dispersals:* none. *Accommodation:* not known.

EAST MOOR, Yorkshire

54°04'10"N/01°05'08"W; SE599641; 88ft asl. 7 miles N of York, E of B1363, SW of Sutton-on-the-Forest

Left: *This interesting overhead view of East Moor shows how the initial triangular runway layout was subsequently developed to enable the north-south runway to become a much-extended main runway. As can be seen, the majority of the dispersals are situated to the south of the airfield.*

On 8 August 1943 no fewer than three Halifaxes suffered accidents, resulting in a major raid being scrubbed.

Constructed during 1941 on the site of East Moor Farm, this airfield was built to typical standards for new bomber airfields, although the presence of local roads and the large area of woodland to the south-east meant that the layout and dimensions of the airfield's runways were somewhat unusual. The main runway was slightly shorter than standard so that it could be positioned between High Carr Wood (used as a bomb dump) and Carr Lane, which ran along the perimeter of the airfield's southern reaches.

The first aircraft to arrive here were the Halifaxes of No 158 Squadron, which moved here from Driffield on 6 June 1942. Having just begun conversion from Wellingtons, most of the unit's initial flying was assigned to training, but by the end of the month their Halifax crews were participating in the last of the famous Thousand Bomber raids, which was directed at Bremen. The squadron's Conversion Flight moved to Rufforth in September 1942 and the rest of the squadron followed in November.

East Moor then became a base for Canadian units, and No 429 'Bison' Squadron brought its Wellingtons here in November as the last of the Halifaxes departed. By January 1943 the squadron was flying operational missions, and operations continued until August when a final mine-laying sortie was completed, after which the squadron transferred to Leeming. From May 1943 No 1679 Heavy Conversion Unit was based at East Moor, tasked with conversion to the Hercules-engined Lancaster, and training activities continued here throughout 1943 until the unit moved to Wombleton in December. By this stage the Wellingtons of No 432 'Leaside' Squadron had arrived, although the unit immediately re-equipped with Lancasters, becoming operational in November and remaining in business at East Moor until May 1945, by which time the unit had re-equipped again, this time with Halifaxes. While operating the mighty Halifax, No 432 Squadron flew no fewer than 2,416 operational sorties, losing forty-one aircraft in the process.

No 415 'Swordfish' Squadron arrived at East Moor in July 1944, swiftly shifting to bomber operations after having been active as a Coastal Command unit for some time. On the 28th of that month the Squadron contributed sixteen aircraft to a raid on Hamburg by seventy-seven Halifaxes. Hamburg was a tempting target for Bomber Command forces as it was a famous shipyard city; the battleship *Bismarck*, now at the bottom of the Atlantic, had been built here, as well as 200 U-boats. For the first time the American Eighth Air Force was invited to join in with a Bomber Command 'battle'. B-17 Fortresses would fly 252 daylight sorties in the two days following the first of four RAF night raids. The American targets were all industrial and included the U-boat yards, but their effort ran into major difficulties mainly due to the fires started by the RAF raids still obscuring their targets. The Americans quickly withdrew from attacking Hamburg and were not keen to follow immediately on the heels of RAF raids in the future because of the smoke problem.

A No 415 Squadron Halifax at East Moor during 1944.

The last aircraft to use East Moor airfield was Jet Provost XN469, which crash-landed here in June 1970 after suffering engine failure.

Sir Arthur Harris directed four major raids against Hamburg in the space of ten nights, known as Operation 'Gommorah'. The most famous of these was on 27/28 July 1943, involving 787 aircraft – 353 Lancasters, 244 Halifaxes, 116 Stirlings, 74 Wellingtons. Seventeen aircraft – eleven Lancasters, four Halifaxes, a Stirling and a Wellington – were lost, only 2.2 per cent of the force. The American commander, Brigadier-General Anderson, again flew in a Lancaster and watched this raid. One of No 415 Squadron's aircraft was lost over Germany on the night of the biggest attack, and another crashed on take-off from East Moor. The centre of the Pathfinder marking – all carried out by H2S on this night – was about 2 miles east of the planned aiming point in the centre of the city, but the marking was particularly well concentrated and the Main Force bombing crept back only slightly. In all, 729 aircraft dropped 2,326 tons of bombs.

This was the night of the firestorm, which started through an unusual and unexpected chain of events. The temperature was particularly high (30°C at 6 o'clock in the evening), and the humidity was only 30 per cent, compared with an average of 40-50 per cent for that time of the year. There had been no rain for some time and everything was very dry. The concentrated bombing caused a large number of fires in the densely built-up working-class districts of Hammerbrook, Hamm and Borgfeld. Most of Hamburg's fire vehicles had been in the western parts of the city, damping down the fires still smouldering there from the raid of three nights earlier, and only a few units were able to pass through roads that were blocked by the rubble of buildings destroyed by high-explosive bombs early in the raid. About halfway through the raid, the fires in Hammerbrook started joining together and competing with each other for the oxygen in the surrounding air. Suddenly the whole area became one big fire, with air being drawn into it with the force of a storm.

The bombing continued for another half-hour, gradually spreading the firestorm area eastwards. It is estimated that 550-600 bomb loads fell into an area measuring only 2 miles by 1 mile. The firestorm raged for about three hours and only subsided when all burnable material was consumed. The burnt-out area was

The crumbling remains of East Moor's control tower.

almost entirely residential. Approximately 16,000 multi-storeyed apartment buildings were destroyed. There were few survivors from the firestorm area and approximately 40,000 people died, most of them by carbon monoxide poisoning when all the air was drawn out of their basement shelters. In the period immediately following this raid, approximately 1,200,000 people – two-thirds of Hamburg's population – fled the city in fear of further raids.

Operational sorties continued until 25 April 1945, when an attack on gun batteries on Wangarooge Island marked the end of East Moor's wartime contributions. The RCAF aircraft soon departed, and No 54 Operational Training Unit arrived in November with a mixed fleet of aircraft comprising mostly Mosquitoes, the unit being tasked with the training of Mosquito night-fighter crews. No 288 Squadron also operated from East Moor for a brief period during the summer of 1946 with its Vengeance aircraft, and other aircraft were also seen here on a sporadic basis, while the airfield remained in use as a satellite field for Leeming. In November 1946 the last aircraft departed and the station closed down.

East Moor remained unused and most of the area was slowly returned to farmland, although the runways and taxiways remained intact. Seventy years later the runways can still be seen, although some agricultural buildings have appeared on them. Much of the perimeter track is still present, as are remains of the bomb dump in the woods, but the dispersals are gone, together with almost all of the former admin and technical site. From the air the unusual layout of this once-busy bomber aircraft is still clearly visible.

Main features:
Runways: 169° 5,700 x 150 feet, 036° 4,300 x 150 feet, 078° 4,200 x 150 feet, concrete surface. *Hangars:* two T2, one B1. *Dispersals:* thirty-six Heavy Bomber. *Accommodation:* RAF: Officers 152, SNCOs 387, ORs 1,555; WAAF: 407.

ECCLESFIELD, Yorkshire

53°26'33"N/01°26'48"W; SK368941; 252ft asl. 3 miles N of Sheffield, between A6135 and M1

The busy outskirts of Sheffield might not seem like an obvious site for military aviation, but Ecclesfield was once a quiet village some distance from the growing city. With the threat of marauding German Zeppelins creating a need for effective air defence, a number of rudimentary sites were selected across Yorkshire from where RFC fighter aircraft could operate when required. Most of these sites were little more than allotted fields with obstacles cleared and grass cut short, and many sites were often left unused for long periods, but they were sometimes vital for the RFC's operations.

No 33 Squadron was tasked with the air defence of the Yorkshire region that included Sheffield (an obvious target for the Germans) and a suitable field was selected at Ecclesfield to accommodate fighter aircraft. It was used sporadically until the end of the Second World War, at which stage it was abandoned. Today, many decades later, there is no evidence of the RFC's presence, and it is difficult to establish precisely where the landing site was; it would appear to have been located in the fields adjacent to what is now the M1, next to Butterthwaite Lane.

Main features:
Runways: none, grass field. *Hangars:* none. *Dispersals:* none. *Accommodation:* not known.

ELSHAM WOLDS, Lincolnshire

53°36'30"N/0°25'25"W; TA043135; 236ft asl. 5 miles NE of Brigg on A15

The flat and largely featureless fields of Elsham Wolds, lying some 9 miles south of Hull on the Lincolnshire side of the Humber, were first utilised as an airfield in 1916 when No 33 Squadron RFC, with its FE.2B and FE.2D biplanes, was deployed between Hull and Lincoln to counter the Zeppelins coming in over the Lincolnshire coast during their night raids on the Midlands. While

Gainsborough served as a headquarters, the limited endurance of the FE.2s necessitated No 33's complement being split into three flights and placed at suitable locations roughly 12 miles apart in a line between the two cities. Elsham Wolds served C Flight, which arrived in December 1916 and stayed until June 1918, flying many sorties to try to counter the Zeppelin raids, but without success. Wooden huts and a small aircraft shed were erected but had been demolished by 1919, when the wold was returned to cattle and sheep.

A wartime aerial view of Elsham Wolds illustrating the typical bomber airfield layout.

In the late 1930s, with the threat of another war, there was a requirement to find new airfield sites for RAF expansion. The 1914-18 locations were some of the first reviewed, but at Elsham Wolds an area to the west of the earlier site was found more suitable. Preparations did not begin until the winter of 1939/40 and were not completed until the summer of 1941 owing to a decision before the station was opened to put down hard runways. These were then extended: the main 14-32 to 2,000 yards and subsidiaries 02-20 to 1,400 yards and 08-26 to 1,600 yards. Initially twenty-seven hardstandings were provided, later increased to thirty-six, comprising three loops and the rest pans. The technical site, with one Type-J hangar and two T2s, was built on the east side of the airfield. Three T2s were erected early in 1944 to serve No 13 Base Maintenance; these were on a spur that ran to the edge of the First World War aerodrome site, south of runway head 28. Domestic sites for up to 2,068 males and 493 females were dispersed in adjoining farmland to the south-east. Bomb stores were located off the north-east side of the airfield.

Approval for the acquisition of land for development on this site was first given in the summer of 1940, initially for construction of an operational bomber base, but subsequently as a satellite airfield for nearby RAF Kirmington. However, the airfield was completed as a fully equipped bomber base with a secondary runway that was slightly longer than standard and a slightly larger than normal complement of hangars.

The site opened as an RAF station on 3 July 1941 as part of No 1 Group, with the Wellingtons of No 103 Squadron arriving during subsequent weeks. This unit remained at Elsham Wolds through the rest of the Second World War (eventually departing in November 1945) and, after replacing Wellingtons with Halifaxes, the unit converted to Lancasters. One of its aircraft was the well-known ED888, which had flown some 140 operational sorties by the time it was retired in December 1944.

Richard Badger was one of many airmen who served with 103 Squadron (and flew in the aforementioned aircraft), and he recalls his experiences at Elsham:

> 'There were a few hairy moments during those operations, like the one where the 2,000lb bomb refused to release over the target and we had to return with it, only to find on landing a slight thump and the groundcrew with their hands on their heads, aghast, as we taxied in. Apparently the bomb had dropped off the bomb rack on landing and was only being held by the bomb bay doors, which had opened about 8 inches or so from the weight! They rushed in with a scissor lift to take the weight before opening the bomb bay doors. Another interesting one was at night, over Germany, when a bomber ahead of us was hit by Flak and exploded. We veered off slightly to the right and felt and heard bits of debris hitting our plane.
>
> There were quite a number of other interesting moments during the tour of operations, one being when we were returning from a low-level raid in France, and had just crossed the coast at Cap Gris Nez, heading out to sea, when the rear gunner shouted, "Christ, we're being fired at!" as tracer started flying past the aircraft. Fortunately we were quickly becoming out of range and breathed a sigh of relief. Just then the mid-upper gunner

shouted, "Blimey, that was a close one!" Apparently a bullet had entered the turret at a bottom Perspex panel and left near the top, leaving quite a hole that he hadn't noticed at the time as both he and the rear gunner were busy firing back.

There was another incident on a night raid and we had just had a fire on No 1 engine (port outer) and had feathered it (to 'feather' an engine involved turning the prop blades edge-on to the slipstream to stop the blade turning – as far as I remember, the port outer engine provided the power to the rear turret) when we were attacked by a night-fighter. I believe it was a Ju 188 from the port side, presumably using cannon from a turret on the top, as we ended up with a huge hole in the starboard wing as we were going through the evasive manoeuvre of 'Corkscrew Port Go' – dive to port, climb to port, climb to starboard, etc. The hole was about 5 feet by 4 feet, and the side of the aircraft was peppered with holes. At the time, I believe, the fuel tanks were of the self-sealing type, thankfully. The miraculous thing was that there was a 2 by 1 inch hole through both sides of the aircraft, passing right through my wireless operator's seat. I had been down the back at the time, pushing leaflets out through the flare chute. Normally we put the leaflets in the bomb bay and they were released when we opened the bomb doors, but this time the ground engineer had asked us to put them out through the flare chute because of some technical problem. This may well have triggered off the attack. I had felt a big bang in the back while down there, and found later that I had been hit by a tiny piece of cannon shell (it felt more like a house brick!), but the parachute harness had absorbed most of the shock. I still have it in me now and I wouldn't be here if I had been sitting in my radio seat. We managed to limp back to the UK and had to make a flapless landing with very little braking, but we made it.

There were many more interesting incidents, like being caught in searchlights and having to twist and dive to evade them, and the night-fighter passing overhead. This was sometimes an Fw 190 or a Ju 188. On the night raids that we did, we flew as individual aircraft in what was known as the bomber stream (not in formation and even a mile or so apart), each with our own navigator, often not seeing another aircraft the whole way there until nearing the target, when the stream began to funnel in towards the target and flares were being dropped. Then you had to really keep your eyes open as there seemed to be Lancasters everywhere – left, right, above and below, some with their bomb doors open. (You had to move a little to one side if they were directly above as they most probably couldn't see you.) As you approached the target area you were listening for the voice of the Master Bomber on the radio (they were the Pathfinder force that had gone way ahead of the main force to mark the target accurately with coloured red and green target indicators), who would advise where to aim, relative to the Target Indicators. The other critical time was over the target when, having dropped your load, you had to hold your course long enough for the camera to take a picture of the result (maybe 10 to 20 seconds, depending on height) before turning off onto the return course home.

The trouble with this was that, with say 200 aircraft all arriving over the target at slightly different times and turning off at different times, the risk of collision was high (and I expect it happened). You were also having to go through quite intense anti-aircraft fire, which was being pumped up over the target area, once they had worked out the altitude you were at. The other thing that we had to contend with was that enemy night-fighters were known to patrol the bomber stream, dropping flares then climbing above us and looking for silhouettes to pick off individually!

As we were returning from a night raid on our last (thirtieth) op, the rear gunner, who was excellent at aircraft recognition, spotted an Me 109 sitting out on our port quarter. We waited, and expected it to start turning in to attack and were ready for it, but it didn't, it just sat there for some time, about 5 or 10 minutes, then turned away. We often wondered, was he on his last mission? Or was he out of ammunition or low on fuel?'

The second unit to arrive at Elsham was No 576 Squadron, which brought its Lancasters to the base in November 1943, staying here until the following October when it moved to Fiskerton. No 100

Squadron arrived in April 1945, and together with No 576 Squadron participated in a final attack on Berchtesgaden on 25 April, after which operations began to wind down at the base. However, some additional activity was provided by the Lancasters and Lincolns of No 57 Squadron, which stayed for a few weeks at the end of 1945, and aircraft operated by Blyton's No 1662 Heavy Conversion Unit, which used Elsham as a Relief Landing Ground. Flying activity ended towards the end of 1945 and the station closed during 1946, after which the airfield and its facilities were abandoned.

The busy A15 now runs across the former airfield and the Elsham Wolds Industrial Estate resides in what was once the main admin and technical site. The single J-Type hangar is still intact together with a couple of the smaller T2s, and traces of the main and secondary runways can still be seen, particularly on the western side of the A15. The loop road that links the industrial estate to the A15 affords an excellent view of the thresholds of both the main and secondary runways, which intersect at this point.

Main features:
Runways: 318° 6,000 x 150 feet, 198° 4,200 x 150 feet, 258° 4,800 x 150 feet, tarmac surface. *Hangars:* two T2, three B1, one Type-J. *Dispersals:* thirty-six Heavy Bomber. *Accommodation:* RAF: Officers 174, SNCOs 480, ORs 1,414; WAAF: 493.

ELVINGTON, Yorkshire

53°55'26"N/0°59'23"W; SE664479; 42.5ft asl. 7 miles SE of York, on B1228

Elvington's airfield was constructed in 1941 as one of many standard bomber airfields in the Yorkshire region. It was intended to become a satellite airfield for nearby Riccall, but even though it was initially expected to be a grass landing site, it was completed to normal bomber standards with three runways in a triangular pattern and the usual array of hangars and support buildings. Constructed on Elvington Common, the land was already ideally suited for development into an airfield and relatively few changes were made to the surrounding agricultural land and road network. Perhaps the only unusual aspect of the airfield's layout was the common intersection shared by all three runways – a design that was dictated by the local topography, which gave the airfield an appearance more typical of a naval air station. Also worthy of note was the rather unusual addition of a railway line that linked the site with the nearby station, enabling construction materials to be brought in with ease, and leading to the line's retention for the delivery of supplies and weapons once the station opened in October 1942.

An evocative view of Elvington during the Second World War, with No 77 Squadron Halifaxes at their dispersals.

Whitleys soon arrived when No 77 Squadron was established here, and flying activity soon developed, a great deal coming from Pocklington – the station for which Elvington acted as satellite. The Whitleys were swiftly replaced by Halifaxes and No 77 Squadron remained as the sole resident unit until May 1944, when it moved to Full Sutton. A very unusual development then took place, and Elvington became a French base as part of No 4 Group, hosting two French units. No

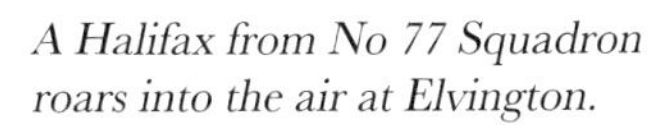

A Halifax from No 77 Squadron roars into the air at Elvington.

Halifax DT807 from No 77 Squadron at Elvington. This aircraft was lost in action during a raid on Kassel in October 1943.

An aerial view of the huge apron at Elvington, with the Yorkshire Air Museum in the distance.

346 'Guyenne' Squadron formed here in May 1944 and No 347 'Tunisie' Squadron followed in June, both equipped with Halifax Mk 5 bombers. The two French Squadrons were soon participating in operational raids over Germany, and despite losses the two Halifax squadrons achieved great success.

One of the worst accidents to affect Elvington was a distinctly home-grown one that occurred on 27 December 1944 when a bomb was dropped during a loading procedure, leading to the detonation of a full bomb load and the loss of thirteen lives. Perhaps even more spectacular (and equally deadly) was an audacious attack performed by the Luftwaffe on 4 March 1945 when a number of enemy fighters followed the French-crewed bombers back to Elvington and waited until the airfield lights were switched on before embarking on a devastating attack on the airfield and the returning Halifaxes. The surprise of this attack shook Bomber Command, but in practical terms it did little to affect the RAF's offensive capability.

The final operational mission flown by the Elvington squadrons took place on 25 April 1945 (a raid on Wangarooge Island), after which both units were transferred to the French Air Force. The only other resident flying unit was No 4 Group's Communications Flight, and this had already left in June 1943. With the Halifaxes gone, the airfield soon fell silent. In January 1946 it became a sub-site for No 14 Maintenance Unit, but by the end of the year the site was again closed and, although the station was retained by the RAF, it was effectively abandoned.

In the summer of 1952 a limited amount of flying returned to Elvington when the site was adopted as a temporary Relief Landing Ground for No 14 FTS based at Holme-on-Spalding-Moor. This activity lasted for a couple of years and parts of the airfield were also used for bomb storage, probably because of the very useful railway spur that enabled weapons to be brought in easily.

By this stage the USAF had expressed an interest in the airfield as a potential site for development into a base for Strategic Air Command (this being a particularly tense period in the Cold War). In June 1953 Elvington was transferred to American control and a drastic reconstruction programme was initiated, which virtually obliterated the former RAF airfield. Two of the bomber runways were removed almost completely, and the main runway was consumed by a huge new 11,000-foot concrete runway laid along its path. This gave Elvington one of the longest runways in the UK, sufficient to support SAC's B-47 Stratojets and B-52 Stratofortresses. However, by the time that the massive construction programme had ended, the USAF had reappraised its plans and decided not to create an SAC base at Elvington. Consequently, even though the airfield had become a very significant site capable of hosting strategic nuclear bombers, it was in effect redundant, and the RAF reacquired it, eventually designating it as a suitable RLG for Church Fenton and (a few years later) Linton-on-Ouse. It was almost comical to witness the diminutive Jet Provosts from No 7 FTS and No 1 FTS seemingly dwarfed by the huge runway that was now at their disposal.

Apart from the daily appearances of the Provosts, very few other aircraft were ever seen here, although Blackburn did transfer some of its Buccaneer trials work here in the early 1960s, including a significant number of RATOG (Rocket-Assisted Take-Off) tests, which saw Buccaneers roaring skywards from Elvington accompanied by two noisy streaks of rocket flames. Ironically, the new-built airfield was never used in any serious way and, although it would seem logical to take advantage of a superb facility that was effectively built for free, the RAF never relocated any units to the site. As flying training requirements wound down, the airfield was used only sporadically, and the RAF finally abandoned it in March 1992. It was sold off by the MoD in January 1999 and since then has been used only for private flying.

There are plans to develop Elvington into a site for limited commercial and business use, but at present the airfield remains mostly unused. However, it is the home of the popular Yorkshire Air Museum, which occupies part of the former admin site, and inside a surviving T2 hangar a beautifully maintained Halifax replica can be seen, creating a sight that is virtually indistinguishable from that which would have been seen seventy years ago. Not far from the museum the monstrous SAC apron is still visible, and the massive runway is still intact and occasionally occupied by the resident Victor bomber, which performs high-speed runs during summer months.

Perhaps less obvious to visitors is the concrete pad close to the eastern runway threshold. This was built to accommodate V-Bombers, which would have been dispersed here in wartime conditions, and Vulcans did occasionally deploy to Elvington to practise their dispersal techniques until the mid-1960s. It is perhaps surprising that jet bombers did not play a far more significant part in Elvington's post-war history.

Main features:
Runways: 263° 6,000 x 150 feet, 203° 4,200 x 150 feet, 143° 4,200 x 150 feet, asphalt surface. *Hangars:* two T2, one B1. *Dispersals:* thirty-six Heavy Bomber. *Accommodation:* RAF: Officers 123, SNCOs 411, ORs 1,696; WAAF: 378.

ESHOTT, Northumberland

55°16'54"N/01°42'38"W; NZ183984. 7 miles N of Morpeth, S of Felton off A1

A site for an airfield at Eshott was first surveyed in early 1941, initially intended as another fighter station for the protection of the North East. Construction began in early 1942 and, as the war progressed, the airfield's role was downgraded to training.

With work still to be carried out on the airfield, the first and main unit to operate from here began to arrive on 10 November 1942 in the shape of No 57 OTU, with a good cross-section of aircraft, including Spitfires, Masters and Battles, the latter being employed for target-towing duties over the local ranges. With more than 100 aircraft on strength, life and operations were quite cramped until Boulmer became the airfield's satellite on 1 March 1943. This not only eased the amount of traffic in the circuit but also meant that certain stages of the training syllabus could be carried out at the satellite.

Because of the rate of development of the Spitfire, there was always a good supply of early marks available for use by the OTUs. Virtually all had seen action, many in the Battle of Britain, and all had served with a multiple of units. Despite being obsolete, the Spitfire Mk I was a perfect introduction to the challenges of flying a single-seat fighter.

On arrival at any fighter OTU, the student pilot spent many weeks undergoing a variety of ground instructions and aviation-

Spitfire Mk IIa P8437 of No 57 OTU is pictured at Eshott. This particular aircraft served with six different units before being struck off charge in November 1944. Via Martyn Chorlton

This modern aerial view of Eshott shows the comparatively pencil-thin modern runways laid on top of the wartime ones. The A1 road runs north-south on the left-hand side of the airfield.
Crown Copyright via Martyn Chorlton

related subjects without stepping into an aeroplane. 'Bull' was high at Eshott, typical of a training station. An average course lasted three months and, although the pilots were introduced to the Spitfire while on the ground, many tests and exams had to be taken before that first magical flight could take place.

Ground studies included exercises in survival and making good your escape if you had to bail out over enemy territory. Physical survival training was carried out at Eshott by a detachment of Commandos, who taught the aircrews the art of survival, camouflage and several ways of crossing a river, the latter making use of the local Longdike Burn.

Before any flight took place, the student pilots had to work through five broad sections of the course: Knowledge of Spitfires, Handling the radio, Cockpit drill and Emergency Procedures, Oxygen system on Spitfires, and Readiness to fly Spitfires. Usually three to four weeks into the course, the trainee pilot was allowed to fly a Spitfire. The final check-out and flying test was carried out with an instructor in one of the unit's Masters. Only once the instructor was satisfied would the student be allowed to take the controls of a Spitfire.

Spitfire Mk IIa P7350 is the oldest airworthy example of its type in the world. No 57 OTU was its last unit, and today she is operated by the Battle of Britain Memorial Flight at Coningsby in Lincolnshire. Crown Copyright via Martyn Chorlton

The course then progressed to actually fighting in a Spitfire, and pairs of students were encouraged to practise dog-fighting tactics both at high and low altitudes between themselves, sometimes resulting in fatal aerial collisions. Aerial gunnery was practised against targets towed by the unit's Battles and Martinets as well as No 4 AGS Lysanders, which provided this service to any unit operating in the area. Bombing skills were also honed on both ground and sea targets; once again, misjudged attacks resulted in many pilots being lost.

With Acklington in such close proximity to the airfield, it was not uncommon to receive visitors, some planned and others mistaking Eshott for its older neighbour. Eshott's emergency services were also called upon to deal with accidents from Acklington as well as their own. No 57 OTU averaged two fatalities a month, with a least four to five aircraft damaged or written off in accidents. The latter also included the loss of six Miles Masters, all of which was described by Training Command as 'acceptable losses'.

Many of the instructors who were posted to Eshott had a great deal of experience, many in combat and several during the Battle of Britain. It has been suggested that many of these individuals resented being on a training unit and were always itching to get back to a front-line squadron. This may have been the case, but their experience and knowledge was invaluable and it was essential that it was passed on to and 'drummed into' these new potential fighter pilots.

The majority of OTUs had a large spread of nationalities, Eshott being no exception. As well as British pilots, others arrived from Canada, Australia, Poland, France, Czechoslovakia and Norway.

From the beginning of 1944 to the end of the war, No 57 OTU's workload was always very high and usually just beyond the resources of the unit, which is how the senior staff preferred it. On 26 January 1944 both Nos 55 and 59 OTUs based at Annan and Milfield respectively were disbanded. Their demise meant that No 57 OTU's throughput of fighter pilots would have to increase. The number of aircraft also rose and even more training exercises were carried out at Boulmer, a role that never decreased until the end of the war.

On 15 May 1945 the hammer fell on No 57 OTU and all training was suspended. Student pilots who had yet to complete their course were transferred to the few surviving OTUs. On 6 June the unit was officially disbanded, and Boulmer ceased to be a satellite of Eshott on the same day. The only task that remained was the ferrying of the OTU's aircraft to other units, although the majority would end up at Maintenance Units for storage and ultimately scrapping.

On 18 May 1945 No 289 Squadron arrived from Acklington with an assortment of aircraft, being equipped with the Spitfire Mk XVI and the Vengeance Mk IV. A few days later the unit was joined by No 291 Squadron, another target-towing unit. Both squadrons had left Eshott by the beginning of June, and by the 26th both had been disbanded.

Between June and October 1945 the rapidly deteriorating Eshott was placed under Care and Maintenance. Activity returned on 1 October with the arrival of No 261 MU, which used Eshott as a sub-site for its main home at Morpeth. The unit was formed as a GED at Long Benton in June 1945, moving to Morpeth not long after. Eshott remained a sub-site until 1 January 1948, bringing an end to its RAF service.

It was 35 years before flying returned to Eshott. The majority of its wartime buildings have disappeared; those that survive are located on the southern side on private property. Large sections of the original runways remain, making the airfield a very attractive proposition for light aviation. It is now the home of Eshott Airfield Flying Club, and this enthusiastically run group will ensure that aviation remains in this area for many years to come.

Main features:
Runways: QDM 264 1,900 yards, 323 1,260 yards, 02 1,250 yards, concrete. *Hangars:* one T1, eight Blister. *Hardstandings:* fifty SE. *Accommodation:* RAF: 1,478; WAAF: 441.

FALDINGWORTH, Lincolnshire

53°21'17"N/0°27'01"W; TF032853; 55.5ft asl. 5 miles SW of Market Rasen, on Spridlington Road off A46

Designed as a satellite field for Ludford Magna, the airfield at Faldingworth took some time to develop, largely as a result of the complicated nature of the site's ownership, which was divided between East Firsby Grange and Faldingworth Grange. The airfield was constructed on an isolated area of farmland covering three parishes south-east of the River Ancholme, 4½ miles from Market Rasen. The contractors involved were Tarmac Ltd and J. Cryer & Sons Ltd, with work totalling £810,000. Site clearance of woodland and hedges began in July 1942, and runway laying was completed by the following summer. Built to Class A standard, the runway lengths were 08-26 at 2,000 yards, and 1,400 yards for each of the subsidiaries, 01-19 and 13-31. Thirty-six hardstandings, all loops, and two T2 hangars and a single B1 were provided. The dispersed camp sites were towards Newton-by-Toft in the north-east, giving accommodation for up to 1,957 males and 281 females. A great deal of work was required to remove field boundaries, communal areas and agricultural buildings, together with a variety of access tracks, and it was not until October 1943 that the station finally opened.

No 1667 Heavy Conversion Unit was the first arrival, its Halifax and Lancaster bombers transferring here from Lindholme. Most of the training activity was conducted on the Halifax (there was a shortage of Lancasters at this time) and the less-active Lancaster element stayed here for only a few weeks before returning to Lindholme as C Flight of No 1 Lancaster Finishing School. The Halifaxes did not stay much longer, and by March 1944 they had departed for Sandtoft.

A 1942 image of Faldingworth with runway construction under way.

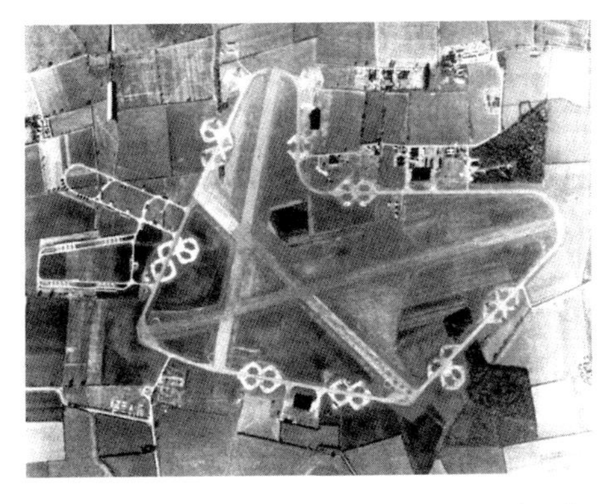

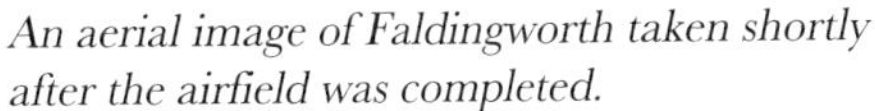

An aerial image of Faldingworth taken shortly after the airfield was completed.

A No 300 Squadron Lancaster at Faldingworth in June 1945.

The airfield then reverted to acting as a satellite for Ludford Magna, although No 300 'Mazowiecki' Squadron soon arrived with Wellingtons, which were swiftly replaced by Lancasters. Operational missions were soon under way for this Polish unit, and some 1,216 sorties were completed with the loss of thirty aircraft. A second Polish unit, No 305 Squadron, arrived here in October 1946, but both units disbanded just a few weeks later. A rather less prominent resident from May 1944 was No 1546 Beam Approach Training Flight, which operated a fleet of Oxfords from Faldingworth until January 1945. With all of these units gone, the airfield was redundant, and by the end of 1948 the site had been reduced to Care and Maintenance status, with all flying activities at an end.

However, the RAF retained the airfield and it was used as a sub-site for Maintenance Units, with No 93 MU (based at Wickenby) using it from 1949 until 1951, and No 93 MU (also at Wickenby) operating here until the mid-1950s. Various parts of the airfield were used to store stocks of HE bombs until they could be disposed of. The MU continued its association with Faldingworth and eventually moved its headquarters here after the site was developed into a storage site for the RAF's nuclear arsenal. Located within easy reach of the V-Bomber bases at Scampton, Waddington and Coningsby, it was an ideal location for the secure storage of atomic and thermonuclear warheads. Faldingworth assumed this role in 1957, with part of the former airfield redeveloped into a secure compound, surrounded by impregnable barbed-wire fences, monitored by security watch towers and patrolled by free-running guard dogs. Inside the compound, a series of concrete storage structures were arranged, each accommodating a nuclear warhead, while other structures were designed to house and load bomb casings. Faldingworth remained active as a nuclear storage facility until 1972, when the RAF finally vacated the site.

Its usefulness as a high-security facility soon saw the arrival of BMARC and the Oerlikon company, which tested and developed its 20mm cannon here. Today the site is still largely occupied by Royal Ordnance (part of BAE Systems), although many buildings have been leased to other companies for secure storage. Most of the nuclear weapons buildings are still visible, although the site is still largely inaccessible and monitored by security personnel. The rest of the former airfield was sold off in 1998 and most of the main and secondary runways can still be seen, although substantial parts of the runways have now been removed. Some of the dispersals are also still visible and one B1 hangar survives. Faldingworth provides a fascinating insight into two distinct periods in the RAF's history, with the former bomber airfield remains laid next to the ominous buildings of an even darker era.

Main features:
Runways: 256° 6,000 x 150 feet, 312° 4,200 x 150 feet, 010° 4,200 x 150 feet, concrete and wood chippings surface. *Hangars:* two T2, one B1. *Dispersals:* thirty-six loop. *Accommodation:* RAF: Officers 87, SNCOs 358, ORs 1,512; WAAF: 281.

FARSLEY, Yorkshire

53°48'46"N/01°41'01"W; SE209352; 472ft asl. 1 mile NW of Pudsey on A6120

Little information is available on this site, but the RFC is known to have established a landing site here early in 1916. This took the form of an area of farmland that was simply cleared of obstructions and made available to the fighter aircraft of No 33 Squadron, which was responsible for air defence of the area. The site would have been used when aircraft were operating in the area, defending air space over the adjacent city of Leeds. The field was abandoned by the end of 1916, and there is no evidence at this location to indicate the RFC's brief presence. Indeed, it is difficult to establish the precise location of the landing field, but it is thought to have been in an area of land next to Priesthorpe School.

Main features:
Runways: none, grass field. *Hangars:* none. *Dispersals:* none. *Accommodation:* not known.

FINNINGLEY, Yorkshire

53°28'46"N/01°0'07"W; SK663985; 19.5ft asl. 7 miles E of Doncaster, off A614

Aerial activity at this site can be traced back to the First World War, when a small area of agricultural land was temporarily used by RFC aircraft that were normally based a few miles away in Doncaster. The RFC site was used only briefly and it was not until 1935 that land was purchased to the south-west of Finningley for development into a bomber airfield. Difficulty in fixing a price for the land caused delays, as did the clearance of buildings and farm tracks, but by the summer of 1936 the airfield was nearing completion and in September the Heyfords of Nos 7 and 102 Squadrons arrived from Worthy Down.

After establishing themselves at Finningley, the two squadrons divided into flights, these becoming separate squadrons. No 76 Squadron formed on 12 April 1937 followed by No 77 Squadron on 14 June. In July Nos 77 and 102 Squadrons departed for Honington, by which time No 7 Squadron had re-equipped with Wellesleys, although these were quickly exchanged for Whitleys, and No 76 Squadron acquired Hampdens. In June 1939 the two units assumed a training role and each squadron was equipped with a mix of Hampdens and Ansons, tasked with the pre-operational training of No 5 Group crews. Just three months later the squadrons moved to Upper Heyford, and in October 1939 No 106 Squadron arrived with more Hampdens, Ansons and (a few weeks later) three Fairey Battle target tugs.

In March 1940 No 98 Squadron arrived from Scampton for a brief stay, and in a matter of weeks the unit's battles were in France. The first wartime operational sorties were flown from Finningley in September 1940, when Hampdens from 106 Squadron flew a mine-laying mission. It was not long before the Luftwaffe turned its attention to Finningley, and a series of fairly low-key raids took place, only one of which caused any significant damage (to one of the hangars). No 106 Squadron transferred to

Finningley in 1936 – Handley Page Heyfords are scattered across the grass apron that would one day house Vulcan bombers.

A No 87 Squadron Wellesley and crews at Finningley in 1937.

A radar-nosed Mosquito on the grass at Finningley, with a visiting Lancaster from St Eval in the background.

Coningsby in February 1941, leaving behind the core of a new training squadron – No 25 Operational Training Unit. As part of No 7 Group, this OTU was equipped with Wellingtons and (eventually) Manchesters. As the unit grew it was divided into flights, which were dispersed to Balderton and Bircotes, but in May 1942 the RAF opted to bring all of the training aircraft up to operational standard. On 31 May some thirty Wellingtons from Finningley participated in the first Thousand Bomber raid, with the loss of only one aircraft (the crew survived).

By the end of 1942 the grass landing field at Finningley had been pounded into a terrible state and the aircraft (including the Oxfords of No 7 BATF, which had arrived in January 1941) were dispersed to other airfields so that some attempt could be made to repair the grass surfaces. Throughout the reconstruction period flying activity continued, however, and training activities were gradually shifted to No 18 OTU – a joint RAF and Polish unit. With the grass field repaired, training activity grew again. No 1507 BATF was replaced by No 1521 BATF, and with a satellite airfield now established at Worksop it was decided that more permanent runways should be laid and flying temporarily stopped until May 1944, when three new tarmac runways were declared

An aerial view of Finningley taken shortly after the RAF's departure, long before so much of the site was redeveloped.

A No 230 OCU Vulcan in company with a civilian-owned Spitfire over Finningley's hangars and V-Bomber pans in 1968.

ready for use. Early in 1942 Finningley passed to No 1 Group, and with no further need for Hampdens or Manchesters No 25 OTU concentrated on Wellingtons, nine of which were lost when the station was called upon to participate in Bomber Command operations. No 25 OTU was disbanded in February 1943, and in March No 18 OTU moved in from Bramcote and began using Bircotes and Worksop as satellites.

In November 1943 the Wellingtons were moved to these satellites so that hard runways could be laid at Finningley. These were put down during the winter of 1943/44, the main 03-21 being 2,000 yards, 07-25 1,400 yards and 12-30 1,400 yards. A concrete perimeter track had been laid in 1942 and asphalt pan-type hardstandings constructed in 1940-41 linked to it, two of the original clusters crossing the A614 road between Finningley village and Bawtry. A single loop-type standing was added to bring the total to thirty-six. Some additional domestic accommodation was provided to cater for a maximum 1,592 males and 459 females. The bomb store was in Finningley Big Wood.

A magnificent line-up of No 6 Flying Training School's Varsities at Finningley in 1975.

The station re-opened for flying in May 1944 when No 18 OTU returned from Bramcote. By the end of the year requirements for operational training had reduced, and in January 1945 the OTU was disbanded and the Wellingtons removed. The Operational Training Unit was replaced in 1945 by the Bomber Command Instructors School with a fleet of Lancasters, Wellingtons, Halifaxes, Spitfires and Hurricanes. This mixed fleet of aircraft was joined by the Mosquitoes of No 616 (South Yorkshire) Squadron when this unit reformed at Finningley in June 1946, and when the Squadron re-equipped with Meteors it brought the first jet aircraft to Finningley. No 21 OTU moved to Finningley in October and became No 202 Advanced Flying School, the BCIS moving to Scampton.

The AFS was replaced by No 1 Refresher Flying Unit in January 1948, this in turn becoming No 1 Refresher Flying School in September 1951 when the unit received its first Meteor jets. Training continued for some time while the longer-term future of the station was considered. It was decided that Finningley would be transferred to operational duties and would become one of the RAF's most important sites, as part of the developing V-Force. In order to accommodate the new (heavy) strategic jet bombers, the existing tarmac runways would be replaced by single new runway some 9,000 feet in length, constructed from concrete 4 feet thick. The creation of this runway, together with new bomber dispersals and associated taxiways, transformed the airfield layout and resulted in expansion to the south, and to the north, which required the closure of the main Doncaster to Finningley road.

After this period of enforced closure, the new airfield reopened for business in June 1957, and in August No 1 Group's Communication Flight arrived with Canberras, Meteors, Ansons and Chipmunks. Finally, in October the first Vulcan bombers arrived and No 101 Squadron reformed here. In December 1958 No 18 Squadron also arrived, operating Valiants in a little-publicised ECM training and development role. In February 1960 the Bomber Command Development Unit was established at Finningley, equipped with Valiants, Vulcans and Canberras. Changes continued, and when No 101 Squadron moved to Waddington in June 1961 the Vulcan training unit (No 230 Operational

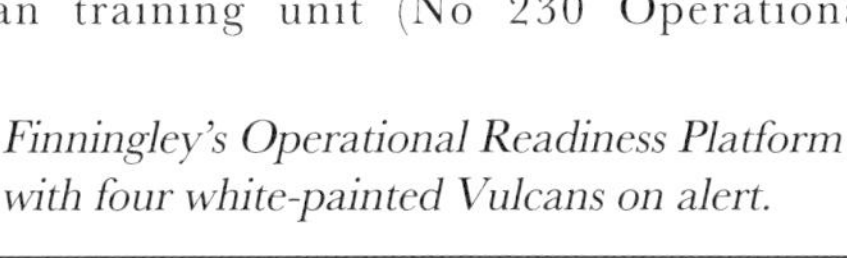

Finningley's Operational Readiness Platform with four white-painted Vulcans on alert.

A 2012 image of Finningley, with the magnificent Vulcan XH558 back at its former home base more than 40 years after it was based here with No 230 OCU.

Conversion Unit) arrived and Finningley settled into a relatively long period as the home of the RAF's Vulcan OCU, the sight and (not inconsiderable) sound of the mighty Vulcan becoming a regular part of daily life as the crews exercised both by day and night. No 18 Squadron disbanded in April 1963 but Vulcan operations continued uninterrupted until the end of 1969, when Finningley's happy association with the V-Force came to an end and the resident units departed.

After a brief period of temporary closure the station reopened back in the training role when No 2 Air Navigation School arrived, becoming No 6 Flying Training School on 6 May, eventually operating a mixed fleet of Varsities, Dominies and Jet Provosts. The Yorkshire Universities Air Squadron's Chipmunks arrived in June 1975 (eventually replaced by Bulldogs) and the venerable Varsity was retired in 1976, marking the end of the RAF's association with this much-loved aircraft.

Helicopter activities came to Finningley in January 1976 when the HQ Flight of No 22 Squadron arrived with Whirlwinds, and No 202 Squadron's HQ Flight arrived in September. Wessex and Sea King helicopters slowly replaced the Whirlwinds, and the skies around the local area were often occupied by the familiar shape of these yellow-painted workhorses, sharing the airfield circuit with the mixed fleet of fixed-wing trainers.

As the RAF's training activities were gradually consolidated, more changes affected Finningley. A small fleet of Hawk jets eventually joined 6 FTS, and the Multi Engine Training School (which arrived in 1979) operated Jetstream aircraft from Finningley, tasked with multi-engine training for students destined for the Nimrod, Hercules, VC10 and other units. Finningley became one of the RAF's busiest and most important stations, and more than £6 million was spent during the early 1980s to improve station and airfield facilities. It was (at least to observers of governmental ineptitude) no surprise that, having invested so much money in Finningley, the MoD then decided to dispose of the station. It is difficult to establish precisely why such a large, well-equipped and important station was abandoned, and there are many conflicting theories as to why this seemingly bizarre decision was taken. Perhaps the most plausible is that the MoD believed that the site could be redeveloped for a new prison, and this plan was indeed proposed, but after encountering huge local opposition it was dropped. But whatever the reason for the decision, the resident units departed during 1995 and early 1996, leading to the station's closure on 31 March 1996 – a desperately sad day that was marked by a final flypast.

The RAF slowly departed, and by the end of 1996 the airfield and facilities were deserted, seemingly destined to remain abandoned. However, thanks to the determination of local councillors, politicians and private enterprise, the airfield reopened for business in 2005 as a new civil airport, with the vulgar name 'Robin Hood Doncaster Sheffield Airport' bestowed upon it. A great deal of the RAF station was bulldozed as part of this transformation, but the mighty hangars survived, joined by a huge new airport terminal constructed on a western portion of the airfield, which enabled most of the existing airfield layout to be retained. Indeed, the only victim of the terminal's appearance was the old bomb dump. Few visitors to the airport car park realise that just yards away from their cars, huge megaton nuclear weapons were once stored.

Thankfully the classic 1960s-style control tower remains in use, the huge runway is still busy, and almost all of the old Vulcan dispersals are still intact. The old Operational Readiness Platform (from where Vulcans and even Victors roared skywards during Finningley's hugely popular annual air display) is gone and the airfield perimeter fence has been replaced by obtrusive chain fencing and a huge earth mound adjacent to Finningley village. But the airfield survives and even hosts occasional visits by RAF aircraft, which inevitably return when their crews cannot resist the urge to revisit such a famous RAF site.

However, perhaps the most remarkable event to have taken place in recent years is the return of the magnificent Vulcan. The sole flying example, safely in civilian hands, returned to its former home and is destined to remain here even when its flying days are over, ensuring that Finningley will for ever be inextricably linked with this iconic example of British expertise. Sadly, the many years when Finningley bustled with military aircraft are gone. The heady days of 1977, when the station hosted what was undoubtedly one of the RAF's proudest weeks (the Jubilee Review of the RAF), are now just a fading memory, and most people who now visit the airfield regard it only as a rather mundane gateway to holiday hotspots. But Finningley has survived and the presence of the mighty Vulcan will ensure that many visitors will continue to remember what this famous base was really all about.

Main features:
Runways: 209° 6,000 x 150 feet, 253° 4,200 x 150 feet, 300° 4,200 x 150 feet, concrete and tarmac surface. *Hangars:* two C-Type (No 1), two C-Type (No 2), one C-Type (Reserve). *Dispersals:* twenty-six circular, one spectacle. *Accommodation:* RAF: Officers 159, SNCOs 473, ORs 1,784; WAAF: 435.

FIRBECK, Yorkshire

53°23'28"N/01°09'47"W; SK557885; 193.5ft asl. 8 miles S of Doncaster, off A634 SE of Maltby

This small site owes its origins to the early development of civil aviation, a small private flying aerodrome having been created here during the 1930s. As the Second World War approached, it was almost inevitable that even this small airfield would be required to play its part in the approaching conflict, and the site was adopted for use by the RAF. The small landing field was enlarged to embrace more of the surrounding farmland, resulting in a substantial site with irregular boundaries.

The first RAF aircraft arrived in September 1940 when Lysanders from No 613 Squadron transferred here from Netherthorpe. Only a small number of the unit's aircraft were usually present, however, as detachments were also maintained at other RAF stations around the country, tasked with Army cooperation duties in association with locally based units. The number of resident Lysanders was increased when No 2 Squadron arrived for a short stay, but by the end of 1941 the units and their Lysanders had gone. Firbeck was also used from the outset as a Relief Landing Ground for No 1 (Polish) Flying Training School based at Hucknall, and a variety of Tiger Moths and Oxfords could be seen here, albeit on a sporadic basis.

The RLG responsibility for the FTS ended in the summer of 1941 and Firbeck became an RLG for No 25 (Polish) Elementary Flying School, resulting in the arrival of many more Tiger Moths from Hucknall, with many aircraft being relocated to Firbeck for a full day's flying before returning to their home base.

A 1994 aerial view of the former airfield site at Firbeck.

No 654 Squadron's Austers came to Firbeck late in 1942 for a brief stay, and more Austers from No 659 Squadron were present from April 1943 for some four months. The Moths from Hucknall's EFTS continued to use Firbeck until the unit disbanded in November 1945, after which only No 28 Gliding School remained at the airfield, having arrived in October 1944. Its Cadet gliders finally transferred to Middleton St George in January 1946, their place being taken by more gliders, this time from No 24 Gliding School.

This was the last unit to use Firbeck, and when it departed early in 1948 all flying ended at the site. With the RAF gone the site was abandoned and no attempt was made to return the site to civilian flying. The hangars were eventually demolished and the entire site was restored to its original status as agricultural land. Today there is very little to indicate the RAF's presence here, although a small farm track off New Road leads to the concrete base of one of Firbeck's two Blister hangars. Firbeck's last significant connection with aviation was the fascinating collection of aircraft and associated exhibits that were on display in the village for many years, but after this collection moved to the site of the former RAF Doncaster, Firbeck's association with aviation finally came to an end.

Main features:
Runways: none, grass surface. *Hangars:* two Blister. *Dispersals:* none. *Accommodation:* RAF: Officers 5, SNCOs 8, ORs 270; WAAF: 0.

FULBECK, Lincolnshire

53°03'14"N/00°39'12"W; SK895510; 50ft asl. 7 miles E of Newark, S of A17

Constructed on land to the east of the small village of Felton, the airfield was eventually named after Fulbeck Grange, which was situated to the eastern side of the new site, thereby avoiding any confusion with RAF Church Fenton in Yorkshire. Designated as a Relief Landing Ground for Cranwell, a grass airfield was to have been laid here, but by 1943 a standard bomber airfield had been built with three full-length runways linked by a perimeter track, and some fifty dispersals for bomber aircraft. Five T2 hangars were also erected at various points around the airfield, and although the technical and admin sites were developed next to the field, the domestic site was built a mile away to the east.

RAF Fulbeck opened for flying in 1940, hosting a variety of training units but primarily acting as an RLG for Cranwell. Airspeed Oxfords became the most common sight here, particularly those operated by No 2 Flying Instructors School. The Bomber Command Air Bomber Training Flight arrived in 1942 with more Oxfords and some Manchesters, and from April 1944 a large number of Horsa gliders came to Fulbeck for storage prior to the D-Day landings.

American units first came here late in 1943 when the 343rd Troop Carrier Group was established at Fulbeck, this unit being replaced after a few months by the 442nd Tactical Carrier Group. This unit's C-47s were very active during the early months of 1944, preparing for the Normandy invasion in which the unit duly participated, some forty-five aircraft flying from here on the first day. The C-47s departed later in 1944, and the RAF had resumed control by the end of the year. No 49 Squadron arrived in October 1944, joined by No 189 Squadron a few weeks later, and their Lancasters remained at Fulbeck until the end of the war.

By June 1945 the flying units had vacated the airfield and No 2 Aircraft Disposal Unit arrived, being duly redesignated as No 255 Maintenance Unit. The unit operated a number of sub-sites but eventually became a sub-site itself after being absorbed into No 93 MU, headquartered at Wickenby. Storage and processing of munitions became the main activity at Fulbeck from the end of 1944, and virtually no further flying too place here from that time. The MU remained here until April 1956, by which time the airfield was being brought back into use as an RLG for Cranwell again. It remained active in this role until the early 1970s, by which stage the familiar Jet Provost had become the most regular user of the site.

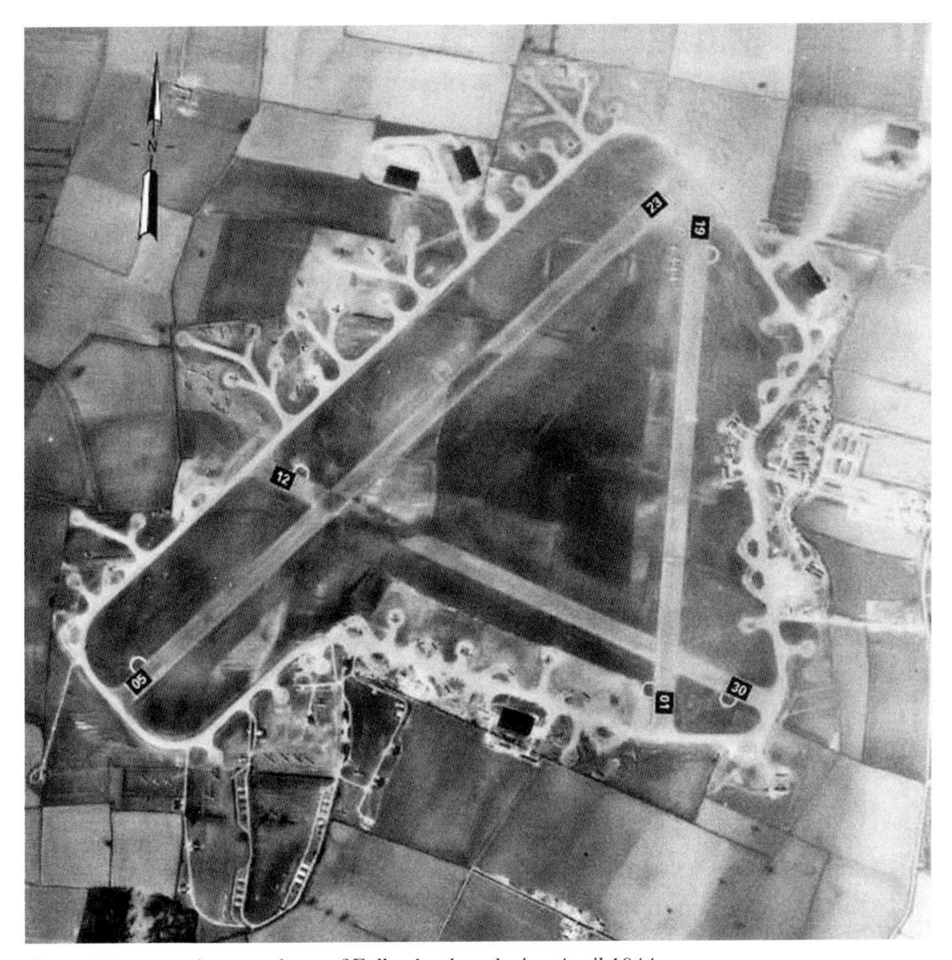

An aerial reconnaissance photo of Fulbeck taken during April 1944.

The RAF had vacated the site by 1976, since when it has remained virtually abandoned, although a small part of the site is used for go-carting. The main runway is gone (although a road runs along its path) and little remains of the secondary runways; most of the perimeter track is still there, together with traces of dispersals and a few crumbling traces of buildings. From a distance there is little to indicate that both the RAF and USAF once had a significant presence here.

Main features:
Runways: 233° 6,000 x 150 feet, 297° 4,200 x 150 feet, 007° 4,200 x 150 feet, concrete and tarmac surface. *Hangars:* two T2, three Glider T2. *Hardstandings:* fifty Heavy Bomber. *Accommodation:* RAF: Officers 517, ORs 2,285.

FULL SUTTON, Yorkshire

53°58'49"N/0°51'53"W; SE745543; 55.5ft asl. 8 miles E of York, off A166

Although the airfield at Full Sutton was created in accordance with standard arrangements for RAF bomber stations, this site was not completed until relatively late in the Second World War, opening as an RAF station in May 1944. As one of many similar airfields in or around the Vale of York, Full Sutton took advantage of the local area's flat features and lack of obstructions, and only a handful of buildings were demolished as part of the airfield's creation. However, the position of the surrounding agricultural land led to two of the three runways being either longer or shorter than standard, but in all other respects the airfield was completed to a standard bomber pattern with runways in a triangular pattern and a set of thirty-six loop dispersals scattered along the perimeter tracks, together with a pair of T2 hangars and a single B1.

A No 77 Squadron Halifax at Full Sutton in May 1944.

A 1972 image of Full Sutton's deteriorating control tower.

Once the airfield opened, the Halifaxes from nearly Pocklington soon arrived, and No 77 Squadron became the main resident unit at Full Sutton for the following year, completing 3,692 operational sorties from Elvington and Full Sutton by the end of the war. The last of these missions took place on 25 April 1945 when the unit attacked gun positions on Wangarooge Island.

With the war at an end, bomber operations were replaced by transport duties, and from 8 May the resident Halifaxes were all allocated to transport missions, although in just a matter of weeks No 77 Squadron exchanged its Halifaxes for Dakotas before moving to Broadwell in August. This might have been the end for Full Sutton but, unlike many other airfields, it remained in use (perhaps because of its short existence of little more than a year). No 231 Squadron established a training flight here on 1 December 1945, tasked with the conversion of Lancastrian crews, becoming No 1699 Heavy Conversion Unit a few weeks later. No 1552 Radio Aid Training Flight was the last unit to operate from Full Sutton during the wartime period, and when its Oxfords and Ansons left in October 1946 the airfield was closed.

It reopened in 1951 when No 103 Refresher Flying School was established here with a fleet of Vampires and Meteors. This became No 207 Advanced Flying School, and flying training became the station's main role until August 1954, when it left and Full Sutton reverted to RLG status for Driffield and Linton-on-Ouse. For a two-year period (until December 1957) the airfield was allocated to the USAF as a forward deployment base for Strategic Air Command; however, it was never used for this purpose and was largely unused through this period.

Flying effectively ended towards the end of 1959 when the RLG commitment was dropped and Full Sutton was designated as a new base for Thor missiles. By the end of 1959 three Thor launch pads and launch facilities had been completed and No 102 Squadron became the resident unit, operating Thor from this site until April 1963, when the force was disbanded. This also marked the end of Full Sutton as an RAF station, and the site was swiftly abandoned. Parts of the airfield were developed for industrial use and in the 1980s a new prison facility was constructed on the northern parts of the old airfield. Because so much development took place here, almost all of the airfield structure was obliterated, but despite this flying returned to the site and a private general aviation centre was opened, using a grass strip adjacent to one of the old runways, which had already been removed. In 2003 a surviving portion of the old main runway was restored for use and some private flying now takes place here. This is in fact the only substantial part of the old airfield that has survived, and to the casual observer the airfield bears very little resemblance to its original condition.

Main features:
Runways: 162° 6,000 x 150 feet, 046° 4,000 x 150 feet, 291° 4,200 x 150 feet, concrete and tarmac surface. *Hangars:* two T2, one B1. *Dispersals:* thirty-six loop type. *Accommodation:* RAF: Officers 183, SNCOs 410, ORs 850; WAAF: 367.

FYLINGDALES, Yorkshire

54°21'34"N/0°40'08"W; SE866967; 839ft asl. On Fylingdales Moor, adjacent to A169

Cold War icons: the BMEWS radomes at Fylingdales pictured during the 1970s.

The story of Britain's 'action stations' inevitably revolves around the history of many active airfields from where military flying has taken place since the First World War. However, it would be foolish to record the history of these sites without making mention of what is perhaps the most significant of all these places – even though it is not (and never has been) an airfield.

RAF Fylingdales is a radar facility that has been at the very forefront of Britain's air defence capabilities since it first opened for business. It enables Britain and the United States to maintain a continual watch over the skies, searching for any sign of what might be the initiation of a nuclear strike against the West. Without it, Britain and NATO would have had no means of launching a retaliatory strike against the former USSR and would therefore have had no credible nuclear deterrent posture. The history of this vital site is outlined by the RAF:

> 'The story of Fylingdales and the Ballistic Missile Early Warning System as a whole began on 4 October 1957. The nation awoke that morning to hear on their early morning news bulletin the "bleep" "bleep" "bleep" of Sputnik I – the first man-made earth satellite to be put into orbit, launched by the Soviet Union. This event had a dramatic effect because it illustrated that the Soviets had the potential to make a missile attack on the West launched from within their own land mass. With the launch of Sputnik I, the threat suddenly and dramatically changed, and because there were no current means of stopping an incoming missile, the Ballistic Missile Early Warning system was conceived to allow retaliatory strikes to be launched. The ability to warn of an attack and to respond rapidly brought peace through deterrence.
>
> The Ballistic Missile Early Warning System (BMEWS) was designed by the USA to give the radar coverage necessary to counter an inter-continental missile attack. In 1960 two stations were completed and became operational at Thule, Greenland, and at Clear, Alaska, both of which were operated by the United States Air Force. To complete the radar coverage for the United States and to give cover to the British Isles it was decided, by joint agreement, to build a third station in Great Britain. A site on the Yorkshire Moors was selected and construction work began in 1960. The site became fully operational in January 1964, and after years of continuous combined operation RAF Fylingdales still monitors for incoming missiles and conducts surveillance of all other objects in low earth orbit around the earth.

A pair of Phantoms from No 56 Squadron streak over the BMEWS radomes at Fylingdales.

The original radar housed in the infamous "golf balls" had a diameter of some 84 feet. The radar was capable of detecting objects out to about 3,000 miles using a peak power of 5 megawatts. Not only was it large and powerful, it was also extremely heavy: each scanner weighed about 112 tons. Nonetheless, these large mechanical radars were successfully maintained for over 28 years with only 13 hours of unscheduled unserviceability during that period.

However, by the mid-1980s the overall radar system at RAF Fylingdales needed modernisation because it was becoming increasingly difficult and expensive to maintain. The UK and US Governments announced their agreement to modernise the mechanical system on 22 May 1986. The US Administration awarded the contract for the radar to the Raytheon Company on 30 June 1988, while the contract for the construction of the buildings was placed in the UK and awarded to John Laing (Yorkshire) Ltd in July 1989. Construction work began the following month and continued apace, helped by the good weather of the 1989/90 winter. Some ninety local firms benefited from this work either as suppliers of goods or as sub-contractors. At the peak of building activity some 350 people swelled the work force at RAF Fylingdales.

The Solid State Phased Array (SSPAR) is very different from the original "golf balls" although the job it does is precisely the same. The current radar has no moving radar dish (no moving parts at all, in fact) and uses changes in electrical phase to steer the radar beam. The SSPAR searches out to some 3,000 miles, continuously looking for missiles. It can track some 800 separate objects simultaneously, thereby greatly enhancing the system's space surveillance capability.'

The US Department of Defense has upgraded the system over recent years so that the tracking capability could be incorporated into US missile defence improvements, and the SSPAR is now integrated into the fourth generation of US missile defence technology. The famous (almost iconic) 'golf balls' are long gone, but the new radar 'pyramid' now sits menacingly atop Fylingdales Moor, reminding passing drivers of Britain's constant state of readiness and the ever-present risks of nuclear war. Fylingdales is a fascinating site about which so little is known, even though it is undoubtedly one of the most important parts of Britain's defence posture, and probably the most obvious potential target for destruction, should the unthinkable ever happen.

Main features:

Runways: none. *Hangars:* none. *Dispersals:* none. *Accommodation:* not known.

GAINSBOROUGH, Lincolnshire

53°23'30"N/0°47'08"W; SK808889; 10ft asl. W outskirts of Gainsborough, adjacent to A631

Drivers approaching the well-known bridge across the Trent as they enter Gainsborough will probably be unaware that adjacent to the busy A631 is the site of a former RFC landing field. The site was adopted for use in 1916 and No 33 Squadron established a presence here in October of that year with BE.2 and FE.2 fighter aircraft. Tasked with the air defence of the local area, the unit operated from a variety of landing sites across the region and the field at Gainsborough was just one of many sites where few facilities were available other than a suitably flat landing field and a scattering of tents and huts.

No 192 Depot Squadron was established here in October 1917, but moved to East Retford just a few weeks after formation, although its headquarters remained at Gainsborough until June 1918, when it transferred to Kirton-in-Lindsey. The headquarters of No 48 Wing was located at North Sandfields and was sometimes also referred to as Gainsborough, but the landing site effectively fell into disuse by the end of 1919 and was returned to agricultural use shortly afterwards. There are now no visible signs of the RFC's presence here other than the general proportions of the landing area, which has remained largely unchanged. However, there are two unusual concrete base structures still to be seen here (albeit crumbling and overgrown), and these may well have been part of the RFC site.

Main features:
Runways: none, grass surface. *Hangars:* none. *Dispersals:* none. *Accommodation:* not known.

GAMSTON, Yorkshire

53°16'42"N/0°57'41"W; SK693761; 72ft asl. Adjacent to Gamston village, off A1(M) and B6387

The history of RAF Gamston can be traced back to the summer of 1941 when approval was given for an airfield to be constructed for bomber operations on an area of land that was declared suitable for purpose, subject to the closure of some small local roads and the demolition of a few buildings. The layout of fields and road networks required the adoption of a slightly unusual airfield layout, but all three runways were completed to standard length, even though it was agreed that there would be no potential for subsequent extensions if required. Likewise, the number of dispersals had to be slightly fewer than usual, although more hangars than usual were provided.

The airfield opened to flying late in 1942 and, as a satellite field for Ossington, No 14 (P)AFU's Oxfords were the first aircraft to operate here. They departed in May 1943 and Gamston was then allocated to Bomber Command's No 5 Group. Vickers Wellingtons and Miles Martinets from No 82 Operational Training Unit arrived in June 1943 and remained at Gamston until June 1944, apart from a few weeks of inactivity during the previous summer when runway work was completed. In that month the OTU was reduced in overall size and there was no longer a requirement for a satellite field.

In November 1944 No 91 Group's Servicing Unit transferred to Gamston, responsible for work on Wellingtons. The next tenant was No 3 Aircrew School, followed by No 93 Group Disposal Unit, but all these units had departed by January 1945, after which No 30 OTU arrived from Hixon, even though its much reduced training commitment meant that it too was redundant by June of that year. This marked the end of Gamston's flying activities, although the station remained very busy as the home of No 9 Air Crew Holding Unit, responsible for the accommodation and 'disposal' of countless RAF personnel who were about to make their way back into civilian life – or even return to their home countries. However, even this important role soon wound down and Gamston was effectively disused by 1946.

It came back into use during 1952 when the newly formed No 211 Advanced Flying School at nearby Worksop required a Relief Landing Ground. The unit's Meteor jets soon became a familiar site on the airfield and Gamston remained active supporting the unit's activities for two years, after which it again became largely unused, although it was not until 1957 that the RAF finally departed.

After a period of abandonment the airfield was adopted for use as a motor-racing circuit, but flying eventually returned when a private general aviation site was established here. As Retford (Gamston) Airfield, the former RAF base has slowly developed into a relatively quiet but successful facility. Even the RAF made a partial return during the 1970s when Finningley-based No 6 FTS occasionally used the airfield as a suitable site for circuit flying (mostly in Bulldogs). Two of the original runways are still intact (most flying takes place on the old main runway), while the third runway is still visible. Gamston has survived and, although only a small amount of flying now takes place here, the old wartime airfield is still very much alive and well.

Main features:
Runways: 216° 6,000 x 150 feet, 270° 4,200 x 150 feet, 332° 4,200 x 150 feet, concrete and tarmac surface. *Hangars:* four T2, one B1. *Dispersals:* thirty Heavy Bomber. *Accommodation:* RAF: Officers 131, SNCOs 240, ORs 594; WAAF: 307.

GILLING, Yorkshire

54°09'8"N/01°04'4"W; SE606739; 515ft asl. NE of Brandby village on B1363 at Well Lane

This virtually unknown site was used only briefly during 1916 when a variety of locations were allocated for military use and occasionally frequented by RFC aircraft. No 33 Squadron's B Flight is believed to have used the site while engaged in air defence of the local region, but no permanent changes to the site were made, nor were any buildings erected. The absence of any permanent structures has resulted in only scant evidence to locate the precise position of the landing field; it may well have been within Gilling Park, but the most likely site appears to have been a cleared field south of Mill Lane, which is now part of the sprawling agricultural landscape.

Main features:
Runways: none, grass field. *Hangars:* none. *Dispersals:* none. *Accommodation:* not known.

GOXHILL, Lincolnshire

53°40.48"N/0°18.56"W; TA113217; 26.5ft asl. 4 miles NW of Immingham, NW of A1077 at Goxhill village

Goxhill is a familiar name to military aviation historians but many people may well be surprised to know that this significant airfield has a location that seems totally at odds with its history. Laid on the shores of the Humber estuary, the industrial town of Immingham is just a few miles away and the city of Kingston-upon-Hull is only a little further away, across the muddy waters of the Humber. Even more unusual is that Goxhill was an all-American facility adopted for use by the USAAF in an area a long way from the more familiar fields of East Anglia, where almost all of America's wartime bases were assembled.

From the outset the airfield was unusual in that the runways (laid in a traditional triangular fashion) were rather shorter than those constructed at most RAF stations. The longest was less than 5,000 feet, and their relative shortness seems to have been due to the surrounding road network, which restricted the size of the airfield. It is not known why any of these roads were not diverted or closed.

A 1946 overhead view of Goxhill. Trees appear to have been planted across the airfield.

Opening as an RAF station on 1 August 1941 the station was part of No 1 Group Bomber Command and, with the airfield capable of supporting only training activities, the first unit to arrive was No 1 Group's Target Towing Flight. The unit's Lysanders stayed for only a few weeks before moving to Binbrook, after which the station transferred to No 12 Group, hosting only temporary detachments from RAF units, including a PAFU.

Early in 1942 the station was relinquished by the RAF and the USAAF assumed control; by June the first of countless American aircraft had arrived here, some of the first being P-38 Lightnings of the 1st Fighter Group. Spitfires (2nd Fighter Group) followed just a few weeks later, followed by P-39s (81st FG), more P-38s (78th FG), and P-47s from the 353rd, 356th and 358th Fighter Groups. All of these units were temporary residents tasked with the ad hoc training of crews assigned to the European Theatre of Operations, but in 1943 a more structured system was introduced and a single Fighter Training Group was established at Goxhill in November. This was the 496th FTG, comprising a P-38 element (554th Fighter Squadron) and a P-51 element (555th FS), although the 496th also acquired P-51s in August of the following year. Tasked with the training of replacement crews, Goxhill was designated as a Combat Crew Replacement Centre, and the skies above the Humber were regularly filled by Mustangs operating from the busy airfield. The 2nd Gunnery & Tow Target Flight was also established here to support the fighter operations, and the 3rd GTTF also operated from Goxhill

The aircrew memorial at Goxhill, with the remains of the airfield in the background. The plaque explains the origin of the propeller blade.

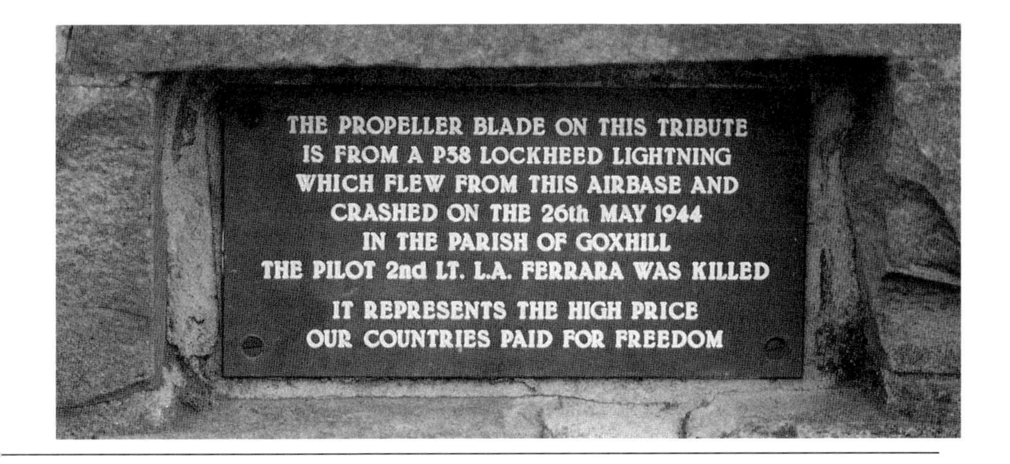

on a temporary basis. Eventually some 2,481 pilots were trained here, which illustrates the importance of this relatively obscure station.

The surviving remains of Goxhill as seen from the air in 1994.

The FTG finally left Goxhill in December 1944 when the last of its aircraft and personnel transferred to Halesworth. Just three weeks later the station was vacated by the USAAF, and the RAF resumed control on 20 January. The busy days when Mustangs, Lightnings and Thunderbolts roared skywards were suddenly gone, and Goxhill fell silent, the only aerial activity being the infrequent movements of aircraft handled by the Maintenance Units that adopted the site once the RAF returned. No 233 MU was the first, followed by No 35 MU in 1947, swiftly followed by No 93 MU. The last of these units to be based here was No 92 MU, which arrived in September 1952, although this unit was concerned primarily with weapons storage rather than aircraft, and by this stage flying activity had effectively ended completely.

When the MU left in December 1953 the station was largely abandoned, and in January 1962 the airfield was sold off, although the hangars were retained for storage use by the Ministry of Supply until the mid-1970s. Today the runways have been removed, although their presence is still clearly visible among the farm fields which now occupy the site. The mighty J-Type hangar and the T2 hangars have survived, and the perimeter track is still largely intact. Some of the old dispersals where Lightnings and Thunderbolts once sat can still be seen, but Goxhill's flying days are long gone.

Main features:
Runways: 070° 4,800 x 150 feet, 010° 3,600 x 150 feet, 130° 3,600 x 150 feet, tarmac surface. *Hangars:* four Blister, two T2, one J-Type. *Dispersals:* fighter pen type and fifty circular. *Accommodation:* Officers 190, ORs 1,519.

GREAT ORTON, Cumbria

54°52'27"N/03°04'35"W; NY310537; 236ft asl. 4 miles W of Carlisle off B5307

Situated in the bleak countryside beyond Carlisle, an area of farmland was first acquired for military use in 1941 when plans were made to create a satellite airfield for nearby RAF Silloth. Its history was short and unremarkable, but the station became familiar to countless RAF personnel who trained here, many referring to the base as RAF Watchtree, although this was only an unofficial name. The airfield was completed with three standard bomber-type runways, which resulted in a number of local roads being closed and the demolition of Watchtree House and farm, which were located in the centre of the new airfield.

Great Orton from the air in 2003.

Opening late in 1943, the first unit to arrive here was No 55 Operational Training Unit, which established a detachment of Hurricanes at the airfield, using Great Orton as a satellite field. By January 1944 this detachment had become B Flight of No 4 Tactical Exercise Unit, tasked with routine flying training for pilots awaiting postings to operational units. Flying activity with Hurricanes continued until March 1944 when the unit acquired Typhoons, operating in the fighter-bomber role; two months later the unit transferred to Acklington. Coastal Command aircraft were next to arrive when No 6 (Coastal) Operational Training Unit established a detachment of Wellingtons at Great Orton in June 1944, staying here until the following April. Air Sea Rescue detachments also arrived, these being from Nos 281 and 282 Squadrons, both equipped with Warwicks.

When these detachments departed in April 1945 the airfield became mostly unused; No 249 Maintenance Unit used it for storage of ammunition and bombs, and little flying activity occurred until the unit departed in 1951, and the RAF vacated the airfield in 1952. Left abandoned for many years, the site was gradually returned to agricultural use and the runways have now been partially removed, although parts do survive and their layout is still very evident. Sadly, monstrous wind turbines are now scattered across the site, but despite the presence of these hideous contraptions the perimeter track survives (much of which now forms part of the local road from Wiganby village) and many of the old dispersals can still be found, but only the ghostly memories of the white Warwicks remain.

Main features:
Runways: 082° 5,900 x 150 feet, 130° 4,250 x 150 feet, 024° 4,200 x 150 feet, tarmac and concrete surface. *Hangars:* four Blister. *Dispersals:* twenty spectacle, one circular. *Accommodation:* RAF: Officers 84, SNCOs 160, ORs 738; WAAF: 102.

GREENLAND TOP, Lincolnshire

53°34'58"N/0°13'22"W; TA177110; 59ft asl. 7 miles NW of Grimsby on Stallingborough Road, adjacent to Keelby village

This small and relatively unknown airfield was in effect little more than an area of farmland cleared for occasional use by the Royal Flying Corps. Opening in 1918, No 251 Squadron established a Flight here with DH.6 aircraft during May 1918, this being designated as No 505 (Special Duties) Flight, assigned to the maritime reconnaissance role and largely tasked with a growing requirement to locate German U-boats in the North Sea and Humber area. DH.9 aircraft were also assigned to the unit, but it is believed that they were not acquired until the unit had moved to West Ayton in November 1918, after which the landing field was abandoned. With no permanent facilities having been constructed, all evidence of the RFC's presence was soon gone and the landing field is now nothing more than a memory, lost among the farmland from which it first emerged.

Main features:
Runways: none, grass surface. *Hangars:* none. *Dispersals:* none. *Accommodation:* not known.

HELPERBY, Yorkshire

54°07'7"N/01°20'4"W; SE426703; 68.5ft asl. W of Brafferton village, 2 miles E of A1(M), 3 miles SE of RAF Dishforth

This First World War site was one of many areas of farmland that was cleared of obstructions and made available for use by the Royal Flying Corps. It was adopted by No 33 Squadron early in 1916, the unit having been established at Tadcaster with a mix of BE.2C, BE.12 and Scout aircraft tasked with air defence of the region. Many sites were scattered around the area and used by the Squadron as required, but unlike many of these rarely used landing sites Helperby was retained when No 33 Squadron left in October 1916 and No 76 Squadron established a Flight here with BE.2C and RE.8 aircraft, together with Bristol F.2Bs from the summer of 1918. Also unlike many of

An RFC aircrew with their RE.8 at Helperby.

these small landing sites, Helperby's landing field was not restricted by roads or boundaries, and (unusually) two hangars were constructed together with a scattering of admin and technical huts. The Squadron transferred its headquarters to Helperby in March 1918 and operated two further Flights at nearby Catterick and Copmanthorpe.

Activity continued at Helperby until the First World War ended, and in May 1919 the unit moved to Tadcaster, having already relinquished its aircraft pending disbandment. The landing field was swiftly returned to agricultural use and the hangars were retained for use as barns. Surprisingly, the site returned to military use during the Second World War when a bomb storage facility was established here, taking advantage of some local woodland for camouflage purposes. All traces of the landing field are now gone, but the fate of the two hangars remains unclear. They may well form part of what is now Cottage Farm.

Main features:
Runways: none, grass surface. *Hangars:* none. *Dispersals:* none. *Accommodation:* not known.

HEMSWELL, Lincolnshire

53°24'12"N/0°35'15"W; SK940905; 213ft asl. 7 miles E of Gainsborough on A631

A wartime aerial image of Hemswell, taken before the concrete runways were laid.

Military aviation first came to this site (then known as Harpswell) in June 1918 when No 199 (Night) Training Squadron transferred here from East Retford. The new site was relatively small and comprised little more than a cleared landing field, although it is believed that temporary hangars and support buildings were also constructed. Operating a variety of aircraft, the unit was joined by No 200 (Night) Training Squadron some five months later, and the two units remained active here until the summer of 1919 when both disbanded and the site was abandoned.

However, during the 1930s the site was identified as being suitable for development into a much larger airfield capable of supporting bomber operations, and construction of a new RAF station began here in 1936. A large, flat landing field was created together with a substantial domestic and technical site, dominated by huge C-Type hangars. A decision was made in 1943 to equip the airfield with concrete runways and three were laid in a standard triangular layout, together with a perimeter track and a scattering of dispersals, some of which were constructed on land on the south side of the adjacent A631 road. No 144 Squadron was the first unit to arrive (in February 1937) with a fleet of Ansons and Audaxes, joined by No 61 Squadron a few weeks later. After re-equipping with Hampdens, the two units departed in July 1941, but by this stage both had been heavily involved in operational duties, their Hampdens having ranged far and wide across occupied Europe; together the two units completed more than 3,000 operational sorties.

When they finally left Hemswell they were replaced by the Wellingtons of Nos 300 'Mazowiecki' and 301 'Pomorski' Squadrons, both Polish units assigned to bomber operations over Germany. These two units remained at Hemswell until the summer of 1943 when the airfield was temporarily closed while the concrete runways were laid. In June No 300 had to switch back to its old base at Ingham, again taking No 305 with it, so that work could begin on laying concrete runways at Hemswell to bring the airfield up to Class A standard. The main runway 17-35 was to be 2,000 yards long, 06-24 1,550 yards and 10-28 1,500 yards. The ends of 24 and 28 were both extended to 1,700 yards, apparently before the station was reopened. Thirty-six asphalt pan hardstandings had been put down in 1940-41, but during runway and perimeter track construction at least four were destroyed. Another six on the south side were compromised by being directly in front of runway 35. Sixteen of the surviving pans were on the other side of the A631, accessed by long taxiways. Evidently several of the original hardstandings were no longer held as suitable for aircraft parking, for seventeen loop-type standings were added along the perimeter track. Also it is evident that some extra work was done

An Avro Lincoln from No 61 Squadron pictured at Hemswell in 1947.

The film crew capture a Wellington T10 taking off from Hemswell during the making of the movie The Dambusters.

to the bomb store situated north-east of the camp. Additional domestic accommodation resulted in a total of 2,807 male and 298 female places at the station. The runway construction was carried out by J. McGeoch & Son Ltd, and the other work by B. Pumfrey & Sons Ltd.

In January 1944 the station reopened for business and No 1 Lancaster Finishing School arrived from Lindholme, its fleet of Lancasters gradually expanding to more than thirty aircraft as its training requirement increased, often using nearby Sturgate as a Relief Landing Ground. The unit remained in business until late 1944, by which stage the training requirement had ended. From 7 October 1944 Hemswell became a satellite for RAF Scampton, and in November No 150 Squadron arrived with its Lancasters, joined by No 170 Squadron later the same month. Both units saw a considerable amount of operational experience, with No 150 Squadron accruing some 840 missions and No 170 Squadron achieving 980 sorties. During this period a Ministry film crew recorded one such mission on colour film, and this rare footage is often seen as part of wartime documentaries even though it is seldom attributed to Hemswell.

The mortal remains of the Bomber Command Aviation Museum shortly before its departure from Hemswell's former technical site.

Flight Lieutenant Owen Scott DFC was a pilot with No 170 Squadron:

'I served in 170 Squadron No 1 Group of Bomber Command and was stationed in Hemswell, Lincolnshire. I joined up at the age of 19 and trained as a pilot. Later in the war I flew Lancaster bombers and served the full term of thirty operations over enemy territory in the period 1944-45. I and my crew were often airborne for 10½ hours. Our take-off time could be at any time of day, depending on the weather, and was a secret to us until we were briefed to report for duty. We might be wakened by our batman and told, "Briefing in 50 minutes, sir!" or receive a radio message while on a training flight. Our aircraft had to be fully operational. Sometimes, with the crew ready in position, the green light that told you to take off was replaced by a red one, telling you that your operation had been cancelled because of the weather or a change in tactics. Then, feeling somewhat relieved, you went back to bed.

Ninety per cent of our operations were at night, flying over enemy territory in the dark, dependent on maps and instruments, aware that enemy fighters would be on our tail. We flew manually and "rolled" the 37-ton loaded plane to see whether any enemy fighters were underneath us. To avoid fighter attack, we performed a "corkscrew" operation. Being caught in a searchlight was a terrifying experience as it made our plane an easy target for enemy shells.

Accuracy was of vital importance. The crew, which included the pilot, flight engineer, wireless operator and bombing team, had to synchronise watches to the second to ensure that every operation went according to plan, and that bombs hit the right target on schedule. The pilot would give the order, "Synchronise watches, 12.05. Hit target 12.10!" In front of every bombing flight flew the "Pathfinders", usually Mosquito aircraft, to drop coloured flares indicating the position of the different targets. This was doubly important when the air was filled with black smoke from exploding shells, spread by the wind into a dark haze, making it impossible to see for any distance. It was terrifying to be flying blind and put us at grave risk of collision with other aircraft. We were never warned of the danger of collision and were not fully informed about the number that took place at the time, though we experienced the enormous explosion that took place when two bomb-loaded planes collided. We were not aware of casualty numbers. All we knew the following day was that a friend or colleague was not to be seen and later heard that he had bought it.

I remember vividly the bombing raids on V-1 (flying bomb) launching pads in France. We had to fly in at a height of 7,000 feet, but the Pathfinders had to fly much lower than that. As we straightened up ready to drop our bombs, we were at grave risk from enemy fighters who knew the exact position of our targets and were waiting to attack. Coming up to the target it was "Left a bit, right a bit, steady, hold it, Skipper!" As soon as our bombs, 7 tons in weight, were gone, the aircraft leapt into the air. We closed the bomb doors, opened up the throttles and went into a shallow dive to increase our air speed to leave the area.

It was not until the war was over that we learned of the massive losses incurred by Lancaster and Halifax bombers and their crews – 7,000 bombers, 7,000 pilots, 55,000 men over all. The chances of survival were one in three. I have vivid memories of the night we bombed Dresden. We had no idea then of the damage our bombs had done, but I shall never forget the sight of the city in flames. It seemed an easier mission than usual owing to the lack of ground fire and enemy fighters.'

No 1687 Flight arrived at Hemswell in April 1945 with Spitfire and Hurricane aircraft assigned to evasion training duties, and when the bomber squadrons departed towards the end of the year the fighters were the sole users of the airfield until the Flight disbanded in October 1946.

Unlike many bomber airfields, Hemswell remained active after the Second World War, and Nos 109 and 139 Squadrons (both with Mosquitoes) were transferred here, after which the Lincolns of Nos 100, 83 and 97 Squadrons arrived. The Mosquito 'target marker' units returned early in 1950, making Hemswell a very busy bomber station, and No 100 Squadron soon departed in order to ease the congestion. The aged Lincolns were soon replaced by the RAF's new Canberra jet bombers, and the Bomber Command Jet Conversion Flight was established here in July 1952.

The Lincoln did not disappear entirely, however, as No 199 Squadron moved to Hemswell in April 1952, tasked with signals and radio counter-measures operations. The Lincoln Conversion Flight also operated here for a year (until 1957), but more changes took place in 1956 when two squadrons disbanded (Nos 83 and 97 became part of the Bomber Command Bombing School at Lindholme) and the two Canberra squadrons transferred to Binbrook. No 199 Squadron took its Canberras to Honington, leaving its Lincolns at Hemswell to become No 1321 Flight and eventually C Flight of No 138 Squadron. This was the last flying unit to operate from Hemswell, and when these Lincolns departed at the end of 1957 the station's flight operations effectively ended.

However, the station was retained by the RAF and after only a year the base was back in business as a site for Thor ICBMs, a cluster of three launch pads having been constructed on the northern side of the airfield. Operated by No 97 Squadron, the missiles were delivered directly to the base thanks to the availability of the main runway, which was still in good condition. USAF C-124 and C-133 transports were seen infrequently here, but the proximity of the Thor site to the runway eventually led to the ending of these flights, and missiles were then routinely taken to and from Scampton by road, for air transportation. This effectively ended all flying activities at Hemswell, and when the Thors were removed in 1964 only glider flying remained, courtesy of No 643 Volunteer Gliding School, which moved to Hemswell in October 1965 and stayed until April 1974.

The RAF's ill-fated TSR2 programme would have seen Hemswell re-emerge in a new role, as the TSR2 Ground School was to have been based here, equipped with a pair of TSR2 instructional airframes. It seems likely that these would have been delivered by air, but with the Thors gone the main runway might well have been put back into use for communication aircraft, or even an occasional visit by operational TSR2 bombers. But when the TSR2 programme abruptly ended the Ground School was closed and the RAF soon vacated the site.

The runways gradually succumbed to the attention of the local farmers (although the northern portions of the main runway survive in surprisingly good condition), but in most places they are now little more than farm tracks. The crumbling remains of the Thor pads are almost gone but the mighty hangars have survived (despite a recent fire) and remain in use for agricultural storage. The majority of the station buildings have also survived, most used for commercial or retail purposes, while some cleared areas of the site are used for weekly car boot sales. Those who watch the final minutes of the famous movie *The Dambusters* will doubtless recall the footage of Guy Gibson (played by Richard Todd) walking down RAF Scampton's main approach road. It is in fact Hemswell, and a surprising amount of the movie's footage was filmed here, the film company's Lancasters (and other aircraft) being based here for the duration of the shoot.

Main features:
Runways: 170° 6,000 x 150 feet, 239° 5,100 x 150 feet, 281° 5,100 x 150 feet, concrete and tarmac surface. *Hangars:* one T2, four C-Type. *Dispersals:* seventeen spectacle, nineteen circular. *Accommodation:* RAF: Officers 240, SNCOs 566, ORs 1,220; WAAF: 256.

HIBALDSTOW, Lincolnshire

53°29'47"N/0°31'18"W; SE981009; 26.5ft asl. 3 miles SW of Brigg on B1206, S of Hibaldstow village

Situated in an area that was literally crowded with military airfields, RAF Hibaldstow was created as a satellite airfield for nearby Kirton-in-Lindsey and land was acquired for this purpose in 1939. Some fairly significant work was required in order to prepare the area, including roads being closed, hedgerows uprooted and some levelling

A line-up of Hurricanes from No 253 Squadron at Hibaldstow.

of the local terrain. By 1941, however, a substantial aerodrome had been created with three runways of fairly standard (if not slightly short) length, together with a perimeter track and a series of dispersals, and some additional parking areas equipped with small Blister hangars.

The station opened in May 1941 as part of No 12 Group, and within days the first Defiants and Hurricanes, of No 255 Squadron, arrived from Kirton-in-Lindsey. The unit continued to use the latter base's hangars for aircraft servicing but daily operations were mounted from Hibaldstow. When the unit received Beaufighters a detachment was set up at Coltishall and the Squadron moved there en masse some two months later. This made way for No 253 Squadron equipped with Hurricanes, destined to operate alongside No 1449 Flight, which moved here in September 1941 with a fleet of Turbinlite Havocs. Having moved to Hibaldstow from the wilds of Skeabrae, No 253 Squadron stayed until November 1942, when operations shifted to North Africa. The Havoc Flight remained at Hibaldstow and, although the combined operations with the Hurricanes had not been a huge success, their night interception capabilities slowly improved and in September 1942 the Flight became No 538 Squadron, joined by No 532 Squadron two months later. However, the Turbinlite Havocs were eventually judged to be insufficiently effective and both units disbanded in January 1943.

After a short period of inactivity the station became home to No 53 Operational Training Unit in May, its headquarters being at Kirton-in-Lindsey, Hibaldstow acting as the main satellite field. The unit's Spitfires became a familiar sight in the local area and with a fleet of more than 100 Spitfires (together with some Masters and Martinets) the two airfields were kept very busy. The Operational Training Unit became part of No 61 OTU (based at Rednal) during May 1945, and training activities at Hibaldstow came to an end apart from a brief three-day stay by Spitfires and Harvards from No 5 (Pilots) Advanced Flying Unit, which went on to become No 7 Flying Training School. The FTS was established at Kirton-in-Lindsey and, although Hibaldstow seems likely to have been designated as a satellite airfield, it was never used by the FTS for this purpose, and flying at Hibaldstow effectively came to an end during the summer of 1946.

The airfield remained unused for many years, but slowly re-emerged as a private airfield and a variety of light aircraft began to appear here. It is now home to a very popular parachute club, and on most days a few aircraft can be seen on the airfield, taking parachute jumpers up to heights of 10,000 feet or more. The three runways have survived in reasonable condition and the perimeter track is still intact too, together with a few traces of some of the dispersals, including an unusual track that leads off to the north-east across South Carr Lane to what may well have been gun butts.

Main features:
Runways: 084° 4,800 x 150 feet, 155° 4,400 x 150 feet, 200° 3,900 x 150 feet, asphalt surface. *Hangars:* eight Over Blister, four Extra Over Blister, one Bellman. *Dispersals:* twelve circular. *Accommodation:* RAF: Officers 112, SNCOs 70, ORs 884; WAAF: 205.

HOLME-ON-SPALDING-MOOR, Yorkshire

53°48'21"N/0°44'40"W; SE827351; 6ft asl. 4 miles SW of Market Weighton, E of A614, SE of Holme-on-Spalding-Moor village

The accepted practice of naming RAF stations after the nearest significant village or parish sometimes created some very unusual results. Holme-on-Spalding-Moor is a classic example and it is little wonder that it was more commonly referred to as 'Holme' or even 'H.O.S.M.' in more recent years. Land was first acquired here in May 1940, and a large area of farmland was cleared for use, all of which lay only a few feet above sea level. The various drainage dykes (combined with local roads) largely dictated the dimensions and layout of the airfield, but despite a slightly unconventional configuration, three full-length bomber runways were eventually constructed together with a perimeter track that curved inwards along its northern boundary to incorporate the station's admin and technical site. The number (and layout) of hangars was also unusual, as was the positioning of dispersals, but despite these quirks a substantial bomber airfield was ready to open in August 1941.

Left: *A No 76 Squadron Halifax pictured at Holme after crash-landing following a mission to Essen in July 1943.*

Below: *An aerial view of the airfield and local area at Holme-on-Spalding-Moor in 1945.*

The first unit to arrive was No 458 Squadron, an Australian unit equipped with Wellingtons. This was swiftly joined by No 20 Blind Approach Training Flight, which brought a small fleet of Oxfords here to conduct night approach training. The unit was renamed as No 1520 Beam Approach Training Flight in October 1941 and stayed here until September 1944, when it transferred to Sturgate. Meanwhile No 458 Squadron had completed sixty-five operational bombing missions by January 1942 when it stood down prior to transferring to the Middle East. The BATF Oxfords became the sole residents of the airfield until August when Halifax bombers arrived from Breighton's Conversion Flight, only staying for a brief six-week period.

It was not until the end of September that operational flying resumed when No 101 Squadron arrived with Wellingtons, which were quickly replaced by Lancasters. The unit stayed until June 1943, when the Lancasters departed for Ludford Magna, replaced just a day later by more Halifaxes, this time from No 76 Squadron. These bombers stayed at Holme until the end of the Second World War, by which stage the Squadron had completed a staggering 5,123 operational sorties (both from here and other bases).

The unit's last operational mission was on 25 April 1945, but the Squadron did not disband until August, by which time the station's only other resident unit had also disbanded. This was No 1689 Bomber Defence Training Flight, formed here in February 1944 equipped with Spitfires and Hurricanes and tasked with dissimilar combat training for No 4 Group's bomber crews. When No 76 Squadron departed, the unit had already re-equipped with Dakotas as part of Transport Command, and earlier in August No 512 Squadron had arrived at Holme-on-Spalding-Moor with its own fleet of Dakotas. However, these were also only short-term residents, and the Squadron left for the Middle East in October, leaving the airfield deserted. It was soon relegated to Care and Maintenance and almost forgotten.

Largely as a part of the Korean War crisis, the RAF increased its training requirements and as part of this expansion No 14 Advanced Flying Training School was allocated to Holme-on-Spalding-Moor in March 1952, equipped with a fleet of Oxfords and Prentices. The unit's stay here was brief, and in January 1953 it left and the station was allocated to the USAF as a standby airfield for Strategic Air Command bombers. No major improvements were made to the airfield and no USAF

Left: *The Blackburn Buccaneer S1 at Holme-on-Spalding-Moor during flight trials.*

Below: *A line-up of Blackburn NA.39 trials aircraft at Holme-on-Spalding-Moor.*

aircraft were ultimately based here. Despite the USAF's presence, very little flying activity took place for some three years, after which the Americans departed and the airfield was again left unused.

It was at this stage that the airfield was taken over by Blackburn Aircraft. This historic and well-known aircraft manufacturer had a manufacturing base many miles away at Brough on the banks of the Humber, but the factory airfield was only small, designed for small, light biplanes. By 1957 Blackburn was working on a jet-powered nuclear strike aircraft, and an airfield with more substantial facilities was

A trials fleet Buccaneer S2 returns from a test flight and makes a low pass for Blackburn's photographer in 1970.

Although the runways are gone, the main site at Holme survives, still dominated by the huge hangars.

urgently required. Consequently, Blackburn set up a flight test operations centre at Holme-on-Spalding-Moor for testing its new NA.39, which became the illustrious Buccaneer. Virtually all of the Buccaneer's flight testing was conducted here, and each aircraft was transported here by road from Brough prior to making its first flight. Test flying on the type continued here long after the aircraft entered service with the Fleet Air Arm and the RAF, and when the services adopted the American F-4 Phantom these too were flight-tested here; Blackburn, by then part of the Hawker Siddeley Group and eventually British Aerospace, had assumed design responsibility for the type in the UK.

RAF activity at the site also included a four-year stay by No 1 Flying Training School, its Jet Provosts sharing the airfield with Blackburn's test fleet until the summer of 1966. The mighty Buccaneer and Phantom jets remained at Holme-on-Spalding-Moor until the end of 1983, when a smaller test facility was established at RAF Scampton – a less expensive option that also avoided the difficulties of transporting aircraft by road across the Humber.

By 1984 all flying had ended at Holme and it was abandoned, although many of the admin and technical buildings eventually found a new role as commercial and industrial facilities. Sadly, the runways were removed with almost indecent haste and today they have been wiped from the landscape with only a narrow farm track revealing the path of the old main runway. Almost all of the airfield is gone, but the huge J-Type hangar and two T-Type hangars remain, together with a stretch of the curved perimeter track that lays in front of them, close to the surviving carcass of the control tower. The roar of the Halifax bombers is long gone, as is the thunder of the Phantoms and Buccaneers. The Holme Industrial Estate is all that remains of this once busy and important airfield.

An unusual view of the approach to what was once Holme's main runway. All that remains is a farm track.

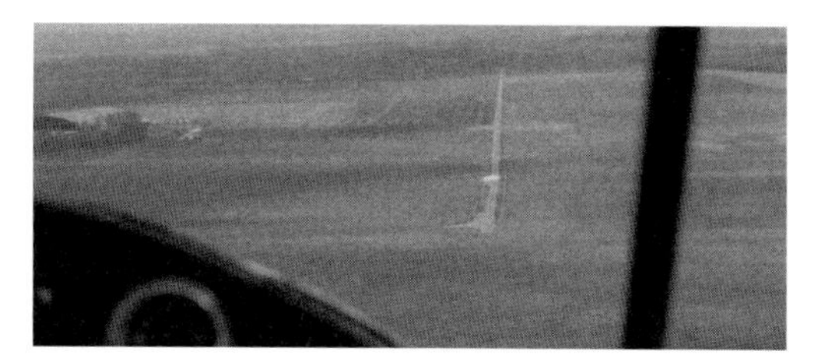

Main features:

Runways: 305° 6,000 x 150 feet, 215° 4,200 x 150 feet, 260° 4,200 x 150 feet, tarmac surface. *Hangars:* five T2, one J-Type. *Dispersals:* thirty-six Heavy Bomber. *Accommodation:* RAF: Officers 134, SNCOs 402, ORs 1,405; WAAF: 301.

HORNBY HALL, Cumbria

54°39'35"N/02°40'07"W; NY569295; 436ft asl. 1 mile E of Penrith, N of A66, adjacent to Barackbank Wood

Although hardly an airfield at all, a site was established here on 1 April 1941 for use by No 22 Maintenance Unit based at Silloth. This site (No 9 Satellite Landing Ground) was created to provide storage space for aircraft awaiting delivery to operational units. Available space at the main MU airfields was often difficult to find at this busy time in Britain's history, and aircraft had to be routinely exported to other sites. Likewise, there was a clear advantage in scattering stored aircraft across a number of dispersed sites where they created a far less obvious target for enemy attention, often concealed in surrounding woodlands.

Hornby Hall was an ideal site for such use with relatively flat and uncluttered fields and a substantial area of woodland that could be used for camouflage. Only temporary structures were built, the most substantial being Robin hangars, and over a four-year period a wide range of aircraft types could be seen here, including Hudsons, Battles, Bothas, Hurricanes and Mustangs. No 12 MU also used the site from January until May 1942, storing Wellington bombers here. The storage of aircraft continued until July 1945, when the last aircraft were removed. After a short period during which German prisoners of war were housed here, all connections with military activity came to and end. There is little evidence of the RAF's presence at this site now, although a few of the scattered farm buildings may well owe their origins to the service's presence during the Second World War.

Main features:
Runways: none, grass field. *Hangars:* none. *Dispersals:* none. *Accommodation:* not known.

HORNSEA MERE, Yorkshire

53°54'21"N/0°11'17"W; TA191470; 0ft asl. W of Hornsea on B1244

The rather rudimentary but meticulously maintained guardroom building at Hornsea. Barry Ketley

This unusual site was created during 1918 as part of the Royal Naval Air Service's efforts to improve its anti-submarine patrol capability in the region. No 251 Squadron formed at nearby Hornsea in May 1918, acting as parent unit for a series of four Flights, each equipped with DH.6 aircraft that were based at other sites in the area. No 79 (Operations) Wing was based at Hornsea and the seaplane base at Hornsea Mere was part of this establishment. No 248 Squadron formed at the base during August 1918, equipped with Sopwith Baby biplanes and a mix of Short 184 and 320 aircraft, the unit having evolved from No 404 (Seaplane) Flight, which had formed at East Halton (Killingholme) some weeks previously. No 248 Squadron joined No 79 Wing at Hornsea Mere, although 404 Flight moved to North Coates with its DH.6 land planes in September.

The base was small, comprising only of a few admin and technical buildings and a couple of Bessonneau hangars, adjacent to two slipways that led into Hornsea Mere's waters. Aircraft operated here until March 1919 when the RNAS vacated the site. Today the area (a promontory of land known as Kirkholme Point) is a haven for recreational boating. Little evidence of the RNAS base remains, although there are some traces of concrete foundations and some of the existing huts and sheds may possibly owe their origins to the Navy's short stay here.

Main features:
Runways: none, lake surface. *Hangars:* two Bessonneau. *Dispersals:* none. *Accommodation:* not known.

Sopwith Baby N1413 pictured at Hornsea after being pushed out from its Bessonneau hangar. Groundcrews are attaching floats before pushing the aircraft to the nearby slipway for launch. Barry Ketley

HOWDEN, Yorkshire

53°47'14"N/0°52'01"W; SE747329; 20ft asl. 5 miles N of Goole off B1228 at Wood Lane

During the dark days of the First World War a large number of small and rudimentary landing fields appeared across the Yorkshire region, mostly used by Royal Flying Corps units engaged in air defence operations. Almost all of these sites were very small and equipped only with temporary structures. By contrast, the Royal Naval Air Service created a far larger site at Howden, which eventually embraced more than 1,000 acres of land, and although both RFC and RNAS fixed-wing aircraft did use the field on a sporadic basis the site was almost exclusively devoted to airship operations.

The site opened in March 1916 and the first airship to arrive here was C11, which transferred from Kingsnorth on 26 June; five of these Coastal types were present by the end of that year. Two large hangars were constructed and the base gradually developed to accommodate more than 1,000 personnel, with admin and technical buildings, domestic facilities and even a railway spur that linked the base to Howden Junction. As the number of airships here increased, a series of out-mooring sites appeared across the region, enabling RNAS crews to operate more effectively in their maritime patrol role, primarily concerned with the location of marauding German submarines.

As the base expanded, two more huge hangars appeared, one of these (No 2 Rigid Hangar) becoming the largest hangar structure anywhere in the world at that time. With a height of 100 feet, a length of 700 feet and a width of 150 feet, it was impressive by any standards, and the hangar complex became the dominating feature of the landscape, visible from a distance of many miles.

Howden's mighty airship hangars under construction.

An aerial view of Howden showing one of the massive airship hangars. Also visible is the huge windbreak structure that enabled airships to enter or emerge from the hangar without any interference from crosswinds. Barry Ketley

Howden's airships completed countless patrols over the grey North Sea, but despite their dedicated activity there is no record of any direct action against the enemy; indeed, the most notable event to have taken place at Howden was the catastrophic loss of five airships in August 1918 when a spark ignited fuel and led to an uncontrollable fire that caused great damage but thankfully resulted in no loss of life. Far more serious was the loss of Howden's R38 in August 1921, which broke up during a familiarisation flight over the Humber. Some forty-four crew lost their lives, many of them Americans, who were due to operate the new airship type in US service.

A look inside one of Howden's airship hangars as R100 nears completion.

The mighty R100 receives attention inside its equally huge hangar at Howden.

By this stage Howden had been transferred to RAF control, although the base's maritime patrol duties remained unchanged after the service came into being in 1918. The RAF finally vacated Howden in September 1921, by which time the base had become the RAF's largest site in the region. However, interest in airship operations had diminished in favour of fixed-wing operations, and the RAF no longer saw any use for the lumbering gasbags. Despite this, civilian interest in airships continued and Howden was sold to the Airship Guarantee Company, part of the Vickers-Armstrong group, and after some reconstruction work the site reopened in 1926.

The company's first (indeed, its only) creation at Howden was the R100, a magnificent and truly massive machine designed by a team led by Barnes Wallis. Completed in 1929, it made its first flight on 16 December when it immediately departed for the Government's Airship Establishment at Cardington, via York. Howden's staff immediately turned their attention to new projects while the R100's sister ship, the R101, was completed at Cardington. When this airship crashed, interest in airship design collapsed immediately, and by the end of 1930 Vickers Armstrong had decided to abandon Howden and the site was vacated during December.

The mortal remains of airship R38 after crashing in the Humber estuary while operating from Howden.

This marked the end of Howden's association with aviation, and the many buildings (including the huge hangars) were demolished. Eighty years later, only a handful of crumbling brick huts remain at this site, and the privately owned land (which cannot be accessed) is just part of the surrounding farmland. The concrete foundations of the hangar door and wind-break structures are still here, and the circular concrete mooring bases can be easily seen from the air. But for anyone driving along the adjacent B1228 there is nothing to indicate that such a large and significant air base was ever here.

The former airship base at Howden pictured in 1994; the few surviving traces of the site are clearly visible.

Main features:

Runways: none, grass surface. *Hangars:* two Coastal Airship Sheds, two Airship Rigid hangars. *Dispersals:* none. *Accommodation:* not known.

The surviving remains of Howden's huge airship hangar.

HUGGATE WOLD, Yorkshire

54°0'14"N/0°41'22"W; SE859572; 708ft asl. 6 miles NE of Pocklington on A166 at Pefham Lane

An unusual image of steel mesh being laid on one of the emerging runways at Huggate Wold.

This very unusual site was not a designated airfield at all, even though it did briefly host a small number of fighter aircraft. The area was surveyed and chosen as a potential location for a bomber airfield but, despite being listed in official documents in 1942, it was never selected for further development. However, the British Government did utilise the site in 1943 for experiments with temporary airfield surfaces, most notably steel mesh planking, which was eventually used extensively at many airfields. Most of the trials work here was geared towards the projected invasion of France and concentrated on the logistical and technical requirements for laying surfaces and operating aircraft from them. As part of these trials, Mustang aircraft were deployed to Huggate Wold in October 1943, being drawn from Nos 168 and 170 Squadrons. These two units sent a large fleet of aircraft over from their base at Hutton Cranswick to conduct a week of intense operations, testing the landing surfaces for durability. When these tests were completed the temporary site was soon abandoned, and returned to agricultural use in 1944. Not surprisingly there is now no trace of the RAF's brief presence here, and visitors can only imagine the short but intense week of operations when Mustangs pounded the field's landing surface.

Main features:
Runways: none, grass surface. *Hangars:* none. *Dispersals:* none. *Accommodation:* not known.

HULL (HEDON), Yorkshire

53°44'42"N/0°13'36"W; TA170290; 6ft asl. 6 miles E of Hull at A1033 junction with Staithes Road

This site was originally a horse-racing course, and like many other similar grounds its flat, uncluttered fields were identified as being ideal for aviation. Racing ended in September 1909 and civil aviation soon took up residence here, the first recorded aircraft being flown by Gustav Hamel in July 1912. The Royal Flying Corps adopted the site (normally known as Hedon) when the First World War began, the former racecourse being controlled by the Yorkshire Regiment.

Although Hedon did have a brief association with military aviation, the site was essentially a civilian airfield as illustrated by these aircraft, a Civilian Coupe and a de Havilland monoplane The proximity of the Hull railway line is clearly visible. Barry Ketley

Like many other sites across the region, Nos 33 and 76 Squadrons used the airfield on an infrequent basis as a landing ground for their fighter aircraft engaged in air defence of the region until the end of the war, when the RFC relinquished any further interest in the site. Attempts were made to reintroduce civil aviation, both for commercial and recreational purposes; Hull Flying Club was established here in 1929, and Hull's Council acquired the site for development into a municipal airport.

1931-vintage Civilian Coupe G-ABNT pictured with its happy pilots at Hedon. Barry Ketley

The former flight shed at Hedon, still standing and in use as a storage facility.

Amy Johnson flew to the airfield for a much-publicised appearance in 1930 just a few weeks after her epic flight to Australia. However, it remained fairly inactive and closed when the Second World War began. Some consideration was given to the possibility of developing the site for military operations, but such ideas were eventually abandoned and scrapped cars were scattered across the field in order to render the area unsuitable for landing and ensure that it remained out of the enemy's reach.

After the war there was another attempt to reintroduce civil flying and the site was cleared of cars and other obstructions ready for a new lease of life, but little interest emerged and after a demonstration flight in 1964 (which almost ended in disaster when the participating aircraft lost its undercarriage) the airfield was soon forgotten. It is now used by motorbike riders and there is little to remind visitors of its links with aviation. Indeed, the only remaining place of significant interest is a mile away at Fort Paull, where the sole surviving example of the mighty Beverley transport continues to brave the salty North Sea air.

Main features:
Runways: none, grass surface. *Hangars:* none. *Dispersals:* none. *Accommodation:* not known.

HUTTON-IN-THE-FOREST, Cumbria

54°42'24"N/02°49'27"W; NY469349; 574ft asl. 3 miles N of Penrith on B5305 at Tollbar Wood

In 1941 this relatively unknown site was allocated to the RAF and No 22 Maintenance Unit, which was headquartered at nearby Silloth. After the area was cleared to create a suitable landing ground, it opened as No 8 Satellite Landing Ground on 1 June of that year. An Avro Anson was the first aircraft to arrive, being used to conduct safety trials before the landing field was approved for regular use. The MU ferried a variety of aircraft here for storage, the first being Blenheims and Bothas, although other types were also present at various times, the largest of which were Wellington bombers. Most of the aircraft were housed off the landing field under the cover of the surrounding woodland, although some Robin hangars were erected. No 12 MU (based at Kirkbride) also used the site, as did No 18 MU from Dumfries. However, by July 1945 the storage facility was no longer needed and the remaining aircraft were flown out, leaving the site to be returned to agriculture. The field is still clearly discernable, and the surrounding woods are still to be seen, where many fighter and bomber aircraft once lurked, hidden away from the enemy's eyes.

Main features:
Runways: none, grass surface. *Hangars:* none. *Dispersals:* none. *Accommodation:* not known.

HUTTON CRANSWICK, Yorkshire

53°56'52"N/0°27'41"W; TA010512; 105ft asl. 4 miles S of Driffield between Hutton Cranswick and Watton on A164

Situated in what is often described as 'Bomber Country', this airfield was in fact designed for fighter operations, even though it was initially selected as a bomber base and built to the appropriate bomber standards with a normal layout of three concrete runways. However, the complex of hangars reflected the airfield's intended use as a fighter base. As part of No 13 Group

A Miles Master from No 1634 AACF, which crashed during July 1943 while based at Hutton Cranswick.

Fighter Command, the station opened in January 1942, and the first aircraft to arrive were Spitfires operated by No 610 Squadron, which relocated here from nearby Leconfield. The unit's four-month stay was followed by more Spitfires, this time from No 19 Squadron, and over the following months the RAF deployed a variety of fighter units to the base to enable each squadron to rotate from operational duties to training and rest periods. Consequently, Hutton Cranswick hosted a long list of units at various periods, most of which stayed for only a matter of weeks before returning to operations.

The Spitfire was undoubtedly the most common type, and among the many Spitfire squadrons based here were Nos 1, 19, 26, 91, 124, 234, 302, 306, 403, 421, 443 and 610. Typhoons also came to Hutton Cranswick, as did Mustangs. Longer-term residents were also here at various periods; No 291 Squadron was established at Hutton Cranswick, combining Anti-Aircraft Cooperation Flights with No 3 Anti-Aircraft Practice Camp to form a new unit tasked with the provision of target tugs and 'adversary' aircraft in the shape of Martinets, Lysanders, Henleys and even Hurricanes. A variety of specialised Flights came here, all tasked with the growing need to provide realistic training facilities for the fighter squadrons; their valuable role enabled countless fighter pilots to hone their skills before returning to operational duties.

The combination of fighter types and the specialised target aircraft ensured that Hutton Cranswick was a busy airfield hosting a variety of aircraft types, although the most unusual were undoubtedly the Supermarine Walrus amphibians operated by No 278 Squadron on Air Sea Rescue duties, from April 1943 until April 1944. No 16 APC (the last of Hutton's specialised target facilities units) disbanded in February 1945, and the last fighter squadron, No 129, departed for Lubeck in May 1946. The last aircraft left Hutton Cranswick later in the same month when No 288 Squadron transferred its Vengeances to East Moor.

The airfield was soon abandoned and remained unused until 27 April 1964 when a No 92 Squadron Lightning made an unscheduled appearance here. The Lightning's pilot declared an emergency and was unable to reach home base at Leconfield, but after attempting to make a forced landing at Hutton Cranswick the aircraft crashed and he was killed. This was a particularly tragic end to this once busy site's valued association with military aviation. The runways are now gone, and only a small part of the perimeter track is still visible. However, the T-Type hangar has survived, hidden among what is now an industrial estate.

Main features:
Runways: 279° 4,800 x 150 feet, 041° 3,300 x 150 feet, 340° 3,300 x 150 feet, tarmac surface. *Hangars:* four Over Blister, four Extra Over Blister, one T1. *Dispersals:* seven fighter type. *Accommodation:* RAF: Officers 122, SNCOs 114, ORs 1,236; WAAF: 275.

INGHAM, Lincolnshire

53°20'26"N/0°33'18"W; SK963835; 210ft asl. 8 miles N of Lincoln on A15, adjacent to Ingham village

Drivers heading north on the A15 cannot fail to notice the huge hangars that dominate RAF Scampton's airfield. The straight Roman road curves neatly around the station's huge V-Bomber runway before resuming its historic path, adjacent to a flat and featureless farm field. Few people realise that this was once another RAF airfield, less than 10,000 feet from RAF Scampton's boundary.

Wellington bombers were the largest and most numerous types to operate from Ingham.

This airfield site was first identified in the 1930s as a potential location for a satellite landing ground, and a grass airfield was established here in 1940, eventually opening for flying as, initially, RAF Ingham. Sandwiched between the A15 and B1398, east of Ingham village, gradients restricted the size of the flying field, making it unsuitable for the hard three-runway configuration then seen as the necessary standard for a bomber airfield. A total of twenty-four pan-type aircraft standings were put down and, in the spring of 1942, a complete perimeter track with twelve extra dispersals. Plans also called for the extension of the grass runway across the road on the southern boundary, but this work does not appear to have been carried out. The technical site, with a T2 hangar, bordered the B1398, and a second T2 was erected near the eastern boundary alongside the A15. Although the landing field was not equipped with concrete runways, a perimeter track was laid and three hangars were constructed and, because the airfield was restricted by a series of roads and farm field boundaries, it was effectively split into two portions with a farmhouse (and a newly constructed control tower) neatly positioned between them.

The first aircraft to arrive here were Oxfords from Shawbury's No 11 Service Flying Training School, which appeared late in 1941. As a designated satellite field for nearby Hemswell, Wellington bombers from No 300 'Mazowiecki' Squadron first appeared at Ingham in May 1942; on the 18th they set up operations here, flying their first mission over Germany the following evening. The Squadron stayed until January 1943, after which it relocated to Hemswell, only to return in June before departing once more in March 1944 to Faldingworth, just a few miles away, in order to re-equip with Lancasters.

While the unit was gone, the Wellingtons of No 199 Squadron arrived in February 1943, departing four months later, and No 305 'Weilkopolski' Squadron arrived in June for a stay of some three months. Through the winter of 1943/44 No 16 Service Flying Training School (a Polish unit based at Newton) used Ingham as a Relief Landing Ground. From March 1943 the Bomber Command Night Bomber Tactical School was based here, staying until May 1944. Spitfires arrived in February 1944 when No 1687 Bomber (Defence) Training Flight moved to Ingham, and just a few weeks later No 1482 (Bombing) Gunnery Flight arrived with a mixed fleet of Martinets and Wellingtons, together with Hurricanes later in the year, although the unit promptly disbanded in December.

One month previously the station was renamed as RAF Cammeringham, in order to avoid confusion with other locations that shared the same name. This seems to have been an odd decision, given that the station was already heading towards closure. The last aircraft (from the BDTF) left the base in December 1944, making the (very) short journey to Scampton. A Polish technical training school remained here until July 1945, when it transferred to Locking, and the RAF's presence here finally ended.

The grass landing strips were quickly returned to agriculture, but traces of the airfield still remain, most notably a strip of the perimeter track leading off the A15. Many commentators have perpetuated a story that a taxi track was constructed to link the airfield with nearby Scampton, but there is no evidence to suggest that this took place, even though Scampton's perimeter track is just a few minutes' walk away. The farmhouse complex that once stood in the middle of the airfield is still here, and although the surrounding fields are still empty it is not difficult to picture the dark shapes of Wellingtons hauling their bomb loads into the night sky.

Main features:
Runways: N/S 4,800ft x 100 feet, SW/NE 4,800ft x 100 feet, E/W 3,600ft x 100 feet, grass surface. *Hangars:* two T2, one B1. *Dispersals:* thirty-six circular. *Accommodation:* RAF: Officers 95, SNCOs 250, ORs 922; WAAF: 134.

INSKIP, Lancashire

53°49'31"N/02°50'05"W; SD451368; 42.5ft asl. 3 miles NE of Kirkham, S of B5269 at Inskip village

Acquired on behalf of the Admiralty in 1942, this site was initially known as Elswick, but by 1943 had been renamed. The flat farmland was deemed suitable for the construction of an airfield, but its poor quality and poor drainage made construction difficult. Nonetheless the station opened on 15 May 1943 as HMS *Nightjar*. Two months later No 747 Naval Air Squadron arrived from Fearn with a mixed aircraft fleet comprising Ansons, Albacores and Barracudas. At Inskip the unit became No 1 Operational Training Unit, embracing No 766 NAS, which arrived with its Swordfish aircraft in July. This element of the OTU specialised in photographic reconnaissance training, whereas the former squadron was concerned with torpedo bombing and radar work.

In August 1943 No 735 NAS was formed at Inskip, equipped with Swordfish and Anson aircraft for ASV (Airborne Surface Vessel) and ASH (Air Search Homing) radar training. Other Fleet Air Arm units also came to Inskip, but most stayed only briefly. No 760 NAS brought its Sea Hurricanes here for six months from May 1944, and No 763 NAS stayed for just over a year from April 1944 with its fleet of Avenger, Swordfish and Anson aircraft. No 766 NAS arrived with Firefly and Swordfish aircraft in July 1943 and stayed until September 1946, then No 787 NAS made a four-month stay from November 1943. Wildcats (and more Swordfishes) arrived in December 1943 when No 811 NAS arrived for a one-month stay, and similar short stays were made by Nos 813 NAS, 838 NAS, 1792 NAS and 816 NAS.

A detachment from No 762 NAS brought a gaggle of Mosquitoes here from June until December 1945, but throughout Inskip's short existence its most numerous aircraft type was the faithful Swordfish, which drifted around the local area in significant numbers on an almost daily basis. Fireflies became an increasingly common sight later in the station's existence, but by the summer of 1946 the station was largely silent, and by the end of the year flying had ended here.

However, the Royal Navy retained the station and it became a tri-service communications centre with a scattering of radio masts emerging across the airfield. It was not until January 2010 that the site was finally abandoned, by which time most of the airfield structures had gone, the runways having been removed in the 1970s in order to provide hardcore for the new M55 motorway. The site's future remains unclear, but the basic proportions of this once busy Fleet Air Arm site are still visible.

Main features:
Runways: 005° 3,000 x 90 feet, 044° 3,000 x 90 feet, 094° 3,600 x 90 feet, concrete surface. *Hangars:* sixteen squadron type, sixteen storage type. *Dispersals:* thirty-six hardstandings. *Accommodation:* FAA: Officers 171, ORs 954; WRNS: 441.

JURBY, Isle of Man

54°21'14"N/04°31'25"W; SC360983; 29.5ft asl. 6 miles W of Ramsey on A10

Unusual by typical standards, Jurby's airfield was constructed in 1939 on an area of land bound by a number of roads, buildings and field boundaries. The result was an unconventional two-runway layout with both runways being somewhat shorter than standard, combined with a large complement of small hangars scattered across the site; some were erected in adjacent fields, requiring a web of linking tracks to connect them to the airfield perimeter track.

No 5 AOS aircraft scattered across the airfield at Jurby.

The monstrous Heyford bomber, basking in the sunlight at Jurby.

No 5 Air Observers School came here from Penrhos in September 1939 with a fleet of Fairey Battles, Bristol Blenheims, Handley Page Hampdens and Hawker Henleys. With more than a hundred aircraft at its disposal, the unit was briefly renamed as No 5 Bombing & Gunnery School before reverting to its previous title, and continued to operate from Jurby until February 1944, providing a valued training facility for countless RAF crews. A variety of RAF units made short stays at Jurby for periods of training, among them Nos 166, 215, 302, 307, 312 and 457 Squadrons, equipped with aircraft such as the Hurricane, Defiant, Whitley and Spitfire.

An unusual image of an air show at Jurby in the 1950s, long before stifling safety laws prevented aircraft from flying over spectators.

When the AOS finally left Jurby, it was replaced by the Air Navigation & Bombing School, equipped with a large fleet of Ansons and Wellingtons. Training activities continued, and were interrupted only by the unscheduled arrival of a Sunderland on 20 May 1945. After making an emergency forced landing on the airfield, the crew escaped before the aircraft's depth-charges detonated, causing significant damage to surrounding buildings.

The ANBS became No 5 Air Navigation School before moving to Topcliffe in September 1946, after which No 11 Air Gunnery School came to Jurby for a one-year stay. When this unit departed, Jurby's flying activities came to an end at the end of 1947, but the RAF maintained a presence with both a ground school and a cadet training unit.

It was not until 1964 that the airfield moved to civilian control as an emergency diversion airfield for the airport at Ronaldsway. Not surprisingly it was only rarely used, and was eventually abandoned, leaving many of the station's buildings to be transferred to commercial use. The RAF did not leave entirely, however, and RAF Jurby survives as a weapons range, with targets off the coast within sight of the now-disused airfield. Most of the RAF's operational aircraft made use of the range over subsequent years, with even the mighty Vulcan making an occasional appearance to deliver inert bombs to the range. Tornado, Jaguar and Harrier aircraft were frequent visitors, and it was not until the 1990s that the range was finally closed down, allowing RAF Jurby to fade into history. The disused airfield is still largely intact, but car-racing is now the only source of noise and spectacle at this once busy site.

Main features:
Runways: 260° 3,600 x 150 feet, 340° 3,000 x 150 feet, tarmac surface. *Hangars:* nineteen Extra Over Blister, four Bellman, two VR, one F-Type. *Dispersals:* none. *Accommodation:* RAF: Officers 141, SNCOs 347, ORs 1,666; WAAF: 194.

KELSTERN, Lincolnshire

53°24'34"N/0°05'55"W; TF264919; 406ft asl. 5 miles NW of Louth off A631, NE of Kelstern village

Although better known as a Second World War bomber base, Kelstern owes its origins to a First World War site created here in 1916 as a suitable landing field for No 33 Squadron RFC, which was responsible for air defence of the region. Although used only on a sporadic basis, the site was retained throughout the war for use by the RFC, and a variety of fighter types were seen here at infrequent intervals. When the war ended the RFC abandoned the site and it was restored to agricultural use.

However, the same area of land was clearly suitable for flight operations, and as the Second World War approached it was selected for development into a bomber airfield. Although roads were closed and field boundaries moved, the land was largely retained in its original condition, and a standard bomber airfield of almost classic proportions was constructed. It had three concrete runways of standard length, a linking perimeter track, bomber dispersals and the usual three hangars. In total, some 400 acres of farmland was acquired and in July 1942 construction commenced on a £810,000 contract to build an airfield to Class A standard. The minor road from Binbrook to South Ellington running across the centre of the site was closed. The three intersecting concrete runways were the main 06-24 at 2,000 yards, and subsidiaries 01-19 and 13-31, both of 1,400 yards. The thirty-six hardstandings were all loop-type. A T2 hangar was placed on the technical site on the north-west side of the airfield between the 13 and 06 runway heads, a second between runway heads 01 and 31, and a third on the north side between 19 and 24. The bomb store was off the south-east side of the airfield, while the camp sites were dispersed in fields to the north-west. The buildings were largely the work of George Wimpey & Co Ltd and provided for 1,585 males and 346 females.

A Lancaster of No 625 Squadron shares airfield space with an A20 from the 409th BG.

The first unit to be based here was No 625 Squadron, which brought its Lancasters to Kelstern when it opened for flying in September 1943. The Squadron was formed from C Flight of No 100 Squadron, and it remained at Kelstern through the war, completing 3,385 operational missions for the loss of more than 400 aircrew. Only one other unit came to Kelstern, No 170 Squadron, which re-equipped with Lancasters here before moving to Dunholme Lodge just one month later. No 625 Squadron stayed until April 1945 and, although Kelstern continued to serve as a satellite airfield for Binbrook through the war, almost all of Kelstern's activity was provided by the sole resident unit and its complement of Lancasters.

The RAF abandoned the station in 1946 and the crumbling remains of the airfield were eventually obliterated by agriculture. Very little of the airfield structure now remains, with only farm tracks tracing the position of the huge runways and only a few scattered remains of the former dispersals being visible. From the air, however, in suitable lighting conditions, the layout of the huge airfield can still be seen, even though the roar of the countless Lancasters has long since faded into memory.

Main features:
Runways: 242° 6,000 x 150 feet, 191° 4,200 x 150 feet, 308° 4,200 x 150 feet, concrete and tarmac surface. *Hangars:* two T2, one B1. *Dispersals:* thirty-six spectacle. *Accommodation:* RAF: Officers 100, SNCOs 296, ORs 1,189; WAAF: 346.

KETTLENESS, Lincolnshire

54°31'40"N/0°42'43"W; NZ834154; 320ft asl. 5 miles NW of Whitby, off A174 adjacent to Kettleness village

Perched on the cliff tops overlooking the stormy North Sea, this bleak and poorly equipped landing field was adopted by the Royal Naval Air Service in 1916 as a suitable site for patrol aircraft operating in the area. The landing field was essentially an area of farmland with obstructions removed; no other modifications were made for aircraft operations. It was used only sporadically by the RNAS, and when the First World War ended it was soon redundant, and was abandoned in 1918. Since then the area has returned to agriculture and there is now no trace of the links with aviation that ended so many decades ago.

Main features:
Runways: none, grass surface. *Hangars:* none. *Dispersals:* none. *Accommodation:* not known.

KINGSTOWN (Carlisle), Cumbria

54°55'22"N/01°57'11"W; NY390595. 2 miles N of Carlisle city centre, now covered by Kingstown industrial estate

Opened in late 1931, this small grass airfield, with just a few wooden huts and a single hangar, had to wait until 23 March 1933 for the first commercial aircraft to land. Another 12 months passed before regular scheduled services began, the bulk of which were to and from Hall Caine and Ronaldsway, both in the Isle of Man.

Locals had the chance to learn to fly for a relatively modest fee when the Border Flying Club was formed at Kingstown in June 1935, mainly with the Moth and Tiger Moth. On 1 July the first military unit was formed on the airfield when No 38 E&RFTS was operated by the Border Flying Club and flew the Tiger Moth, Audax and Hind. The airfield's facilities were expanded in response, with new hangars and technical buildings, some accommodation and a concrete hardstanding, although the total useable area of the airfield was still only 90 acres.

With the outbreak of war all civilian-operated flying schools were closed, including No 38 E&RFTS. The airfield fell silent until activity returned on 24 November 1939 with the arrival of twelve Ansons of No 3 AONS from Desford. A reshuffle of units meant that No 3 AONS was

Magister I R1851 crashed at Gates Hill on 11 June 1942 while taking part in a searchlight exercise. Note the civilian 'prop swinger' by the aircraft's nose. Via Martyn Chorlton

moved to Weston-super-Mare on 3 June 1940, as space was needed for another training unit.

The South East of England in May 1940 was not an ideal location for an EFTS to operate, so No 15 EFTS, based at Redhill, was ordered to move to the relatively safe skies of north Cumbria. The Chief Instructor and OC 15 EFTS, Wg Cdr F. S. Homersham, was summoned without any warning to HQ Flying Training Command, where he was instructed to move, complete with all civilian staff, to Kingstown and told that a Fighter Command Unit would take over Redhill on 3 June.

In a somewhat dazed condition, the first batch of one course of forty pupils were sent by train, under three Officers and two Ground Instructors, to reach Kingstown on 3 June with instructions to start work on lectures on Monday morning, which they duly did without losing any training time whatsoever. Fortunately, Monday was a fine day and the fleet of forty Magisters, with the exception of one under repair, flew up with the remaining instructors and the senior course of pupils who had only had a few weeks training left.

On arrival at Kingstown, all stores and records had to be moved up from Redhill. New defence schemes and regulations had to

Despite having been established since 1939, it was not until August 1946 that No 15 EFTS received an official badge. Via Martyn Chorlton

be prepared and conditions were made very difficult by the fact that the buildings had been laid out for a different type of flying school and were not adequate or suitably arranged for No 15 EFTS – the departing No 3 AONS, with half the number of pupils, had already complained that the buildings were inadequate for them. This overcrowding included the fact that the Army was sleeping in what should have been the airmen's and pupils' recreation rooms. There was no Sergeants' Mess and no Officers' Ante-room, among other shortages. The cooking apparatus might have been sufficient for approximately 100, had the stove been in a satisfactory condition.

To complicate matters still further, the English courses were suddenly reduced from eight to seven weeks, and the Polish course from three to two weeks, with the same amount of flying per course as previously asked for in each case. The Instructors flew approximately 80 hours per month on average, and had less than one day in seven off. The civilian staff worked even harder in dealing with the increase in routine work due to the shorter courses. Shortly after reaching Kingstown, the Polish course, which had been slowing down as the supply of Polish airmen in England reduced, suddenly accelerated, and the school started courses of thirty pilots lasting a fortnight instead of the previous twenty pilots on three-week courses – this was due to the influx of Polish airmen from France.

A new RLG became available at Burnfoot on 5 July, taking considerable pressure off Kingstown's small landing ground. Built on a clay sub-soil, drainage was a major problem, and any alternative airfield was being viewed at the time. A second RLG was also made available at Kirkpatrick.

On 26 August the length of the elementary flying training course was reduced by a further five weeks, and at the same time a special request was received to hurry the 'testing and grading' of Polish pilots to maximum capacity.

All hopes of maintaining the high safety standards that had been set in pre-war days rapidly disappeared from December 1940 onwards. In the first three months of 1941 alone, the EFTS had lost ten 'Maggies' in a variety of accidents, the majority on or near the airfield.

The Tiger Moth was introduced into the inventory of No 15 EFTS from February 1941. Embarrassingly, however, the first Tiger Moth to arrive, T7055, with Plt Off S. P. Richards at the controls, overturned on landing in snowy conditions after running into one of the many soft spots on the airfield.

The Magister's time with No 15 EFTS was almost at an end by May 1942. On 31 May the first of many ferrying flights were flown from Kingstown, via Clyffe Pypard in Wiltshire, Peterborough and back to Kingstown. Magisters were flown out of the airfield and Tiger Moths were returned in their place.

Harry Wappler, the crewman of an He 111, and Heinz Schnabel, who was among more than eighty pilots shot down in September 1940, were both being held in Camp 13 at Shap Fells Hotel in Penrith (known locally as the 'U-Boat Hotel' because of the amount of German Naval Officers held). On 13 November 1942 they both managed to escape from the camp by hiding in laundry baskets, then successfully stowed away on a train heading for Carlisle. Posing as Dutch airmen, without difficulty they gained access to Kingstown airfield and with even less trouble persuaded a pair of civilian groundcrew to start a Magister for them.

The intrepid duo flew south with the intention of refuelling at an airfield in East Anglia and continuing to the continent. It is rumoured that Wappler and Schnabel landed at Hucknall, but wherever it was they managed to persuade another unsuspecting group of groundcrew to refuel them and send them on their way. On the next leg, bad weather closed in and the German airmen became lost. Low on fuel, they landed in a meadow 5 miles north of Norwich. Despite their escape being successfully concealed at Camp 13, news of the missing No 15 EFTS Magister was sent around the country and all airfields in Norfolk were on the lookout. Still claiming to be Dutch airmen, Wappler and Schnabel were apprehended by personnel from Horsham St Faith and eventually returned to a POW camp. The RAF officer fraternity, who were all encouraged to escape if caught by the enemy, thought the German pair had put on a 'jolly good show'. However

One of the first military aircraft to arrive at Kingstown was the Ansons of No 3 Air Observer & Navigator School in November 1939. Via Martyn Chorlton

this did little to detract from the embarrassment caused to the CO of No 15 EFTS and the staf involved in 'handing over' the aircraft.

With the invasion of Europe on 6 June 1944, the need for pilots was greater than ever. The wa was still far from being won, and at any one time at least eighty Tiger Moths could be seen on th airfield during that summer.

At the beginning of 1945 the strength of the unit had peaked, with 108 aircraft being flown b more than seventy flying instructors. When the war in Europe came to an end, the throughput o pupil pilots began to decline, as did the size of the EFTS, while still maintaining a high standard o training. By 15 August 1945 the unit had been reduced to forty-five aircraft.

Kingstown became the home of No 274 MU from Great Orton from 28 May, while bot RLGs at Burnfoot and Kirkpatrick were now closed to flying.

By 30 June 1946 the strength of No 15 EFTS was reduced still further, with only forty-fiv flying instructors teaching a group of thirty-five pupil pilots.

After almost seven years as CO, and the unit's Chief Flying Instructor, Wg Cdr C. M Homersham AFC MM was posted on 1 April 1947. Wg Cdr T. M. Scott AFC, who was seconde from the RAF for duty with British Air Transport Ltd, replaced him.

By 1 October the unit had a mere thirty aircraft on strength, and the last full flying course t pass through Kingstown began on 13 October 1947, with four Burmese, two Iraqi officers and tw Belgian NCOs attending.

The remarkable history of No 15 EFTS came to an end on 31 December 1947 when the uni was disbanded. Since its formation at Redhill, the unit had flown a total of more than 270,000 flyin hours and trained more than 12,000 pupil pilots, but not without cost. The EFTS lost eighty-fou 'Maggie's' and thirty-one Tiger Moths in a variety of flying accidents at Kingstown and th supporting RLGs. This still equates to an excellent safety record of one accident for every 235 flyin hours. The unit was reformed a few months later at its original home at Redhill as No 15 RFS, stil flying the Tiger Moth.

Kingstown came under the complete control of No 14 MU Carlisle on 20 April 1948 transferring from No 23 Group to No 40 Group Maintenance Command, whose HQ was a

Bicester. No 14 MU made use of the airfield until 24 June, when all key buildings situated on the airfield were handed over to Carlisle Corporation.

A small amount of civilian aviation returned to the airfield during the post-war years. Manx Airlines, flying the Rapide, flew a limited service to the Isle of Man until the airfield's closure in 1957. Airlines were expanding, aircraft were getting bigger, and Kingstown was simply not large enough to cope. It was impossible to extend the airfield because of the vast No 14 MU site to the north and a railway line running down the western side.

Sadly, today nothing remains of this once active training airfield. It has been swallowed up by the demand for industrial estates, and it is a shame that the local council has not named a road Magister Close or Tiger Moth Drive in its memory.

Main features:
Runways: QDM NW-SE 1,000 yards, NE-SW 1,000 yards, grass. *Hangars:* seven various. *Hardstandings:* none. *Accommodation:* RAF: 385.

KIRKBRIDE, Cumbria

54°52'57"N/03°12'12"W; NY229548; 30ft asl. 10 miles W of Carlisle off B5307, S of Kirkbride village

Designed and constructed for use as a Maintenance Unit storage facility, land was acquired in 1937 and the station opened as RAF Kirkbride on 5 June 1939. Unlike many airfields that were often allocated to roles for which they had not been designed, Kirkbride remained allocated to the MU role and was created in an area that was already deemed suitable for storage of aircraft, not only because of the flat and uncluttered nature of the local terrain, but also because Cumbria was commendably distant from mainland Europe, from where the Luftwaffe ventured in search of targets. The need for suitable storage space resulted in an interesting airfield layout, which embraced more than 100 dispersals of various types (including standard spectacle loops), and a wide range of hangars of different types, including one large C-Type building. The various component parts of the airfield were littered across a large area, which ran across many fields and roads; indeed, one part of the dispersed parking ran for more than a mile away from the main airfield. The three runways were constructed in 1940, and although they are somewhat shorter than those provided for bomber airfields they were more than adequate for the routine ferry flights that would be necessary.

No 12 Maintenance Unit was established here and three satellite fields were allocated to it, at Wath Head (10 SLG), Brayton Park (39 SLG) and Hornby Hall (9 SLG). Within a matter of weeks the MU had begun its unglamorous but vital task of storing aircraft, and a wide variety of types were soon visible on the airfield. Spitfires were the most numerous, but these were later joined by many other types including Wellingtons, B-17 Fortresses, B-25 Mitchells, Hurricanes and others. No 16 Auxiliary Ferry Pilots School was formed at Kirkbride in July 1941 and became responsible for the delivery of aircraft to and from the station. By 1945 the site was literally crammed full of aircraft – more than 1,000 were recorded as being present at one stage.

A 1942 vertical view of Kirkbride illustrating the surprisingly effective camouflage measures employed by the RAF.

Part of the main site at Kirkbride with a visiting Anson parked in front of one of the airfield's many hangars.

An RAF reconnaissance photo of Kirkbride in 1946.

Right: *Another overhead photograph of Kirkbride showing the airfield layout. The picture appears to have been taken after the site's closure.*

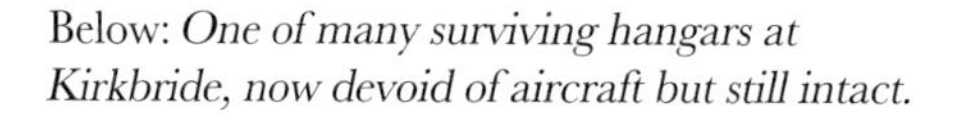

Below: *One of many surviving hangars at Kirkbride, now devoid of aircraft but still intact.*

Such was the MU's importance that it remained in business long after the Second World War, and many post-war aircraft types also found their way to this airfield. Most notably, a huge number of Meteor jets were stored and eventually scrapped here, and it is perhaps sad that for many people Kirkbride will for ever be associated with the sorry sight of countless Meteors slowly falling apart on its dispersals. The RAF finally left the base during the summer of 1960 and, after a period of inactivity, the airfield was used for private flying.

The airfield is now mostly active at weekends, and is often almost deserted, but most of the original RAF site has survived, and although the runways are in a rather sorry state they are still present, as are many of the hangars. Kirkbride never earned much publicity thanks to its rather unglamorous role, but it was undoubtedly a vital, busy and significant site that deserves a place in the Royal Air Force's history.

Main features:
Runways: 281° 4,200 x 150 feet, 165° 3,800 x 150 feet, 232° 3,000 x 150 feet, concrete and tarmac surface. *Hangars:* six E-Type, four D-Type, four L-Type, one C-Type. *Dispersals:* more than 100, various. *Accommodation:* RAF: Officers 0, SNCOs 10, ORs 226; WAAF: 0.

KIRKLEATHAM, Yorkshire

54°35'46"N/01°05'21"W; NZ589227; 36ft asl. At Redcar on A1042 on Kirkleatham Lane

Known both as Kirkleatham and Redcar, this site was (like many others) established on a horse-racing course, but instead of creating a landing strip on the actual course, an area of land adjacent to the course was cleared and prepared for use in 1918. The Royal Naval Air Service operated here while mounting Home Defence duties, although much of the flying was concerned with maritime patrol and training. Various aircraft types were present, including HP.100s from No 7N Squadron, although the first flying machines to appear here were actually airships, as the site was allocated for use as an out-mooring station for Howden; Zero Class airships SSZ 54, 55 and 62 were the first to be seen here.

No 510 Special Duty Flight formed here on 7 June 1918, and became part of No 68 (Operations) Wing; the unit transferred to West Ayton in November. The North Eastern Area Flying Instructors School was also established here in the summer of 1918, and in January 1919 it was joined by the North Western Area FIS, the two units combining during the following May. No 53 Training Squadron arrived in October 1918 and the airfield remained busy until the end of 1919, by which time the requirement for aircrew had rapidly diminished. By October the airfield was largely dormant, the RAF vacated the site at the end of the year, and it was abandoned. The few airfield structures were eventually demolished and the landing field was returned to farming, the racecourse also disappearing in the process. Today the fields are still largely unchanged and Cleveland Police maintain a facility on what was once the southern boundary of the landing field.

Main features:
Runways: none, grass surface. *Hangars:* four flight sheds. *Dispersals:* none. *Accommodation:* not known.

KIRMINGTON, Lincolnshire

53°34'41"N/0°20'43"W; TA096103; 72ft asl. 6 miles NE of Brigg on A18 (Humberside Airport)

Land was acquired here by the Air Ministry in 1940 and construction took place in 1941. Located directly south-west of Kirmington village on the A18 Scunthorpe to Grimsby road, it was a standard bomber airfield for the RAF. The local terrain and road network enabled an airfield of almost perfect proportions to be built, with three runways of full length (04-22 originally 1,450 yards long, 15-33 1,150 yards, and 09-27 1,100 yards), a full complement of spectacle-type dispersals, and the standard three hangars, all linked by a perimeter track. The result was perhaps one of the best examples of the standard bomber airfield. However, it appears that extension of the runways was

An airman adds a mission marking to record a Lancaster's return from Berlin during February 1944.

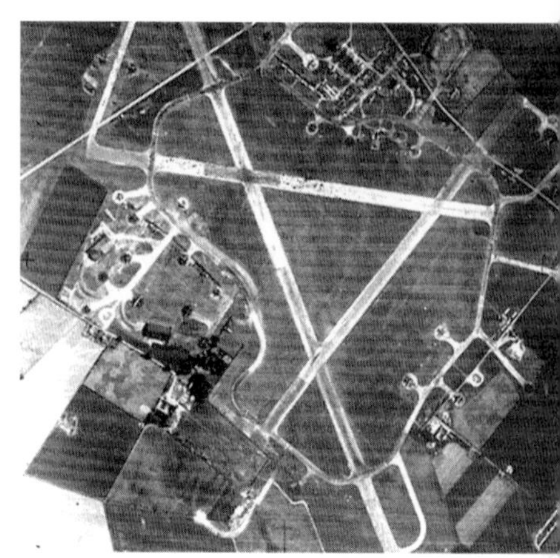

An RAF reconnaissance photograph of Kirmington taken in 1945.

The ubiquitous Airspeed Oxford was a familiar sight at Kirmington, as it was at many RAF stations across the region.

carried out before the station was opened, and this involved the closure and diversion of the A18 jus west of Kirmington village, as well as the closure of the minor road running north to the village on th east side of the airfield. When extended, the runway lengths were 04-22 at 2,000 yards and the other both 1,400 yards. Of the original thirty-six pan-type hardstandings, two were lost to the perimete track and hangar taxiway extensions, and two loop-type standings were added as replacements.

A T2 hangar was located on the main technical site between runway heads 22 and 27 an another on the maintenance site near Kirmington Villa. Another B1 was built a little to the north Two Blister hangars were erected on pan hardstandings adjacent to the technical site T2. The bom store was off the south side of the airfield, and eleven camp sites, of which seven were domestic, wer dispersed around Kirmington village and further to the east, allowing for 2,177 males and 345 females

Kirmington opened during October 1942 as part of No 1 Group Bomber Command, althoug flying had already commenced some months previously when No 15 Pilots Advanced Flying Uni began operating its Oxfords here from its home base at Leconfield. The first proper residents were th Wellingtons of No 150 Squadron, although most of this unit's personnel and aircraft soon departed fo the Middle East. The remaining cadre of aircraft and crews were soon joined by those from No 14

Squadron, and these were merged to create No 166 Squadron in January 1943. By the end of that month the unit was already flying operational missions over Germany. The Wellingtons were gradually replaced by Lancasters, but the unit stayed at Kirmington until the end of the Second World War.

Late in 1942 the station was designated as a satellite for Elsham Wolds, and in October 1944 part of No 166 Squadron was re-roled to become the nucleus of No 153 Squadron, which moved to Scampton once established. No 166 eventually completed 789 Wellington and a staggering 4,279 Lancaster missions from Kirmington before disbanding in November 1945.

The RAF vacated Kirmington at the end of that year and the site lay abandoned for some time until it was purchased for civilian use in 1966. Still in good condition, the airfield reopened in 1967 for commercial and private flying, and in 1970 the site was purchased by Lindsey County Council for development as a regional airport. The basic airfield layout was retained, but the runways were repaired and resurfaced and a new airport terminal was constructed, opening as Humberside Airport in 1974. The airport has been a success and continues to thrive; air shows have even been staged here, attracting modern aircraft such as the Harrier and F-16 and of course the world-famous Red Arrows. Two of the original runways remain in daily use (the main runway having been extended considerably), but virtually all other parts of the wartime airfield are gone, and even the perimeter track and dispersals have disappeared. This is a great pity considering the almost classic proportions of this once busy bomber base.

Main features:
Runways: 217° 6,000 x 150 feet, 326° 4,200 x 150 feet, 272° 4,200 x 150 feet, concrete and tarmac surface. *Hangars:* two T2, one B1. *Dispersals:* thirty-six Heavy Bomber. *Accommodation:* RAF: Officers 162, SNCOs 332, ORs 940; WAAF: 210.

KIRTON-IN-LINDSEY, Lincolnshire

53°27'44"N/0°34'37"W; SK945970; 183.5ft asl. 16 miles north of Lincoln off A15 on B1398 S of Kirton-in-Lindsey village

The origins of this airfield can be found in the First World War when, like many such sites around this region, it was allocated for use as a landing field for the Royal Flying Corps. No 33 Squadron used it on a sporadic basis while engaged on air defence duties in the area. However, the Squadron moved its headquarters here in June 1918 and, in addition to its existing compliment of BE and FE fighters, acquired its first Bristol Fighters, followed by Avro 504Ks from August 1918. The unit continued to operate the 504 from Kirton until June 1919, when the unit moved to Harpswell (Hemswell) and disbanded.

The airfield at Kirton was soon abandoned and it quickly reverted to farmland, but it remained of interest to the Air Ministry and as the Second World War approached it was acquired for development into a more substantial aerodrome. Developed as a fighter base, two grass runways were laid within the

A Spitfire from No 452 Squadron returns to Kirton's grass landing field.

No 264 Squadron Defiants line up at Kirton-in-Lindsey.

Above: *A visiting civil Tiger Moth stands on the grass landing field at Kirton in front of the distinctive control tower.*

Right: *Thanks to the Army's presence, the airfield and main site at Kirton remain largely unchanged since the days of the Second World War.*

confines of the two roads that pass to each side of the site. A concrete perimeter track was laid, whic linked the runways to the technical and hangar site, and some dispersals were constructed, some o them being across the public roads; it would appear that the roads may have actually been used a taxi tracks in parts, presumably closed to the public when required by the RAF.

The first aircraft arrived in May 1940 when Nos 65 and 22 Squadrons flew in with Spitfires together with No 253 Squadron with Hurricanes. However, the station hosted most squadrons on rotational basis and few stayed for very long. Consequently a wide variety of Spitfires, Hurricane and Defiants could be seen here as each unit came and went. In June 1942 the 94th Fighte Squadron (1st Fighter Group) USAAF moved to Kirton from Goxhill, and P-38s became a familia sight as the American crews honed their piloting skills. The 91st Fighter Squadron (91st Fighte Group) also came to Kirton to train with its fleet of P-39 aircraft in October 1942, but both unit had gone by the end of the year to engage in operational duties.

Situated far away from the RAF's main defensive region, Kirton's use as a fighter base slowl diminished and in April 1943 the Polish Nos 302 and 317 Squadrons were the last fighter units t leave, together with the Lysander-equipped No 1489 Target Towing Flight detachment. Kirton wa now very much a training base, and a mix of Spitfires and Masters could be seen on the airfield these belonged to No 53 Operational Training Unit, which moved here in May 1943. During th summer the training syllabus was widened to include low-level bombing, and a range was laid ou on the airfield; however, bomb splinters created a danger to both property and personnel, so range was set up at Middle Farm, Manton. The unit was a particularly busy one and at one stag boasted more than ninety Spitfires together with a fleet of Masters and Martinets, some of whic operated from the unit's nearby satellite field at Hibaldstow. Training continued until May 194 when the unit moved to Keevil.

No 3 Aircraft Preparation Unit formed at Kirton in July 1944, tasked with the handling o various types but most notably the Wellington and Warwick. The unit merged with No 16 Ferr Unit a few weeks later, although much of Kirton's activity now surrounded the ground-base Aircrew Holding Unit, which was established here in May 1945. The next significant developmen vas the arrival of No 7 Flying Training School with Harvards and Oxfords in April 1946, forming he Central Synthetic Training Establishment (which became the Link Trainer School). RAF 'inningley also adopted the airfield as a satellite, and the Flying Refresher School's Spitfires, Vellingtons, Oxfords and Harvards became familiar sights here until April 1950, after which No .01 FRS adopted Kirton as a satellite. This was the last unit to use Kirton in any significant way and nost of the station's activities had by now shifted to ground-based units.

By the end of 1951, when No 101 FRS departed, the airfield was virtually silent, with only glider flying taking place courtesy of No 22 Gliding School, which arrived in August 1957, becoming No 643 VGS in September 1955. No 2 Gliding Centre arrived from Newton in September 1959. After both glider units left in October 1965 the RAF vacated the airfield at the end of that year. Kirton was then transferred to Army control and the Royal Artillery arrived.

The RAF did not disappear entirely, however, as the Army conducted a great deal of air lefence training from Kirton and the RAF's Canberra target facilities squadrons sometimes flew attack' missions on the airfield to aid with this. The station was returned to the RAF in 2004 when No 1 Air Control Centre was established here, but flying activity has not returned. The airfield is used by a civilian glider club, but military flying here is largely confined to occasional visits by communications aircraft and helicopters. Because the station has remained in military hands for so ong it remains in excellent condition, and it seems difficult to believe that so many years have passed since Spitfires and Defiants sat on this now deserted airfield.

Main features:
Runways: N/S 3,300 x 100 feet, E/W 3,300 x 100 feet, grass surface. *Hangars:* four Over Blister, three C-Type, one B1. *Dispersals:* ten, various. *Accommodation:* RAF: Officers 89, SNCOs 102, ORs 1,146; WAAF: 531.

KNAVESMIRE, Yorkshire

53°56'14"N/01°05'52"W; SE593493; 29.5ft asl. At York (Racecourse) off A64 at Dringhouses

York's racecourse was an obvious choice for adoption as a landing field, being flat and clear of all obstructions. It was used for flying from 1912 with civil aviation sharing the site with horse-racing activities. Some early Blackburn types are known to have been tested here, but it was not until the outbreak of the First World War that the site became associated with military aviation. The Royal Flying Corps arrived and No 33 Squadron deployed its BE.2C aircraft here in 1916, while mounting air defence sorties around York. The unit's stay was short, however, and it had departed by the summer, leaving the site abandoned. After the war the area returned to horse-racing, and aviation was destined never to return.

Main features:
Runways: none, grass surface. *Hangars:* none. *Dispersals:* none. *Accommodation:* not known.

KNOWSLEY PARK, Merseyside

53°27'27"N/02°48'52"W; SJ460959; 197ft asl. 3 miles W of St Helens S of A580, W of B5283 adjacent to Pony Coppice

Situated in the relative safety of Merseyside, Knowsley Park was one of many sites adopted by the RAF for aircraft storage, being many miles from the industrial heart of the country and far from the South East, where the Luftwaffe could easily search for such facilities – and destroy them. Various airfields were allocated as storage bases, but the sheer number of aircraft (and the obvious need to spread them around in order to present smaller potential targets to the enemy) meant that many small areas of farmland were often adopted for use, cleared of all obstructions and prepared for use by aircraft. Knowsley Park became No 49 Satellite Landing Ground, and was opened in

October 1941 on an area of land owned by Lord Derby. However, a spell of poor weather soo rendered the site unusable and it was not until May 1942 that it was again capable of receivin aircraft. No 37 Maintenance Unit at Burtonwood duly dispatched its aircraft here and th surrounding woods were soon occupied by a wide variety of types.

In July 1942 the site was transferred to No 48 MU (based at Hawarden), and a large number o the MU's Wellingtons were soon ferried to Knowsley Park, followed in 1943 by the even large Halifax bombers, which could comfortably land and take off from this relatively large landing field When the requirement for so many storage facilities gradually diminished, No 49 SLG was n longer needed and No 100 SLG at Hooton Park took over, enabling the RAF to abandon its site a Knowsley Park. The temporary Robin hangars were soon demolished and in a matter of just a fe years there was no trace of the RAF's presence here. However, a careful search of Pony Coppic and Chain Acre Wood may well reveal a few traces of the areas where so many aircraft once lurked

Main features:
Runways: none, grass surface. *Hangars:* none. *Dispersals:* none. *Accommodation:* not known.

LECONFIELD, Yorkshire

53°52'37"N/0°26'07"W; TA029434; 20ft asl. 2 miles N of Beverley off A614

This well-known airfield site first emerged during the RAF's expansion period in the 1930s, established on an area of low-lying land close to the notorious Vale of York, where low cloud and fog were often a daily problem, but sufficiently close to the East Coast to counter this difficulty. Opening in December 1936, the airfield comprised a grass landing field flanked by five huge C-Type hangars.

A Wellington undergoes maintenance at Leconfield during the summer of 1943.

During January 1937 the first aircraft arrived in the shape of No 166 Squadron's lumbering Heyford bombers, which transferred here from Boscombe Down. No 97 Squadron (from which the former unit had been formed) arrived a few days later. The two units remained in business for some time and in June 1938 became the Temporary Air Observers School, but by December the units had returned to their more conventional roles as bomber squadrons. Early in 1939 No 97 Squadron re-equipped with Ansons, reflecting its increasing training role, while No 166 kept its Heyfords until June, when it received Whitleys. Both

A panoramic view of RAF Leconfield and the local area in 1945.

units then moved to Abingdon in September and Leconfield fell silent, Fighter Command taking control of the airfield a few weeks later.

On 23 October 1939 the first fighter unit arrived in the shape of No 616 (South Yorkshire) Squadron, which worked up here with its first Spitfires. By January 1940 the unit was able to mount its first operational patrol. Nos 234 and 245 Squadrons had also arrived some weeks previously with Blenheims, and over the next few years a large number of fighter units came and went, most staying for only a matter of weeks for rest or training. No 616 Squadron engaged the enemy on 15 August when a large Luftwaffe raid was mounted on northern England using bases in Norway and Denmark. Leconfield's Spitfires intercepted and No 616 Squadron downed no fewer than eight Luftwaffe bombers; the unit later moved to Kenley.

Through 1940 and 1941 fighter squadrons came and went, and No 60 Operational Training Unit was established here in April 1941, equipped with a variety of aircraft tasked with the training of crews destined for Blenheim and Defiant night-fighters. The unit stayed only briefly, as did Nos 81 and 134 Squadrons, both equipped with Hurricanes. Organised as 'Force Benedict', they later deployed to Russia, based at Venga to defend the Russian's convoy route. The last of the many fighter units to leave Leconfield was No 610 Squadron, and when its Spitfires departed Leconfield returned to Bomber Command. Indeed, Halifax and Stirling bombers had already arrived towards the end of 1941 to begin conversion training, but they left again after only a few weeks so that the airfield could be modified to accommodate the concrete runways that were deemed necessary for the heavy bombers. No 15 P(A)FU continued to use parts of the airfield while it was closed

On reopening in December 1942 Leconfield boasted three new runways linked by a concrete perimeter track, littered with dispersals. The main runway was (unusually) not aligned with the prevailing wind direction but was placed in the most convenient location between the confines of the roads and railway line that surrounded the airfield boundary. Nos 196 and 466 Squadrons duly arrived with Wellingtons, and by January the units were embarking on operational missions over Germany. No 196 Squadron moved to Witchford in July 1943 (having flown 517 missions from Leconfield), and No 466 Squadron acquired Halifaxes in September. No 640 Squadron arrived with more Halifaxes in January 1944 and stayed until May 1945, by which stage some 2,400 missions had been completed. The Australian No 466 Squadron departed for Driffield in June, but in the opposite direction came the Oxfords of No 1520 Flight. No 51 Squadron was the last bomber squadron to arrive, making a brief stay from April until August 1945. By this time more changes were becoming apparent, and No 96 Squadron was formed in December 1944 with a fleet of Halifax transport aircraft. The unit soon departed for the Middle East, but laid the way for Leconfield's transfer to Transport Command.

However, it was a distinctly non-transport unit that first arrived here in the shape of the Central Gunnery School with an unusual mix of aircraft types including the Wellington, Mosquito, Meteor, Spitfire, Martinet and Master. The unit became a long-term resident, staying here until 1957, by which time it had become the RAF Fighter Weapons School. In that year the first helicopters arrived at Leconfield when No 275 Squadron's HQ was established here with a fleet of Sycamores. It heralded the beginning of a long association with Search and Rescue (SAR) helicopters that continues to this day.

In 1959 fighter aircraft returned and No 19 Squadron arrived with Hunters, followed by No 72 Squadron with Javelins. In order to accommodate these jet fighters the main runway was extended and Operational Readiness Platforms were constructed adjacent to each threshold. The noisy Javelins shared ramp space with the sleek Hunters, and when No 92 Squadron arrived with yet more Hunters

Lightnings from Nos 19 and 92 Squadrons undergo deep maintenance at Leconfield.

A line-up of No 19 Squadron Lightnings at Leconfield.

many aircraft soon appeared in an eye-catching overall blue paint scheme, indicative of the unit's brief role as the Royal Air Force's Aerobatic Team, the 'Blue Diamonds' (a forerunner of the famous Red Arrows). In the 1960s both Nos 19 and 92 Squadrons acquired the fast (and even more noisy) Lightning, and Leconfield became one of the RAF's busiest fighter stations with numerous Lightnings pounding the main runway both day and night.

When the units eventually departed for Germany in 1965, Leconfield's future suddenly looked uncertain, but the arrival of No 60 Maintenance Unit ensured that the station's link with the mighty Lightning continued, and the RAF's Lightning fleet continued to be rotated through the base for repair, storage or (eventually) scrapping. With the RAF's main Lightning base at Binbrook only a few minutes' flying time away across the Humber, Leconfield was also used regularly for Lightning flying, and hosted all of Binbrook's units when the latter base was closed for runway work.

No 72 Squadron's Javelins pose for inspection at Leconfield during 1972, then start up in unison.

A 1980 aerial image of Leconfield showing the airfield layout before the Army's presence changed it for ever.

Right: *A aerial view of Leconfield taken during 1994, the Army's presence already revealing many changes to the airfield layout.*

Below: *Leconfield in 2011, the huge airfield now obscured by trees.*

However, in 1977 the RAF finally handed control of the base to the Army, and the airfield was closed. The School of Mechanical Transport arrived and over successive years an extensive series of driver training facilities appeared, trees were planted, and the airfield was transformed. The RAF maintained a presence, however, with an SAR helicopter facility on the southern part of the now disused airfield. Sea King helicopters from No 202 Squadron continue to provide SAR cover from this base, although the demise of the RAF's SAR helicopters is fast approaching and the RAF's final foothold at Leconfield will doubtless be gone in a few years.

From outside what is now Normandy Barracks, only the huge hangars indicate that the RAF was once present here. However, closer examination reveals that the 1960s-era control tower is still here, looking out over a rather bizarre landscape that is now dominated by trees. But between the trees and the mounds of rubble, the main runway is still intact. As the muddy Army trucks go about their daily business, it is difficult to imagine that less than 40 years ago Lightning interceptors were streaming along the very same stretch of concrete.

Main features:
Runways: 190° 6,000 x 150 feet, 050° 4,200 x 150 feet, 140° 4,200 x 150 feet, concrete surface. *Hangars:* five C-Type. *Dispersals:* thirty-six Heavy Bomber. *Accommodation:* RAF: Officers 120, SNCOs 392, ORs 1,231; WAAF: 249.

LEEMING, Yorkshire

54°18'01"N/01°32'09"W; SE303895; 95ft asl. 4 miles SW of Northallerton off A1(M) adjacent to Leeming Bar

The precise date when flying commenced at this site is unclear, but it is known that by the mid-1930s an area of land was in use here for commercial and private flying, and by 1938 it was to be used more extensively by Yorkshire Air Services. However, by the end of that year the site had been allocated for development into a military airfield and a much larger stretch of land was requisitioned embracing the original airfield

A Halifax from No 35 Squadron pictured during a sortie from Leeming.

A No 7 Squadron Stirling at Leeming.

Suitably decorated in preparation for a Christmas mission, this mine awaits its Halifax carrier aircraft at Leeming late in 1944. A Halifax from No 424 Squadron is visible in the distance. Ken Cothliff

together with surrounding farmland out to the Great North Road to the west and a river to the east. These boundaries (together with various roads) resulted in a bomber station with an unusual layout.

By the summer of 1940 it was ready to receive its first aircraft and No 219 Squadron established a detachment of Blenheims here, although the first proper resident was No 10 Squadron, which arrived in July 1940 from Dishforth with its Whitleys. Within a matter of days the unit was flying operational missions and in August it was joined by No 7 Squadron, with new Stirling bombers, with which the Squadron trained for some two months before moving to Oakington. In November No 35 Squadron arrived to work up in similar fashion on the Halifax, and in December the unit left for Linton-on-Ouse. No 77 Squadron brought its Whitleys to Leeming in September 1941 and No 7 Squadron soon converted onto the Halifax, but both of Leeming's squadrons moved out in the summer of 1942, to be replaced by No 419 Squadron with Wellingtons and No 408 'Goose' Squadron with Hampdens (although the unit was already converting onto the Halifax). Both units had their own Conversion Flight and these were combined to form No 1659 Heavy Conversion Unit on 7 October, this unit then moving to Topcliffe in March 1943.

A rare view of groundcrew loading mines into a Halifax prior to a 'Gardening' mission at Leeming during 1945. Ken Cothliff

An aerial image of Leeming, showing the partially completed runway extension that now forms part of the airfield's main runway.

No 427 'Lion' Squadron came to Leeming in May 1943 with Halifaxes and No 429 'Bison' Squadron arrived in August, also with Halifaxes. Both units kept their Halifaxes until the later stages of the Second World War, when Lancasters started to arrive. Ultimately No 427 Squadron flew some 2,800 Halifax sorties and 239 Lancaster missions, while No 429 flew 2,519 Halifax sorties and 114 on the Lancaster. Both squadrons disbanded in May 1946 and Leeming then became the home of No 54 Operational Training Unit, tasked with Mosquito night-fighter training. In May 1947 it was merged with No 13 Operational Training Unit (which had moved to Leeming four weeks previously) to become No 228 Operational Conversion Unit, and with a fleet of Mosquitoes (and some Wellington 'flying classrooms') the OCU was divided into three component squadrons, each tasked with a different aspect of the unit's training task. Hornets and Meteors gradually joined the OCU.

The advent of new high-performance jets led to a decision to improve Leeming's runways, so the airfield closed in April 1956 for ten months while the main runway was extended, crossing what was formerly the main road link between the station and the Great North Road. A barrier was installed so that the road could still be used by the public, although it is not known when the crossing was closed more permanently. Eventually the OCU's main equipment was the Canberra T11 and the Javelin fighter, together with a handful of Meteors. The unit remained in business at Leeming until September 1961. By this stage No 33 Squadron had arrived (in September 1957)

Right: *Jetstreams operated for a short period at Leeming before eventually relocating to Finningley.*

Below: *The Central Flying School's 'Vintage Pair' Vampire and Meteor are seen during a training sortie from Leeming.*

with Meteor night-fighters and (eventually) Javelins, but it left Leeming again in September 1958.

With the OCU gone, Leeming was unused for a while, but became active again late in 1961 when No 3 Flying Training School began training operations here with a large fleet of Jet Provosts. Vampire jets also arrived in 1966 courtesy of the Vampire Advanced Training Unit (part of No 7 FTS). When Manby's operations wound down, the School of Refresher Flying was also relocated to Leeming, in 1973, and a few months later it was combined with the Royal Navy Elementary Flying Training School, which moved here from Church Fenton. Now a very busy training base, even more aircraft arrived when the Multi Engine Training Squadron was established here in May 1977 with a fleet of Jetstreams. The Central Flying School also arrived, and as part of this move the unique 'Vintage Pair' team (comprising a Vampire T11 and a Meteor T7) arrived and operated from Leeming over successive years, appearing at air show venues around the country. Flying training continued until April 1984 when the last of the resident units left and Leeming closed down.

The airfield and support buildings were refurbished and the base reopened in January 1987 as a fully equipped fighter base, ready to received the Tornado F3 interceptor. By this stage the airfield had been littered with concrete Hardened Aircraft Shelters and the old road crossing the main runway had been closed for good. Nos 11, 23 and 25 Squadrons duly reformed at Leeming on the Tornado F3 and once again the skies around the region reverberated to the thunderous roar of jet fighters. They were joined by the Hawks of No 100 Squadron in 1995 and operations continued apace until a slow wind-down began in March 1994 when No 23 Squadron departed (to reform on the E-3 Sentry). No 11 Squadron left in October 2005 and the last Tornado unit (No 25 Squadron) said goodbye in April 2008.

An unusual formation of aircraft types from Leeming – Javelin fighters are accompanied by Canberra and Meteor radar trainers.

Today only the Hawks of No 100 Squadron remain here, together with the Tutors of the Northumbrian Universities Air Squadron. Leeming's main role is now as Headquarters for No 135 Expeditionary Air Wing, but as such it does not operate any aircraft. No 100 Squadron is expected to continue flying its Hawk aircraft for a few years, but the future of Leeming as a flying base is still uncertain.

Above: *An aerial view of RAF Leeming captured from one of the station's Hawks during 2006.*

Right: *A Tornado pilot's view of one of Leeming's HAS sites in 2006.*

Main features:
Runways: 347° 5,850 x 150 feet, 220° 4,950 x 150 feet, 305° 4,200 x 150 feet, concrete and tarmac surface. *Hangars:* Five C-Type. *Dispersals:* thirty-six circular. *Accommodation:* RAF: Officers 194, SNCOs 664, ORs 1,534; WAAF: 317.

LINDHOLME, Yorkshire

53°33'08"N/0°58'02"W; SE685066; 23ft asl. 2 miles SE of Hatfield Woodhouse on A614 (HMP Lindholme)

An unusual reconnaissance image of Hatfield Moors with Lindholme airfield to the left. A number of aircraft crashed on the moors during Lindholme's operational years.

Originally known as RAF Hatfield Woodhouse, the airfield here was completed in 1940, the station opening as part of No 5 Group on 3 June of that year. However, within a couple of months a decision had been made to change the station's name to Lindholme in order to avoid confusion with nearby Hatfield Woodhouse and Hatfield, which caused logistical problems for deliveries of stores and mail. The large expanse of flat land seemed ideal for the construction of a bomber base, but the area was in effect a large peat bog and the construction of the grass landing field and station buildings was by no means easy; it took two years to complete the site to an acceptable standard. Work began in the spring of 1938, taking in approximately 250 acres of pasture for the airfield itself and a further 150 for the camp and support facilities. Three C-Type hangars fronted the south-west side of the bombing circle, with a fourth and fifth behind the two outer hangars. The administration, technical and barrack area lay alongside the A614. As was common with these Expansion Scheme airfields, the construction of buildings took place over several months and the pace was only quickened with the outbreak of war. However, the finished station was of a typical layout with admin and technical buildings all dominated by a group of huge C-Type hangars.

On 10 July 1940 the first aircraft arrived in the shape of Hampdens from No 50 Squadron, which transferred from Waddington. The unit was already operational and within days its aircraft were back on bombing missions, remaining active here until July 1941. No 408 'Goose' Squadron RCAF had formed here during the previous month, but it swiftly moved to Syerston. This allowed Lindholme to transfer to No 1 Group, accepting Nos 304 'Silesian' and No 305 'Ziema Wielposka' Squadrons from Syerston in exchange, both Polish units equipped with Wellingtons. Other units also made brief stays here including No 11 SFTA with Oxfords and a detachment from No 110 Squadron with Blenheims. In May 1942 the busy No 304 Squadron departed to become part of Coastal Command while No 305 Squadron made the short move to Hemswell a few weeks later, enabling Lindholme to close temporarily while two concrete runways were laid.

The condition of the surrounding land and the adjacent boundaries made a third runway impractical, but the main runway was completed to a full 6,000-foot length and a large collection of dispersals appeared, most connected to the perimeter track but with additional dispersals attached to spurs that reached out into the surrounding fields, some crossing the main public road. When the station reopened in June 1942 it was assigned to training activities and No 1656 Heavy Conversion

A Bomber Command Bombing School Varsity pictured at Lindholme in 1961.

Unit arrived with a mixed fleet of Manchesters and Lancasters, although a shortage of the latter meant that Halifaxes were also acquired to undertake the bulk of the multi-engine training. The unit was briefly joined by No 1667 HCU for a few weeks in 1943, and relatively short stays were also made by the ABTF, 1481 Flight, 1503 Flight and the BDU, resulting in the regular presence of smaller aircraft types such as the Oxford and Martinet.

Right: *A farewell flypast by a gaggle of Bulldogs to mark the end of flying activities at Lindholme.*

Below: *The former Northern Radar tower, still intact at Lindholme.*

In November 1943 the resident bombers were grouped into the new No 1 Lancaster Finishing School, which finally transferred to Hemswell over a five-month period beginning in January 1944. By this stage Lindholme was part of No 7 Group with satellite fields at Blyton, Sandtoft and Sturgate. Training activities swiftly wound down in 1945, but Lindholme remained active and hosted a variety of bomber units including No 1653 HCU, which became No 230 OCU. Nos 57 and 100 Squadrons arrived in May 1946 with Lincolns, departing again after just five months, although the OCU's Lancasters were present here until February 1949, after which Lindholme became part of Technical Training Command.

The Central Bombing Establishment brought a variety of types to the airfield from April 1949, staying until December, and No 5 Air Navigation School arrived in March 1951, staying until November 1952. It was at this stage that the Bomber Command Bombing School arrived to begin a 20-year stay at Lindholme, initially equipped with a variety of types but mostly Lincolns. The BCBS was renamed as the Strike Command Bombing School in 1968 when Bomber Command finally stood down, and the unit eventually acquired Varsity and Hastings aircraft, tasked with the training of crews destined for the V-Bomber force. The Lincolns were slowly phased out and the pugnacious Varsities became the most numerous residents, in company with the much larger Hastings transports, which were kitted out for rear crew radar training. The SCBS stayed at Lindholme until September 1972 when its remaining Hastings aircraft transferred to Scampton to become part of No 230 OCU.

The airfield was then effectively abandoned, although it remained in RAF hands with Northern Radar (a regional radar unit) situated in its own complex on the old bomber dispersals across the A614. Flying was restricted to occasional use by RAF Finningley's Varsities, Dominies and Jet Provosts, the last significant event taking place here being a services cadets show in the late 1970s when a Harrier operated from the old main runway, becoming in the process the last combat aircraft to visit Lindholme. Helicopters were positioned here during the Queen's Review of the RAF held at Finningley in 1977, and Finningley continued to shift its aircraft here during its annual air show week, but by the 1980s the RAF was gone and the station was modified to become a prison.

The runways were eventually removed and the former airfield is now virtually obliterated, with the prison complex occupying the south-western area, a large sports field situated at the very point where the countless bombers and post-war aircraft once touched down. The station's buildings have mostly survived and are now used for a variety of commercial purposes, and Northern Radar's mortal remains can still be found across the A614.

Main features:
Runways: 210° 6,000 x 150 feet, 132° 4,200 x 150 feet, concrete and tarmac surface. *Hangars:* five C-Type. *Dispersals:* thirty-six Heavy Bomber. *Accommodation:* RAF: Officers 204, SNCOs 681, ORs 1,307; WAAF: 365.

LINTON-ON-OUSE, Yorkshire

54°02'56"N/01°15'13"W; SE489616; 46ft asl. 9 miles NW of York off B6265, E of Little Ouseburn

Situated in a remote area of flat land in the misty (and often foggy) Vale of York, work on creating a bomber airfield adjacent to the village of Linton-on-Ouse began in the early 1930s, leading to the opening of an RAF station in May 1937, although it was still only partially completed by this stage. No 4 Group took control and Nos 51 and 58 Squadrons arrived in April 1938 with a large fleet of Whitley bombers. Training was geared towards preparation for war and, when hostilities began, Linton's bombers were involved on the very first night, embarking on leaflet-dropping flights over Germany, delivering some 5.4 million propaganda papers.

Donald J. R. Wilson was with No 51 Squadron at this time and recalls his experiences:

'First operational flights by 51 Squadron from RAF Linton-on-Ouse, September 1939. The first pre-operational briefing was obviously a major milestone of the war years. Seven crews from 58 Squadron and three from 51 Squadron foregathered in Station Headquarters,

Linton-on-Ouse is seen from a passing Hawk during 2006. The original bomber airfield layout is mostly unchanged, although a large flight line apron is now visible in front of the hangars, and is occupied by Tucano trainers during weekdays.

Linton-on-Ouse, to learn that they were to transport and distribute 5.4 million leaflets to enlighten German citizens of some Ruhr towns; my destination, however, was Frankfurt-on-Maine. The atmosphere was tense, and excitement tempered with anxiety showed on the faces of the assembled crews. This was a great adventure, the first incursion into enemy territory with the probability of a hostile reception across the frontier. Being a married man of some two days standing, with no honeymoon or reception, motivation was the strongest for a speedy and safe return from this first operational sortie. The round trip was sufficiently lengthy to necessitate a refuelling stop at the French Air Force base at Villeneuve.

The British Air Force component of the Expeditionary Force had not as yet been sent to France. Weather conditions were good, and map-reading down the prearranged corridor to the South Coast was relatively easy. An alteration of course took us to the point of entry at Le Treport, where we circled a hilltop fortress, awaiting permission to proceed to Villeneuve. The passage to Riems was made easy for the navigator by a string of flashing beacons, and on landing refuelling commenced. Aerodrome defence, like all the other services provided, was antiquated, and First World War Hodgkiss machine guns were installed in perimeter gun pits. As darkness approached the ten aircraft took off for their respective dropping zones, and the leaflet bundles, secured by elastic bands, were prepared for launching down the parachute flare chutes.

A wartime image of Linton-on-Ouse with bombers occupying many of the airfield dispersals.

The border between Holland and Germany was clearly defined – Holland was a blaze of lights whereas Germany was in complete darkness. Opposition was very light, the searchlights causing more anxiety than the distant gunfire. Having completed the task of distributing what was described as free toilet paper, we set course from Frankfurt to the Cherbourg Peninsula. The coastline was bathed in bright moonlight with sparkling sea and silver sand. We had just altered course over

A No 78 Squadron Halifax airborne on a mission from Linton-on-Ouse.

the Channel, heading for Shoreham, when the tranquillity of the scene was rudely disturbed by a frantic voice shouting over the intercom, yelling, "Christ, we're out of fuel!" The Whitley banked sharply, turning through 180 degrees, and made for the French coast.

The port engine was already spluttering when we scraped over a high sand dune; with the undercarriage down we made a remarkably good landing between high and low water lines. Within minutes the aircraft was surrounded by French Marines, who, when they had established that we were not members of the Luftwaffe, conducted us through a gap in the sand dunes to a fisherman's hut serving as a Guard Room. With uncharacteristic haste, they arranged for a caterpillar tractor to retrieve the Whitley from the incoming tide. While this was being accomplished, we had a visit from a local farmer carrying a small barrel. He dispensed unpleasant and very dry cider in badly stained coffee mugs, but our crew was fortunately able to hide their true feelings about this kind, if unappreciated gesture.

The arrival of another dawn initiated great local activity; inhabitants with horses and tractors paraded on the foreshore, and as the tide receded all manpower (and women) was engaged in removing rocks, stones and other obstacles to construct a makeshift runway. The French Naval Air Service arrived from the naval base at Cherbourg with supplies of aviation fuel in dozens of small containers. Sufficient fuel for take-off and a flight to Cherbourg base was transferred, and in a storm of flying sand the Whitley took to the air. Having completed the refuelling we had an uneventful flight back to Linton. Subsequently we found that three

De Havilland Hornets from No 64 Squadron on the flight line at Linton.

out of the ten aircraft taking part in the operation had to make unscheduled stops in France during the return journey. Was it due to inefficient ground servicing, or could it have been failure to carry out pre-flight cockpit checks? Undoubtedly great pilot skill ensured a successful emergency landing without damage, but avoidance of so many rocks, boulders and debris was nothing short of a miracle. However, an enquiry and reprimand was not avoided.'

No 58 Squadron remained at Linton until April 1942, and although No 51 Squadron's stay was rather shorter other bomber units transferred here, including No 35 Squadron, which brought its Halifaxes from Leeming in December 1940, working up on what was then a new bomber type for the RAF. The unit then spawned another Halifax squadron, No 76, which duly moved to Middleton St George.

Linton's contribution to the air offensive over Germany was considerable and it was no surprise when the Luftwaffe retaliated in May 1941 and considerable damage was done to the station facilities, resulting in the death of the Station Commander. Nos 35 and 58 Squadrons continued to provide Linton's main striking force until 1943, when they departed for Breighton. Linton then became part of No 6 (RCAF) Group and Wellingtons from No 426 'Thunderbird' Squadron arrived from Dishforth on 18 June; on arrival they were swiftly replaced by Lancasters. By August the Squadron was flying operational missions and joined a mass attack on Peenemunde during the night of the 17/18th. This was the first raid in which No 6 (Canadian) Group operated Lancaster aircraft. No 426 Squadron dispatched nine Mark II Lancasters, losing two including that of the Squadron Commander, Wing Commander L. Crooks DSO DFC, an Englishman, who was killed.

This was a special raid that Bomber Command was ordered to carry out against the German research establishment on the Baltic coast where V-2 and V-1 rockets were being built and tested. The raid was carried out in moonlight to increase the chances of success. There were several novel features. It was the only occasion in the second half of the war when the whole of Bomber Command attempted a precision raid by night on such a small target. For the first time there was a Master Bomber controlling a full-scale Bomber Command raid; Group Captain J. H. Searby, of No 83 Squadron, 8 Group, carried out this task. There were three aiming points – the scientists' and workers' living quarters, the rocket factory, and the experimental station – and the Pathfinders

The sleek de Havilland Hornet pictured on a sortie from Linton-on-Ouse.

A No 92 Squadron F-86 Sabre gets airborne at Linton-on-Ouse.

employed a special plan with crews designated as shifters, who attempted to move the marking from one part of the target to another as the raid progressed. Crews of 5 Group, bombing in the last wave of the attack, had practised the time-and-distance bombing method as an alternative method for their part in the raid. The Pathfinders found Peenemunde without difficulty in the moonlight and the Master Bomber controlled the raid successfully throughout.

A Mosquito diversion to Berlin drew off most of the German night-fighters for the first two of the raid's three phases. Unfortunately, the initial marking and bombing fell on a labour camp for forced workers, situated 1½ miles south of the first aiming point, but the Master Bomber and the Pathfinders quickly brought the bombing back to the main targets, which were all bombed successfully. In all, 560 aircraft dropped nearly 1,800 tons of bombs, 85 per cent of which was high-explosive. An estimate has appeared in many sources that this raid set back the V-2 experimental programme by at least two months and reduced the scale of the eventual rocket attack. Approximately 180 Germans were killed at Peenemunde, nearly all in the workers' housing estate, and 500-600 foreigners, mostly Polish, were killed in the workers' camp, where there were only flimsy wooden barracks and no proper air-raid shelters.

Bomber Command's losses were forty aircraft – twenty-three Lancasters, fifteen Halifaxes and two Stirlings. This represents 6.7 per cent of the force dispatched, but was judged an acceptable cost for the successful attack on this important target on a moonlit night. Most of the casualties were suffered by the aircraft of the last wave when the German night-fighters arrived in force; the groups

A Short Tucano from No 1 FTS pictured during a sortie from Linton.

involved in this were 5 Group, which lost seventeen of its 109 aircraft on the raid (14.5 per cent) and the Canadian 6 Group, which lost twelve out of fifty-seven aircraft (19.7 per cent). This was the first night on which the Germans used their new Schrage Musik weapons; these were twin upward-firing cannons fitted in the cockpit of Me 110s. Two Schrage Musik aircraft found the bomber stream flying home from Peenemunde and are believed to have shot down six of the bombers lost on the raid.

The second Canadian unit to arrive at Linton was No 408 Squadron, which brought its Halifaxes from Leeming a few weeks later and converted onto Lancasters, returning to operations a couple of months later. Both units eventually returned to the Halifax during the course of 1944 and remained in business at Linton until the end of the war. The units then returned to Canada in preparation for their participation in the Tiger Force, and Linton was briefly silent, the mighty Halifaxes and Lancasters that had lumbered around the airfield for so long having all gone. Indeed, during the war years very few non-bomber units had spent any time here, the only significant exception being No 6 Flying Training School, which had first arrived with a mix of Ansons, Harvards and Oxfords in August 1938, staying until April 1942.

With the last of the Canadians having gone in June 1945, Linton was dormant for many weeks until bombers returned in November. This time they were equipped as transports, Linton having been returned to No 4 Group, now in Transport Command. No 1665 (Heavy Transport) Conversion Unit arrived from Marston Moor with Halifaxes and Stirlings, but its stay was brief and in July 1946 it amalgamated with No 1332 at Dishforth. Linton was then transferred to Fighter Command and in July 1946 No 264 Squadron transferred here, joined by Nos 64 and 65 Squadrons in August, bringing their Hornets and Mosquitoes to the base. No 6 FTS had briefly returned in December 1945 for a stay of just five months, but Linton was now destined for the jet age and both the Mosquitoes and the Hornets were soon replaced by the Meteor.

By 1951 the Linton Wing had been formed, comprising Nos 66 and 92 Squadrons (from Duxford) and No 264 Squadron. From 1954 the sleek, swept-wing F-86 Sabre replaced many of the Meteors, but the type's troublesome period of service with the RAF did not last too long and after

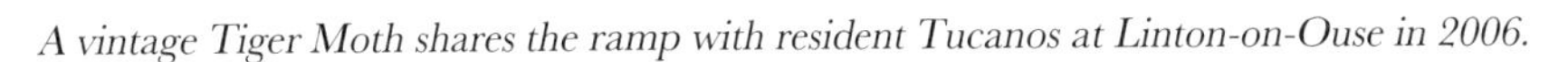

A vintage Tiger Moth shares the ramp with resident Tucanos at Linton-on-Ouse in 2006.

only a couple of years they were replaced by even more capable Hunters. Linton thrived as a very significant and busy fighter base until 1957, when another major change took place and the station shifted back to training activities. The Hunters departed and in November 1957 No 1 Flying Training School arrived from Syerston, initially flying a mixed fleet of aircraft including the Provost and Vampire, but eventually standardising on the Jet Provost. This marked the beginning of a long and continuous stay for No 1 FTS, which saw the unit expand to embrace training that had hitherto been performed by other units. The much-loved Jet Provost was ultimately replaced by the turboprop-powered Tucano, and it is this type that still operates from Linton, equipping Nos 72 and 207 Squadrons, the component elements of No 1 FTS.

The days of the huge bombers and the noisy jet fighters are long gone, but Linton is still a very active base that is now responsible for providing all of the RAF's basic flying training. The black-painted Tucanos buzz around the airfield on a daily basis and occasional graduation ceremonies often see appearances from other RAF aircraft, including the Red Arrows. The longer-term future of Linton is uncertain, however, and with the RAF's force structure continually being reduced it seems likely that all flying training may eventually be concentrated at RAF Valley, rendering Linton redundant. But for the time being Linton is still very much in business, the station's buildings virtually unchanged from 1940 and the runways (the main runway extended for jet operations), perimeter track and dispersals all still in place.

Main features:
Runways: 223° 6,000 x 150 feet, 179° 4,200 x 150 feet, 287° 4,200 x 150 feet, concrete surface. *Hangars:* five C-Type. *Dispersals:* thirty-six Heavy Bomber. *Accommodation:* RAF: Officers 258, SNCOs 552, ORs 1,596; WAAF: 406.

LISSETT, Yorkshire

54°0'22"N/0°16'11"W; TA134580; 16.5ft asl. 6 miles S of Bridlington off A615 at Lissett

This airfield was constructed from 1940 onwards as part of the RAF's expansion plans, being one of many sites in the region used to create a bomber airfield. However, the site chosen was far from ideal and a great deal of work was required to close roads, flatten fields and demolish buildings before a suitable area was established. Despite this, a fairly standard airfield was eventually created with three

This Liberator from the 448th BG skidded off the runway after landing at Lissett during November 1944.

A Halifax from No 158 Squadron has returned safely to Lissett after suffering severe damage from a falling bomb during a raid on Cologne.

full-length runways in a triangular layout, the longest fitting neatly between drainage channels and surrounding roads. Bomber 'frying pan' dispersals were constructed on a series of spurs running from the perimeter track. Initially, many of the admin and technical buildings were distinctly temporary in nature, yet quite a few remained in use until the station was abandoned after the end of the Second World War.

In February 1943 the station finally opened while some construction work was still being undertaken, and No 158 Squadron arrived from Rufforth on the 28th of that month. After settling in, the Squadron embarked on its first operational mission on 11/12 March attacking targets in Stuttgart, ten of its aircraft joining a force of 314 aircraft. Among the unit's Halifaxes was LV907, which eventually amassed some 128 operational missions. Painted as 'Friday the Thirteenth', this aircraft is now represented by the Halifax replica/restoration at the Yorkshire Air Museum.

Arnold Hawthorn was engineer on LV907:

'Returning from a target in Chemnitz the bomber "Friday the Thirteenth", of which I was the engineer, lost the starboard engine, which therefore had to be feathered, while further problems due to heavy icing only made matters worse. The decision was made to rise above the clouds. Although the tops of the clouds could be seen, it wasn't possible to climb above them, so we tried to go below. The pilot asked the navigator for the highest ground on the route home, the answer being 1,500 feet. The skipper decided to go down to 2,000 feet to clear the ice from the aeroplane. Then, as we were flying happily along, the rear gunner complained that he could see telegraph poles and houses flashing past! We re-entered the clouds as quickly as possible, and after discussion the skipper decided to slowly descend until clear of the clouds to verify the altitude. After a few minutes the same sight was seen again, but this time it was seen not only by the gunner but the whole crew! So once again we had no alternative but to re-enter the clouds.

A third attempt was made to escape the icy clouds, and this time the skipper saw a light shining ahead and decided to take a fix on this point. As it came closer it turned out to be a cyclist, who quickly jumped off his bike into a nearby ditch! The resulting tactic was to fly just below the clouds while the whole crew kept a sharp eye out for any obstructions. Eventually the clouds became less icy and intense, so the plane could rise to a reasonable height on the way to the French coast. Before reaching this destination I noticed that the fuel gauges were reading almost zero; not wanting to bail out on such a night, I carefully avoided

A 1945 aerial view of the local area around RAF Lissett.

mentioning it to the skipper, the reason being that I thought my own fuel calculations were accurate and that there was sufficient fuel to return home.

Halfway across the channel I decided to tell the skipper. As I was casually eating an orange, I remarked, "This is a smashing old kite," to which he replied, "Yes, this is a grand old – what do you mean?" Then I told him, "Well, we've been flying for the last few minutes with empty tanks!" He immediately erupted and cursed me, then asked the navigator for the location of the nearest airfield. He was told Blackbushe near Reading. So on reaching the English coast we turned west for Blackbushe. The weather conditions were still atrocious with mist and rain – nothing could be seen. The skipper chose to use a short-wave radio to contact the airfield at Blackbushe, but all we heard was German, much to the consternation of the whole crew! We tried again, but with the same result. By this time even I was getting worried about the fuel situation.

Eventually, after trying Blackbushe yet again, we heard the voice of a female operator still asking for "Aircraft calling Blackbushe say again please." The skipper tried again a number of times with the same results. After some minutes we saw a glow in the sky, which we hoped was Blackbushe, and as we got closer we realised with relief that it was the lights of the airfield runway. And the operator was still calling, "Aircraft calling Blackbushe say again please." With numerous curses that only a Canadian skipper could come out with, he replied that he was coming in anyway as there was no fuel left to circle the airfield. Finally we landed after a sharp manoeuvre and taxied to a dispersal area just in time as the engines cut out from lack of fuel!

The crew were invited to go to the mess but, as I was in the doghouse, I was angrily turned away. The following morning I was told to service the plane, and when asked if I could drive I replied "Yes", although I was not a driver. A petrol bowser and trailer were pointed out to me and I was told to refuel the aircraft. After sorting out the ten gears I managed to drive to the plane without accident but with a few close shaves. The rest of the crew were sent back to the squadron while I was left to see to a new engine that was flown down and mounted, then I was allowed back to the squadron.'

RAF Lissett was part of No 43 Base (headquartered at Driffield) and, as a relatively late arrival, continued to host just the one bomber squadron, although No 1484 Flight also came here in January 1944 for a stay of just a few weeks with a fleet of Defiants, Whitleys and Martinets. By the end of hostilities the Squadron had amassed some 5,400 missions and new Mk VI Halifaxes were slowly being introduced. However, No 4 Group was now transferred to Transport Command, which resulted in No 158 Squadron re-equipping with Stirlings. Once reformed on the new type, the unit moved to Stradishall and RAF Lissett's relatively short existence came to an end.

The area quickly returned to agricultural use and it has remained as such until the present day, the runways and most of the perimeter track being removed over a period of some years. Some traces of the runways remain, as do many of the old dispersals, but most of the admin and technical site is now gone. Lissett's history is unremarkable but the base was undoubtedly significant and played a key part in the RAF's bomber offensive.

Main features:
Runways: 092° 6,000 x 150 feet, 028° 4,200 x 150 feet, 150° 4,200 x 150 feet, concrete surface. *Hangars:* two T2. *Dispersals:* thirty-six Heavy Bomber. *Accommodation:* RAF: Officers 110, SNCOs 310, ORs 1,310; WAAF: 364.

The aircrew memorial at Lissett, and the ghastly wind turbines that now dominate the former airfield.

LONGTOWN, Cumbria

55°00'22"N/02°55'16"W; NY411683; 108ft asl. 7 miles N of Carlisle on A6071, SE of Longtown village

Created as a satellite field for Crosby-on-Eden, the site that became RAF Longtown was established on an area of land adjacent to Hall Burn, where little modification to field boundaries or roads was needed. From the outset a series of three runways was laid together with a connecting perimeter track and a series of dispersals designed for fighter-type aircraft.

Opening in July 1941, the Hurricanes of No 59 Operational Training Unit were the first aircraft to arrive, and these stayed here until August 1942. No 41 Squadron brought its Spitfires here for a brief stay from the beginning of August 1941, but it was the OTU that became the station's main tenant with its mixed fleet of Hurricanes, Martinets, Beaufighters and Beauforts, as well as examples of other types. No 9 Operational Training Unit came here in September 1942 and stayed for a year, and No 55 Operational Training Unit also stayed for approximately a year from April 1942.

A significant change occurred late in 1943 when the fighter types departed and the station became home to No 1674 Heavy Conversion Unit, which brought its Halifax and Fortress aircraft here in October, tasked with the training of Coastal Command crews. The bombers soon became a familiar sight in the area, and training for the Halifax's meteorological role continued until January 1944, after which the whole unit was re-established back at its former home at Aldergove in Northern Ireland. No 6 (Coastal) Operational Training Unit arrived for a three-month stay from October 1943.

When the HCU's Halifaxes departed, the airfield was stood down for a year until No 1521 Flight arrived in October 1945 with its Oxfords. The unit's Beam Approach training kept the airfield busy until April of the following year, when the Flight disbanded and Longtown was no longer required. This long-abandoned airfield is still recognisable even though it has been used only for agriculture for more than 70 years. Most of the main runway is still intact, as is most of a secondary runway and parts of the perimeter track, but apart from a handful of old concrete buildings still be found among the farms, there is little else.

Main features:
Runways: 267° 6,000 x 150 feet, 209° 4,140 x 150 feet, 330° 3,450 x 150 feet, concrete surface. *Hangars:* one T2, two Blister. *Dispersals:* twenty-eight fighter type, five loop type, two circular. *Accommodation:* RAF: Officers 62, SNCOs 142, ORs 693; WAAF: 95.

LOWTHORPE, Yorkshire

54°02'02"N/0°22'16"W; TA067609; 98.5ft asl. 8 miles SW of Bridlington on A614, 1 mile NE of Nafferton

After the RNAS established a large airship station at Howden, plans were soon made to acquire further land where some of the station's airships could be positioned. This would relieve Howden of some of its growing fleet and also enable RNAS crews to be scattered more strategically around the region. A variety of mooring-out sites emerged, and Lowthorpe was one such facility. It opened in April 1918 and soon accommodated three SSZ-type airships in the shape of SSZ 23, 32 and 38 (32 was damaged in an accident and subsequently replaced by 63). These were eventually replaced by other ships, and from this location daily reconnaissance flights were mounted.

Eventually an area of the acquired land was also used for fixed-wing flying, and No 251 Squadron brought its DH.6 aircraft here late in the summer of 1918 to conduct maritime reconnaissance flights for a few weeks. However, when the First World War ended there was no longer any perceived maritime threat and the airships and DH.6s were no longer needed. Lowthorpe was duly handed back to its former owners and the scattering of temporary huts and tents was soon gone. Today there is no evidence of the old airship site or landing field, but it is possible that the farm buildings at Airey Hill are built on concrete foundations that were first laid for the RNAS so many years ago.

Main features:
Runways: none, airship base. *Hangars:* none. *Dispersals:* none. *Accommodation:* not known.

LUDFORD MAGNA, Lincolnshire

53°22'23"N/0°11'45E; TF201877; 443ft asl. 6 miles E of Louth on B1225, off A631 at Ludford Magna

A No 101 Squadron Lancaster has made a crash landing at Ludford Magna in February 1944.

The tiny village of Ludford Magna lays on the A631 road, which climbs out of Market Rasen across the rolling hills towards the East Coast. It is a bleak area of land full of beauty in the summer months, but often almost inaccessible in winter when the area can disappear under feet of snow. It was here, high on the hills, that an airfield was commenced in June 1942 and completed early in 1943. A standard bomber airfield was constructed with three runways of standard length, although more hangars were built than was usual for such sites. Dispersals were scattered around the airfield and some were laid across the B1225, which skirts the western boundary of the airfield. A wartime airfield built to Class A standard, George Wimpey was the main contractor on this £803,000 project. Located 5 miles east of Market Rasen directly south of the A631 and the village after which it was named, the site took over land in two other parishes.

The three concrete runways were 02-20 at 1,950 yards, 11-29 at 1,430 yards and 15-33 at 1,400 yards. The usual thirty-six hardstandings adjoined the perimeter track and all but one were pans. Two other pans were lost due to the erection of four T2 hangars off the south-west perimeter in the technical area. A lone T2 was located on the east side of the airfield and a B1 and T2 on the station technical site near Ludford Magna village, between the heads of runways 15 and 20. A road on the west side from the A631 to the A157 was crossed by dispersals in this area. Domestic and communal sites were dispersed in farmland to the north of the airfield. There were seven domestic sites, two messes, a communal site and sick quarters. Maximum accommodation was given as 1,953 male and 305 female.

The first aircraft arrived in June 1943 when No 101 Squadron's Lancasters appeared. This 'Airborne Cigar' unit stayed at Ludford until the end of the Second World War, although other units were also seen here on a temporary basis. Jack Squire recalls his uncle's experience of life with No 101 Squadron at Ludford:

> 'On Thursday 1 February 1945 at 15.44hrs, Avro Lancaster "C for Charlie" of 101 Squadron RAF took off from the Lincolnshire Bomber Command base of Ludford Magna, near Louth. On board was its crew of eight, captained by Flight Lieutenant Robert Harrison, on his twentieth mission. My uncle, Jack Squire, aged 20, was rear gunner that night, flying as a stand-in for a sick member of Flt Lt Harrison's regular crew. Thirteen minutes behind them, Lancaster "K for Kilo" with Captain Robert Boyd at the controls climbed into the dusk. Just over 3 hours later, the two aircraft and sixteen young men were destined to meet again in the sky over northern France. "C for Charlie" was one of twenty-five 101 Squadron Lancasters that night, on their way to bomb Ludwigshafen in Germany. H-hour was 19.15hrs.
>
> By the time they had all reached the necessary height and settled on course, darkness had fallen. Unseen in the pitch black night around them, hundreds of other blacked-out Lancasters from Nos 1 and 6 Groups converged into a bomber stream, all heading for the same target, the idea being to get as many of them across the target in the shortest possible time. All had the same aiming point, adjusted only by the Master Bomber, who would circle the area giving instructions over the radio as he watched the bombs explode far below.

Like all Bomber Command aircrew, Sergeant Jack Squire was a volunteer. One of thirteen children from Hammersmith, West London, he had already completed more than twenty missions with his regular crew, mainly as mid-upper gunner. Tonight's raid, as tail-end Charlie, would be his last. No 101 Squadron was a Special Duties Squadron operating top secret radio-jamming and communication equipment. The aircraft carried an additional crew member, a German-speaking Special Duties operator whose job it was to confuse the German night-fighter defences by giving false orders over the radio, and interrupting their broadcasts. However, this meant that the Squadron flew an above average number of missions.

At approximately 19.00hrs and 14,000 feet, with 10/10 cloud at 5-6,000 feet, while the aircraft was climbing headed approximately south-eastwards, Flt Lt Harrison saw another Lancaster at the same height, about 100 yards to starboard on a course directly at right angles to 101C. At that moment a shout was heard and the aircraft went into a steep dive that Captain Harrison could not control. He was rolled over, his face forced against something, and his helmet ripped off. The next instant he was falling and, pulling the ripcord on his seat-type parachute, he landed safely in a field near the village of Bezang-Le-Grand. During his descent, while above cloud, the Captain saw two large flashes from below cloud, followed by the sound and shock of explosions, and pieces of aircraft falling around him.

At the time of the collision Sergeant Robert Whiteford, the wireless operator, was out of communication with the rest of the crew, as he was tuning in to the Master Bomber to receive any orders concerning bombing procedure. As he did so he heard a shout from a member of the crew and the aircraft went into a steep dive. Sgt Whiteford hit his head on the fuselage roof, and vaguely remembers several miscellaneous articles swirling around inside the aircraft. There was a large fire that appeared to come from the pilot's position, where there was also a large hole directly in front of the pilot. Although the Captain had his parachute attached to his seat, the rest of the crew were unable to wear theirs; the gunners hung them up outside the doors of their gun turrets due to lack of space.

By chance, Sgt Whiteford saw a parachute swirling around with all the books, etc, and, hanging on with one hand, managed to catch it and put it on against the rush of air through the hole in the nose of the Lancaster. He jumped, and made a safe landing in a ploughed field about 5 miles from where Captain Harrison had landed. The two fortunate crew members were picked up by an advance party of the US Army and taken to their Divisional HQ. While there, various articles were brought in, together with a list of four names, two of which the Captain recognised as being members of No 101 Squadron.

The mighty Thor missile erected in launch position during a training exercise at Ludford Magna.

Eventually Flt Lt Harrison and Sgt Whiteford were returned to the UK, and went on to safely complete their tours of duty. I made contact with Robert Harrison, who finished his RAF time as a Squadron Leader, in 1983, and also with Robert Whiteford, in 1993. I am indebted to them for taking the trouble to write to me then, with such memories as they could recall of what happened that night. My

This aerial view of Ludford Magna from 1980 shows the traces of the runways and the remains of the three Thor launch pads.

> Uncle Jack is buried, together with most of the other crew from the two aircraft, in Choloy War Cemetery, about 3 miles from the city of Toul in north-eastern France. I was able to visit the graves in 1997.'

As Headquarters No 14 Base, the station parented sites at Wickenby and Faldingworth, and aircraft from these stations were sometimes seen at Ludford. The importance of 101 Squadron's activities led to Ludford's main runway being equipped with FIDO fog dispersal gear, and this inevitably led to the appearance of aircraft from other units when weather conditions dictated a diversion here.

The Luftwaffe turned its attention on the station on the night of 3 March 1945 when German aircraft followed British bombers back to their bases and wreaked havoc at various locations. However, Ludford survived with little damage and No 101 Squadron flew its last operational mission on 25 April, having amassed 2,477 sorties at Ludford, from an overall total of 4,895.

When Ludford's Lancasters departed in October 1945 the airfield fell silent and it was reduced to Care and Maintenance status as a sub-site for No 61 Maintenance Unit at Handforth, and subsequently No 93 MU at Wickenby. But the station was to re-emerge a few years later when three launch pads were constructed on the airfield to accommodate the Thor missiles of No 104 Squadron. This unit occupied a high-security compound on the site until May 1963, when the Thor force stood down and the unit disbanded.

Flying activities had effectively ended in the early 1950s when even the occasional flights associated with the MUs ended, but it was not until the Thor's departure that the RAF finally said goodbye to Ludford Magna, and in 1965 the last portions of the site were sold off. Almost half a century later the concrete perimeter track is the only clear evidence of the airfield's existence, almost everything else having been obliterated. However, a farm track traces the path of the airfield's secondary runway, and leads to the site of the former Thor complex, where the launch pads can still be found, slowly decaying under the weeds. From the air these are instantly visible, as is the path of the old main runway, from where the mighty Lancasters once became airborne.

Main features:
Runways: 196° 6,000 x 150 feet, 286° 4,200 x 150 feet, 329° 4,200 x 150 feet, concrete and wood chippings surface. *Hangars:* six T2, one B1. *Dispersals:* thirty-six spectacle and circular. *Accommodation:* RAF: Officers 167, SNCOs 450, ORs 1,336: WAAF: 305.

MANBY, Lincolnshire

53°21'30"N/0°04'53W; TF386866; 56ft asl. 4 miles E of Louth on B1200 at Manby village

Opening as a Royal Air Force station in 1938, this airfield was somewhat unusual in that it was confined between existing roads and fields. This created few difficulties when the landing field was first established, but it led to an unusual layout when concrete runways were subsequently laid, there being room for only two, of modest length. However, five large hangars were erected and numerous dispersals appeared, many in fields across the surrounding roads. Another very unusual aspect of Manby's development was the erection of a huge steel mesh wall some 50 feet high and 1,500 feet in length; this was part of a series of experiments into ways in which the effects of crosswinds could be negated. The trials yielded some encouraging results, but ultimately it was accepted that the creation of sufficient runways in a variety of directions provided a cheaper and more effective solution, and the huge screen was removed in February 1938.

A fascinating line-up of Manby's resident aircraft types, probably taken around 1945.

No 1 Air Armament School came to Manby in August 1938 with a varied fleet of aircraft that included Wallaces, Hinds, Furies, Harts and Overstrands. Other aircraft joined the unit in subsequent years, including Hampdens, Blenheims, Henleys, Herefords, Hurricanes, Lysanders, Wellingtons and Whitleys. The unit operated on nearby ranges, Wainfleet in the Wash being the first, but nearby Theddlethorpe becoming the preferred site. The Armament School kept Manby very busy, and in December 1942 Caistor was adopted as a satellite airfield. A great deal of test flying was conducted at Manby and many new items of equipment were flown on trials here, which inevitably meant that the already mixed fleet of aircraft included many 'one-off' specialised trials variants.

The Central Flying School formed a Handling Flight at Manby in 1949, and as part of its duties a number of new aircraft types were evaluated so that 'Pilots Notes' documents could be created for almost every aircraft type that was in service, or due to enter service. The Fleet Air Arm also operated a Handling Flight, and as part of its work a number of FAA types were test-flown at Manby, including the Attacker and Sea Venom. The bizarre contra-propped Wyvern was also evaluated here, and the unique sound of its mighty Python engine became familiar around this region of Lincolnshire for many weeks.

The Armament Synthetic Development Unit arrived in June 1943 with Wellingtons, and the Empire Central Armament School formed at Manby in April 1944, before becoming the Empire Air Armament School in October. It eventually became part of the RAF Flying College, which formed at Manby in June 1949 (as mentioned previously), divided into component squadrons devoted to fighters, bombers, development and handling. Strubby was adopted as a satellite field for the unit, situated just a few miles to the south.

Jet Provosts get airborne from Manby in 1971.

Flying training first came to the station in March 1951 when No 25 (Flying Training) Grou arrived here, and the Group's Communications Flight operated Chipmunks, Ansons and Meteor from Manby. The College of Air Warfare was formed in July 1962, and as part of this new school th School of Refresher Flying was established, equipped with Varsities and Jet Provosts, joined b Dominies a few years later. By the 1960s Manby's flying activities were almost exclusively share between the RAF College of Air Warfare and the School of Refresher Flying. Meteors (and eventuall Jet Provosts) were the most frequent users of the airfield, although Varsities were often seen chuggin around the circuit together with a handful of Canberras. However, a great deal of the Varsity an Canberra flying was exported to Strubby and almost all of the Dominie flying was performed there Manby's runways being rather short for the Dominie's luxurious take-off performance.

This publicity photograph at Manby in 1971 shows the wartime control tower, which was replaced by a new post-war tower some distance away.

Like many other Jet Provost bases, Manby eventually created its own aerobatic team. Led by Brian Hoskins (who went on to lead the Red Arrows), the Macaws team became a popular attraction at shows across the UK and occasionally in Europe. The name 'Macaws' was derived from a combination of Manby and the CAW. Although based at Manby, much of the formation flying practice was performed over Strubby. The CAW transferred to Cranwell in 1974 and this marked the end of Manby's long and fascinating association with aviation.

The runways were eventually removed, but most of the perimeter track remains, together with almost all of the station architecture, which is recognised by English Heritage as being a classic example of Expansion Period design. Local authority offices now occupy many of these buildings, which will hopefully mean that they are preserved into the future. The post-war control tower (situated on a site separate from the technical site) is in poor condition but appears to have finally been saved from demolition, at least for the time being. Manby is a small and seemingly insignificant site, but it was in fact a very active station that hosted a wide range of unusual and significant units. It is all the more gratifying, therefore, that much of the station has survived.

Main features:
Runways: 290° 4,200 x 150 feet, 220° 3,750 x 150 feet, concrete and wood chippings surface. *Hangars:* four C-Type, seventeen Extra Over Blister, one Special. *Dispersals:* fifty-one circular. *Accommodation:* RAF: Officers 193, SNCOs 254, ORs 1,062; WAAF: 602.

MARKET STAINTON, Yorkshire

53°18'23"N/0°09'59"W; TF222803; 282ft asl. 1 mile N of Ranby, south of A157 on B1225 at Stainton Covert

Although this site was not developed into an airfield, it was an important part of the Royal Air Force's wartime capabilities. Opened in January 1943, No 233 Maintenance Unit established a base here. It was of a distinctly temporary nature, primarily tasked with the storage of bombs, including gas bombs. Documentation suggests that the bombs were stacked in long rows along each side of the B1225, transported here by road from a nearby railway link. This 2-mile stretch of road was closed off to the public with armed guards at each end, guarding not only the bombs but also large stocks of gas stored in 4-gallon drums. Eventually, even Tallboy bombs were stored here and more than 250 personnel were assigned to the unit, most living in temporary accommodation.

The precise location of this unusual site is now difficult to establish, but the old (and now disused) railway track crosses under the B1225 in a tunnel just north of Market Stainton, therefore it seems likely that the temporary rail terminal was at this point. An adjacent patch of shrubbery in the farmland (which has the appearance of a disused and filled-in storage pit) may well be one relic of the RAF's presence here, which ended in 1948.

Main features:
Runways: none, grass field. *Hangars:* none. *Dispersals:* none. *Accommodation:* not known.

MARSKE, Yorkshire

53°52'47"N/01°07'10"W; SE579429; 36ft asl. 1 mile SE of Redcar on A1085 coastal road at Marske-by-the-Sea

Aviation first came to this site in 1910 when Robert Blackburn established a small presence here, working on new monoplane designs. A maintenance shed was built to house the aircraft, which made a short hop to the nearby beaches for most test flying. In July 1910 a flying school was opened, and the site was used by civilian flyers until the outbreak of the First World War, when the site was identified as suitable for military use. The RFC soon arrived and a new station (Marske-by-

An aerial view of Marske and No 2 Fighting School's Bessonneau hangars.

The row of RFC hangars at Marske.

The surviving First World War-era hangars at Marske.

the-Sea) was opened in November 1917, the first resident unit being No 4 (Auxiliary) School o Aerial Gunnery, operating a varied fleet of aircraft including Camels and FK.8s. A large number o canvas hangars soon appeared, and eventually four larger brick-walled Belfast-truss hangars wer built, together with many technical and admin buildings. The station grew in size and eventuall embraced more than fifty aircraft, most of which were used to hone gunnery skills on the nearb weapons ranges near Zetland Park.

Interestingly, one of the station's instructors was a Second Lieutenant W. E. Johns who, despit suffering a series of flying mishaps, maintained his passion for aviation and eventually becam author of the famous 'Biggles' books.

When the RAF was formed the site shifted to RAF control and after a series of name change the resident unit became No 2 Fighting School. Training activities continued through until 1919 some time after the war had ended. The RAF then abandoned the site and it remained unused unti the Second World War, when Army units moved into the four large hangars, but no more flying ever took place here. The historic hangars were sadly demolished during the 1990s.

Main features:

Runways: none, grass surface. *Hangars:* eighteen Bessonneau. *Dispersals:* none. *Accommodation:* not known.

/IARSTON MOOR, Yorkshire

3°57'47"N/01°17'58"W; SE460521; 69ft asl. 4 miles NE of Wetherby off B1224, adjacent to 'ockwith village

Marston Moor is notorious as the site of the largest battle ever to have been fought on :nglish soil, way back in 1644. No stranger to varfare, the area became an important part of a nuch later conflict when an area of land was cquired here in 1940. This was developed into bomber airfield, and RAF Marston Moor pened in November 1941. Originally the tation was to have been called RAF Tockwith a name still used by some locals), although it vas changed in order to avoid confusion with 'opcliffe – or perhaps someone felt it appropriate to perpetuate a name that was already so well ssociated with war. As one of many airfields in the local area, it was positioned on a very suitable tretch of land, although various roads had to be closed and a number of buildings demolished in rder to make way for the landing field, comprising three full-length concrete runways laid in a tandard triangular fashion. Unusually, no fewer than seven hangars were erected, but in all other espects the airfield was of standard bomber design.

This Halifax from No 1662 HCU was written off in a crash during April 1943.

No 1652 Conversion Unit formed here in January 1942, combining the elements of Nos 28 and 07 Squadrons' Conversion Flights, which had arrived a few weeks previously. Its fleet of Halifax ombers eventually grew to more than thirty, but strength varied because of difficulties with ccommodation, runway condition and other problems. However, in October the unit absorbed No 35 quadron's Conversion Flight and another, by which time it had been renamed as a Heavy Conversion Jnit. One element soon moved to Pocklington because of a lack of available space at Marston Moor.

The station eventually assumed control of three others in the area, Acaster Malbis, Riccall and Rufforth. The number of resident Halifaxes continued to grow and the station became extremely usy. The HCU also participated in operational missions from time to time, although its vital raining task remained as its main role, continuing until the end of the Second World War, when he HCU disbanded in June 1945.

Marston Moor then became part of Transport Command, and No 1665 HCU arrived from 'ilstock during August 1945 with a mixed fleet of Halifaxes and Stirlings. Upon arrival it became No 1665 (Heavy Transport) Conversion Unit and training again resumed. However, the unit's stay vas short and in November it was transferred to Linton-on-Ouse. The RAF vacated the station a ew months later, and by 1946 the airfield and its facilities were abandoned.

Today a great deal of the station survives, with many of the hangars and support buildings being art of the local industrial network. The eastern portion of the airfield has been swallowed up by ousing development and the runways have largely been modified for driver training use. A small air trip is still present on a small portion of a secondary runway (but is rarely used), and most flying here ow comprises radio-controlled model aviation. The historic battlefield area lies just to the south-east f the airfield and has attracted many visitors, as have local breweries set up in the former RAF site. Roman Road, on the airfield's western boundary, neatly crosses part of the old main runway, and rom here it is possible to picture the monstrous Halifax bombers thundering skywards.

Main features:

Runways: 226° 6,000 x 150 feet, 353° 4,200 x 150 feet, 286° 4,200 x 150 feet, concrete surface. *Hangars:* six T2, one B1. *Dispersals:* thirty-six Heavy Bomber. *Accommodation:* RAF: Officers 206, SNCOs 744, ORs 1,164; WAAF: 358.

MELBOURNE, Yorkshire

53°52'11"N/0°50'25"W; SE763421; 26.5ft asl. 5 miles SW of Pocklington off B1228, south o Melbourne village

Created as a satellite for Church Fenton, land here was developed into an airfield in 1940. Th nearest village was Seaton Ross, but the more distant Melbourne was chosen to provide th station's name and a grass landing field was laid with little alteration to the existing countrysid When the airfield opened towards the end of the year it was designated as a satellite for Leeming and No 10 Squadron's Whitleys soon began to make use of the facility. However, the site was soo closed again while more substantial construction work was completed, which included the laying c three concrete runways. This required some changes to the area, but whenever possible existin buildings were retained and many of the new bomber dispersals were initially laid across the loca roads in adjacent fields, although this arrangement appears to have been abandoned fairly swiftly.

By August 1942 the airfield was reopened and No 10 Squadron returned. Now operational o the Halifax, the Squadron immediately embarked upon operational missions and stayed a Melbourne through the rest of the Second World War, providing regular contributions to Bombe Command's nightly offensives. The unit's last such sortie took place on 25 April 1945 when some c the unit's aircraft joined an attack on Wangarooge Island. No 10 Squadron amassed some 4,80 operational sorties, the highest total within No 4 Group.

The airfield was unique in being the only facility in the region equipped with FIDO fo dispersal gear, and this attracted many aircraft from other units, which diverted here in ba weather. In May 1945 the station transferred to Transport Command, and No 575 Squadro arrived with Dakotas in August 1945, just days after No 10 Squadron left. However, the Dakota

Left: *A Halifax from No 10 Squadron returns from a mission to Germany.*

Below: *An aerial view of Melbourne, illustrating how the airfield layout is still largely intact.*

ayed only briefly and in November the unit transferred to Blakehill Farm. A series of short stays ere then made by Nos 1510, 1552, 1553 and 1554 Flights (operating Oxfords and Ansons), but by ugust 1946 all of these had also gone and Melbourne was no longer required by the RAF.

Following closure later in 1946 the site lay abandoned. The airfield has since slowly decayed, ut although many of the buildings have gone the main and secondary runway are still largely ntact. Drag racers now roar along the runway where the mighty Halifaxes once became airborne.

Main features:
Runways: 242° 5,700 x 150 feet, 326° 4,200 x 150 feet, 193° 4,100 x 150 feet, concrete and tarmac surface. *Hangars:* two T2, one B1. *Dispersals:* thirty-six Heavy Bomber. *Accommodation:* RAF: Officers 155, SNCOs 287, ORs 1,337; WAAF: 288.

IENTHORPE GATE, Yorkshire

3°48'10"N/0°57'09"W; SE690345; 19.5ft asl. 5 miles NE of Selby, 1 mile W of Breighton airfield n Dyon Lane

Like many other such sites in the region, this small landing field was little more than an area of cleared farmland, secured for use by the Royal Flying Corps from 1916. No permanent tructures were built here and only tents and huts were erected to support occasional deployments rom No 76 Squadron, which was responsible for air defence of the local area. Avro 504 aircraft requented the area for some time during 1916, but by the end of the year the site appears to have een abandoned and it was immediately returned to agricultural use. Today there is no evidence of he RFC's brief presence and the only local link with aviation is a mile away, where the far more ubstantial remains of RAF Breighton can be seen.

Main features:
Runways: none, grass field. *Hangars:* none. *Dispersals:* none. *Accommodation:* not known.

IIDDLETON, Yorkshire

3°44'52"N/01°32'04"W; SE307280; 407ft asl. 3 miles S of central Leeds off A654 on Middleton ing Road

Although this site is now engulfed by the sprawling city of Leeds, in 1916 it was a rural area, and part was cleared for temporary use as a landing field. The RFC established three such sites round Leeds in order to provide landing grounds for No 33 Squadron, which was responsible for the ir defence of the city and region. From March 1916 the landing field was occupied (usually for night perations) by B Flight of No 33 Squadron, protecting the city and local area from the attentions of narauding German airships. It was abandoned in October and no further flying activity took place ere. With no permanent structures (only tents and huts) any evidence of RFC operations was soon one and today it is difficult to establish precisely where the landing field would have been situated, lthough two grass sports pitches may well indicate where the field could once have been found.

Main features:
Runways: none, grass field. *Hangars:* none. *Dispersals:* none. *Accommodation:* not known.

IIDDLETON ST GEORGE, Yorkshire

4°30'41"N/01°25'20"W; NZ375130; 121ft asl. 5 miles SW of Darlington on A67 (Teesside Airport)

First surveyed in the mid-1930s, land was acquired here in 1939 for development into an airfield. The original airfield site was constructed without any significant changes to the local area and only minimal disruption. When the runways were laid they were positioned neatly within the existin road structure, but when the main runway was subsequently extended it required the closure of on of these roads. The same runway was extended yet again some years later when jet aircraft arrived.

Opening on 5 January 1941 the first aircraft to arrive were the Whitleys of No 78 Squadror which transferred here from Dishforth in April. The unit immediately resumed operational dutie and more aircraft arrived in June the following year when No 76 Squadron brought its Halifa bombers from Linton-on-Ouse. The former unit also gradually converted onto Halifaxes, an Middleton soon became a very active Halifax base, joined by the diminutive Oxfords of No 151 Beam Approach Training Flight, followed by No 1535 Flight from December 1942. Trainin became a major task for both squadrons and each spawned a Conversion Flight. These eventuall amalgamated and moved to Riccall, joining No 1658 Heavy Conversion Unit.

The existing squadrons left for Linton in September 1942 and Middleton then became home to Nos 420 and 419 Squadrons, both Canadian units equipped with Wellingtons. RAF Croft was adopted as a satellite field, and Wellingtons soon became a familiar sight around the region, but in May 1943 No 420 Squadron departed for the North African Theatre, replaced by No 428 Squadron flying Halifaxes. Early in 1944 both resident squadrons converted onto the Lancaster and remained active at Middleton with that aircraft until the end of the Second World War. No 419 Squadron completed 1,616 Halifax missions and 2,029 Lancaster sorties, while No 428 completed 1,406 missions on the Halifax and 1,677 on Lancasters. In June 1945 the Canadians returned home in preparation for joining the Pacific war, and Middleton returned to RAF activities.

A wartime aerial reconnaissance image of Middleton St George.

Fighter Command assumed control and No 13 Operational Training Unit moved here from Harwell with its Mosquitoes, although they had left again by April 1947. In October of that year No 2 Air Navigation School arrived from Bishops Court with Ansons and

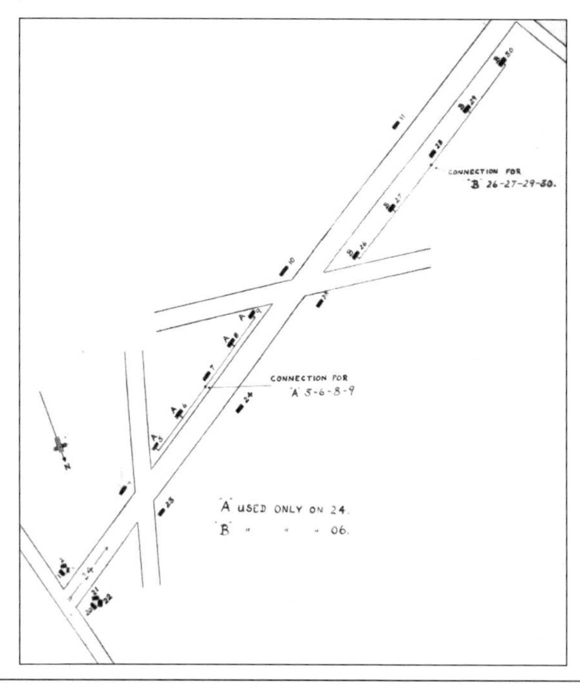

A rare drawing illustrating the flare path layout designed for Middleton St George's main runway. The flare installation was a vital asset for bomber crews when many of the RAF's airfields were shrouded in fog and mist. Ken Cothliff

Above: *Halifax KB762 is pictured between sorties at Middleton St George.*

Above right: *A No 419 Squadron Halifax at Middleton St George.*

Right: *No 419 Squadron Lancasters at Middleton St George in 1944.*

Below left: *Personnel watch Halifaxes embarking on a mission to Cologne during October 1944.*

Below right: *A Lancaster and crews pose for the camera at Middleton St George in 1944.*

Vellingtons. They operated as 'flying classrooms' for some three years before heading south to 'horney Island. The jet age arrived at Middleton when No 205 Advanced Flying School was stablished here in September 1950, equipped with Meteors. The unit moved to Croft early in 1953 'hile the main runway was improved, but returned some months later and became No 4 Flying 'raining School, acquiring some Vampires to join its fleet of single- and twin-seat Meteors.

No 264 Squadron came to Middleton with Meteor night-fighters in February 1957, and a few 'eeks later No 92 Squadron arrived from Linton-on-Ouse, flying the swept-wing Hunter. After only even months the night-fighter Meteors departed and in September 1958 the mighty Javelin all-'eather fighter arrived when No 33 Squadron ıoved here from Leeming. The Javelin's erformance required the main runway to be xtended to 7,500 feet, and the awe-inspiring ınd very noisy) jet became another familiar ıape in the region, joined by more Javelins 'om West Raynham's Javelin Instrument .ating Squadron from August 1961. However, ighter Command's modernisation was far 'om complete and in August 1961 the RAF's ll-new Lightning interceptor arrived at Iiddleton, the Lightning Conversion Squadron

Lancaster KB731 prepares to depart on a mission from Middleton St George during April 1944.

This magnificent view from under the wing of a Lancaster shows a line-up of No 428 Squadron's aircraft at Middleton St George during June 1945. Ken Cothliff

(part of the Central Fighter Establishment) forming here. In June 1962 this became No 22 Operational Conversion Unit, given a shadow designation as No 145 Squadron. The skies aroun Middleton thundered to the sound of the magnificent Lightning for some time as the RAF fighte crews honed their skills on this phenomenal supersonic aircraft, but in April 1964 the OCU heade south to Coltishall and Middleton suddenly fell silent.

In contrast to the awesome Lightning, glider units had already been present here, but from earl 1964 the last such, No 645 Volunteer Gliding School, vacated the station. With the Lightnings gon Middleton was redundant, and in 1964 the site was vacated. It was quickly adopted as the site for a ne regional airport, and in November 1966 it reopened as Teesside Airport. It has continued to develo since the 1960s and it is now a thriving facility, renamed as Durham Tees Valley Airport in 2005.

The OCU Lightnings at Middleton St George were undoubtedly the most colourful examples of the type, adorned with white, red and fluorescent orange trim.

Above: *A Lightning F1A on the flight line at Middleton St George. Javelin fighters are visible in the distance.*

Right: *Lightning mishaps were common at Middleton St George. XM933 has suffered a landing gear collapse while landing in December 1962.*

Below: *Another hapless Lightning on the runway at Middleton St George after suffering a nose-wheel collapse.*

The RAF returned when No 11 Air Experience Flight and the Northumbrian Universities Ai Squadron relocated here from 1985 (with Chipmunks and Bulldogs), but their stay was brief an from 1988 the site has been solely used by commercial traffic, apart from occasional air displays an some visiting military aircraft. Most of the original RAF airfield has survived in remarkably goo condition and the main runway is now used by airliners instead of Lightnings. The old jet apron i still in situ, as are the two main hangars, and of particular interest is the airport's fire dump, whic sits on what was once a V-Bomber Operational Readiness Platform, occasionally frequented b Vulcans and Victors. Middleton St George is a thriving and busy centre of aviation, but the glor days of the mighty Halifax and the awe-inspiring Lightning are little more than memories.

Main features:
Runways: 195° 4,200 x 150 feet, 059° 6,000 x 150 feet, 283° 4,200 x 150 feet, tarmac surface. *Hangars:* two T2, one B1, one C-Type. *Dispersals:* thirty-six Heavy Bomber. *Accommodation:* RAF: Officers 140, SNCOs 504, ORs 1,728; WAAF: 301.

MILFIELD, Northumberland

55°35'26"N/02°05'14"W; NT945329; 131ft asl. 4 miles NW of Wooler on A697, S of Milfield village

This site owes its origins to the Royal Flying Corps, which established a base here in 1917. The known as Woodbridge (the name was later changed so as not to be confused with the much bigge airfield in Suffolk), No 77 Squadron operated its biplane fighters here on defence duties. By 1918 th RFC had abandoned the landing field, but in 1941 the area was again selected for use and a mor substantial airfield was created on a suitable stretch of land that embraced the original RFC site and wider area bounded by the River Till and the A697 road. With high ground only a couple of miles away it was not the ideal location for an airfield, but by 1942 a fighter station had been created with thre concrete runways of moderate length, and a pair of standard T2 hangars, together with other facilities.

No 59 Operational Training Unit arrived here in August 1952 and quickly expanded t considerable size, with more than 120 Hurricanes, Typhoons and Austers at its disposal. The trainin of fighter pilots was a vital and demanding job, and Milfield was a busy airfield throughout the rest o the Second World War. The OTU disbanded in January 1944 when it amalgamated with th Specialised Low Attack Instruction School to become the Fighter Leaders School. The FLS continue the training task (extended to embrace tactical training) and acquired Spitfires to add to the huge flee at Milfield. However, at the end of 1944 the unit moved to Wittering to join the Central Fighte Establishment. No 53 Operational Training Unit arrived in December and resumed the training task equipped with Typhoons, Tempests, Masters and Martinets. As the RAF's training requirement wound down, the OTU was effectively redundant by February 1946 and disbanded, the RAF leavin Milfield shortly afterwards.

The airfield lay disused for many years and eventually the runways, perimeter track and dispersals were removed. Today only a small portion of the northern runway thresholds can be seen, together with a small cluster of dispersals to the south-east. But despite the airfield's destruction, flying still takes place on the site and a gliding club operates from a series of three grass strips laid directly on top of the old airfield layout, albeit on different headings. Flying is certainly still to be seen here, but the roar of piston engines is long gone.

A reconnaissance photo of Milfield taken during 1941.

Above: *No 1 Specialised AIS Hurricane pictured at Milfield.*

Right: *A 1948 aerial view of Milfield illustrating how only portions of two runways have been resurfaced with tarmac.*

Main features:
Runways: 058° 3,300 x 150 feet, 357° 4,200 x 150 feet, 298° 3,300 x 150 feet, concrete and wood chippings surface. *Hangars:* two T2, eight Blister. *Dispersals:* forty-four fighter type, six various. *Accommodation:* RAF: Officers 110, SNCOs 265, ORs 762; WAAF: 521.

MILLOM, Cumbria

54°12'06"N/03°19'22"W; SD138792; 36ft asl. 7 miles N of Barrow-in-Furness, 2 miles W of Millom on A5093 (HMP Haverigg)

Plans to construct a fighter airfield on this site began in 1940, and by February of the following year RAF Millom was ready to open, named after a town some 2 miles away, rather than the nearest village of Haverigg, which should have been adopted, according to standard nomenclature practice.

By mid-February 1941 the first aircraft were starting to arrive here, these being Bothas (sixty-seven aircraft) and Battles (eighteen) from No 2 Bombing & Gunnery School. It took some time for the entire fleet to be built up at Millom, and with the station only partially completed flying activity was not without problems. The local terrain (with high ground very close) and poor weather contributed to these problems, but by June the School was up to strength and was renamed as No 2 Air Observers School, some Ansons joining the aircraft fleet to provide facilities for additional aircrew training.

Towards the end of the year (but only for a stay of a few months) No 1 Anti-Aircraft Cooperation Unit established a presence at Millom with Henleys and Defiants. In June 1942 the AOS became No 2 (Observers) Advanced Flying Unit and, after transferring from No 25 Group to No 29 (Training Group), activities continued, with more Ansons and Bothas supplementing or replacing earlier aircraft, and some Lysanders joining the large fleet. The unit's primary task was to train students who had only overseas experience to become familiar with the European environment, particularly the fickle weather conditions – something that Millom could replicate easily. Training wound down as the end of the Second World War approached, and the unit finally disbanded in January 1945. By this stage the only other flying unit at Millom, No 776 Naval Air Squadron's detachment, had ended its two-year stay in December 1944 and, with the AFU gone, Millom fell silent.

An Officer Cadet Training Unit kept the station active until September 1946, but by then the airfield was abandoned and the RAF station closed at the end of that year. Army units used the airfield on a sporadic basis through the 1950s, but in 1966 the site was transferred to HM Prison Service and a prison was constructed on the site of the RAF station. This facility is still in use today and the adjacent airfield can still be seen, even though it is blighted by ghastly wind turbines. The old runways are overgrown but are still here, as are many of the dispersals and the perimeter track. If the presence of the turbines can be ignored, it is still possible to imagine the countless fighter types heading in from the stormy Irish Sea, just off the end of the old main runway.

Main features:
Runways: 124° 3,000 x 150 feet, 169° 4,200 x 150 feet, 079° 4,200 x 150 feet, tarmac and hardcore surface. *Hangars:* thirteen Extra Over Blister, eight Bellman. *Dispersals:* fifty finger type. *Accommodation:* RAF: Officers 179, SNCOs 550, ORs 1,209; WAAF: 413.

MORPETH, Northumberland

55°07'48"N/01°43'59"W; NZ171817; 318ft asl. 3 miles SW of Morpeth off B6524, N of Shilvington village

Designed originally as a base for a fighter Operational Training Unit, construction of an airfield began here in 1941, but when the station opened in February 1942 it was occupied by an Air Gunnery School, No 17 AGS forming here on 17 March. As part of No 29 (Training) Group, the airfield had a fairly conventional layout with modestly sized runways, but with a loop of dispersals running out into nearby Cockhill Plantation and a variety of buildings and other facilities within the woodland that skirted the airfield perimeter track.

Equipped with a fleet of Bothas and Lysanders, Martinets and Ansons were added to the fleet in 1943 and the AGS was often extremely busy with a steady stream of students passing through. The airfield was not ideal for flying, situated in an area prone to poor weather, and the surrounding terrain and woodlands often created turbulent conditions that made circuit flying hazardous. On one occasion two of the Blister hangars were ripped up and turned over by the unpredictable wind conditions. Despite this, the station remained very active, and in addition to the School's intensive flying other units often deployed here for brief stays, although the only recognised deployment of an operational unit was for one week in August 1942 when Spitfires of No 72 Squadron were based here. The nearby ranges at Otterburn attracted a variety of visiting aircraft, including those of No 652 Squadron, which flew its Austers from Morpeth during June 1943. By the end of 1944 the Gunnery School's activities had wound down, and the airfield was reduced to Care and Maintenance status.

However, in April 1945 it reopened when No 80 Operational Training Unit arrived here, a French unit within No 12 Group equipped with Spitfires, Masters and Martinets. Tasked with the training of pilots for the four French squadrons in Europe, the unit's Spitfires and support aircraft stayed until July, when they moved to Ouston and flying operations finally ended. The station remained in RAF hands and No 261 Maintenance Unit moved here, operating sub-sites at Eshott, Holme-on-Spalding-Moor, Riccall and Wombleton. The MU closed in May 1948 and by the early 1950s the RAF had abandoned the site.

The airfield slowly returned to farmland, but even today its basic structure can still be seen with the paths of the three runways still visible and parts of the concrete still intact. Parts of the perimeter track are still to be found and some dispersals have survived, including the unusual loop of dispersals running into the woodland across the Shilvington road. Many other crumbling buildings can also be seen, hidden in the surrounding woods.

Main features:
Runways: 230° 4,200 x 150 feet, 290° 3,300 x 150 feet, 350° 3,300 x 150 feet, tarmac surface. *Hangars:* three T2, seventeen circular. *Dispersals:* thirty-six spectacle. *Accommodation:* RAF: Officers 58, SNCOs 153, ORs 1,038; WAAF: 437.

MURTON, Yorkshire

53°57'53"N/01°0'23"W; SE652525; 56ft asl. 1 mile E of York off A64 at Murton village, E of Murton Lane

One of many sites adopted by the Royal Flying Corps, this area of farmland was requisitioned in 1916 and towards the end of the year was used on a sporadic basis by the Avro 504s of No 76 Squadron. This unit was tasked with air defence of the local area, and the landing field at Murton was used to support air defence patrols over York, protecting the region from the attentions of marauding German airships. The landing field had few facilities apart from scattered tents and huts, and was only used by the one RFC unit. By 1917 the RFC no longer had any need for the site and it was swiftly returned to agricultural use, the huts and tents quickly disappearing. Having spent so little time at this site, no permanent evidence of the RFC's presence survived, and the area is now no more than an empty field, adjacent to the Yorkshire Museum of Farming.

Main features:
Runways: none, grass field. *Hangars:* none. *Dispersals:* none. *Accommodation:* not known.

NETHERTHORPE, Yorkshire

53°18'60"N/01°11'51"W; SK535802; 240ft asl. 3 miles NW of Worksop, SW of Shireoaks on Thorpe Lane

This small grass landing field was first used in the early 1930s for civil flying, and in 1940 was identified as a suitable area for development into a much larger and more substantial RAF airfield. However, the proposal was never pursued and the existing

Above: *Lysanders taking cover in the trees at Netherthorpe in 1940.*

Right: *Although military flying is long gone, Netherthorpe is still an active site for private flying.*

airfield remained unaltered. Operated at that time by M&H Mining (this being a site in the very heart of the coal-mining community), Netherthorpe was requisitioned by the RAF during the Second World War, although control of the airfield remained with the mining company.

No 613 Squadron came here in 1940 with its Army Cooperation Lysanders, but by the summer of 1941 they had gone. No 24 Gliding School formed here in September 1943 and its Cadet gliders became a common sight in the skies around Worksop, but these also departed in May 1946 to nearby Firbeck, and Netherthorpe resumed its pre-war status as an all-civil commercial and recreational flying site. The airfield is still very active although it remains small, and capable of accepting only light aircraft. Two grass runways are now laid within the confines of the original landing field, but all of the flying here is now civilian orientated, apart from occasional appearances by former military aircraft that survive as 'warbirds'.

Main features:
Runways: none, grass surface. *Hangars:* various (civil flight sheds). *Dispersals:* none. *Accommodation:* not known.

NEWCASTLE-UPON-TYNE: GOSFORTH (Dukes Moor), Tyne & Wear

54°59'25"N/01°37'32"W; NZ240662; 240ft asl. Central Newcastle between A617 and B1318 at West Jesmond

The aircraft manufacturer Armstrong-Whitworth constructed a production factory at Dukes Moor on the (then) outskirts of Newcastle during the summer of 1913, in response to a variety of contracts available from the War Office. Two BE.2A aircraft (Nos 383 and 385) were the first to be built here, and led to contracts for more BE.2A, 2B and 2C aircraft. The new factory was developed from what had once been the grandstand of the original Newcastle racecourse, which had been abandoned when racing had shifted to Gosforth Park in 1882. The new factory (known as the Gosforth Works) was expanded in 1914 when the War Office requested that production capacity be increased, leading to the manufacture of 250 BE.2C aircraft here.

A two-bay hangar was created adjacent to the small landing field and the site was soon occupied by C Flight of No 1 Squadron RNAS, which stayed here with its Bristol TB.8s until

A rare image of workers at Dukes Moor in front of one of the aircraft sheds.

One of few surviving images of Town Moor, the hangars visible in the background.

March 1915. Aircraft production continued here with various FK-series machines emerging from the factory, but in October 1919 the site was closed after officials had declared it unsafe. By this stage more than a thousand aircraft had been manufactured here, but Armstrong-Whitworth shifted its activities to a new site at Town Moor, leaving Dukes Moor abandoned.

The new site was literally adjacent to the old landing field but much larger; acquired in June 1916, Armstrong-Whitworth slowly moved its activities here from Dukes Moor. No 9 Aircraft Acceptance Park was established here in 1917 and handled types such as the FK.8, Bristol F.2B and Sopwith Cuckoo. A 1917-pattern flight shed was erected here together with three Bessonneau hangars, and activities continued until January 1920, when the site was finally abandoned. Today the area is still open moorland and the Dukes Moor site can be found on the northern side of Grandstand Road. The larger Town Moor site lies to the south side of this road. There is little evidence of the area's links with aviation other than a few small patches of concrete to be found along the perimeter of the site.

Main features:
Runways: none, grass field. *Hangars:* various flight sheds. *Dispersals:* none. *Accommodation:* not known.

NORTH COATES, Lincolnshire

53°29'50"N/0°03'57"W; TA371020; 13ft asl. 5 miles SE of Grimsby off A1031, E of North Coates village on Sea Lane

Perched on the very edge of the East Coast, this site became associated with aviation during the First World War when the War Office requisitioned a number of fields here in 1918 to create space for a simple rectangular landing field. A hangar and a few support buildings were also constructed, but the airfield remained very basic, yet immensely useful for the fledgling RAF, which required a suitable base in the area to enable aircraft to patrol the North Sea where German submarines were posing a major threat to shipping.

No 404 (Seaplane) Flight of No 248 Squadron arrived here in September 1918 with DH.6 aircraft, and although other aircraft undoubtedly used the site it was this unit that became the only resident here until it departed in March of the following year. The short-lived landing field was then left unused for some time until activity began to increase on the nearby weapons range at Donna Nook. In 1928 the original landing field was reacquired and brought back into use for occasional Armament Training Camp detachments. In 1 January 1932 No 2 Armament Training Camp was officially established here at what was then known as RAF North Coates Fitties. The unit subsequently changed its title to Temporary Armament Training Camp, then to Temporary Armament Training Station, but its role remained much the same, hosting detachments from various RAF and Fleet Air Arm units.

A 10 Squadron Vickers Virginia at North Coates in 1935.

The Air Observers School formed here in January 1936 with a fleet of Wallaces, and in November 1937 the TATS became No 2 Air Armament School, changing yet again in March 1938 to No 1 Air Observers School. Both resident units moved out in September 1939, as the airfield's location on the coast was judged to be perilously close to the newly declared enemy across the North Sea. Spitfires from No 611 Squadron operated here late in the year. No 1 Ground Defence Gunners School was formed in November, flying Wallaces and Harts as targets until July 1940, by which stage the station had been renamed as RAF North Coates and had become part of Coastal Command.

In February 1940 the first torpedo bombers arrived when No 235 Squadron moved here with Blenheims for a two-month stay, and Nos 236 and 248 Squadrons also came here for a similar duration. No 22 Squadron transferred here from Thorney Island in April with its Beauforts and stayed until June 1941, operating detachments at various other airfields; it also spawned what became No 431 Flight, operating Marylands on strategic reconnaissance duties, eventually leaving for the Mediterranean. No 812 Squadron FAA brought its Swordfish aircraft to North Coates in May 1940 and operated with the RAF for the rest of the year on mine-laying and bombing operations. No 816 Squadron did likewise during the following year.

As one of the RAF's main anti-shipping bases, North Coates became the home of No 2 Mobile Torpedo Servicing Unit in December 1940, becoming No 21 Mobile Servicing Unit (Torpedo) in September 1942, and also operating from Donna Nook. No 42 Squadron sent a detachment of Beauforts to North Coates early in 1941, as did No 86 Squadron in May, after which No 22 Squadron moved to Thorney Island. In the opposite direction, the Hudsons of No 407 'Demon' Squadron RCAF arrived in June, and this aircraft type became the most numerous at North Coates over the next year. When No 86 Squadron moved to St Eval in January 1942, No 59 Squadron arrived with more Hudsons, followed by No 53 Squadron's Hudsons a few weeks later, although this unit only stayed until May. Hampdens arrived here when No 415 'Swordfish' Squadron came in June for a two-month period.

No 59 Squadron left in August to be replaced by No 143 Squadron, bringing the first Beaufighters to North Coates. This immensely successful torpedo bomber led to the creation of the North Coates Wing, which was established by the end of 1942 and comprised Nos 236 (which arrived in September), 245 (arriving in November) and 59 Squadrons, all equipped with Beaufighters. The first operational mission took place on 20 November, but it was unsuccessful and a tactical reappraisal was conducted before

This Beaufort from No 22 Squadron crashed shortly after this photo was taken at North Coates.

operations got truly under way in 1943. The Strike Wing became a vital part of the RAF's offensive capability and North Coates remained busy, hosting countless Beaufighters as well as No 278 Squadron's Air Sea Rescue detachment with Walruses and Lysanders. This unit arrived in October 1941 and stayed until early 1944. Detachments from No 6 AACU and No 7 AACU were also present in 1942 and 1943 respectively.

In August 1943 the Strike Wing lost No 143 Squadron, although it briefly returned twice, having re-equipped with Mosquitoes by the time it made its final appearance here in October. By March 1945 No 254 Squadron had also begun to receive Mosquitoes, but by this stage operations were winding down, the North Coates Strike Wing having been responsible for the destruction of more than a hundred enemy ships. By June 1945 the Beaufighters and Mosquitoes had gone and North Coates became virtually silent, the station being transferred to No 61 Maintenance Unit until May 1950. No 15 School of Technical Training was also based here from October 1948 until September 1953.

After many years of inactivity helicopters came to North Coates in November 1954 when B Flight of No 275 Squadron was established here with a gaggle of yellow-painted Sycamores. They stayed until October 1957, and their departure marked the end of flight operations here. However, the RAF stayed and in December 1958 No 264 Squadron reformed here as a Bloodhound Surface to Air Missile unit. A large missile battery was constructed on the south side of the airfield during that year and the North Coates squadron contributed towards the network of Bloodhound batteries that emerged around Lincolnshire, tasked with the protection of the Thor ICBM sites that lay further inland. When Thor was withdrawn the Bloodhounds were no longer needed and 264 Squadron disbanded in November 1962.

The Surface to Air Missile Operational Training School formed at North Coates in October 1963 and stayed until the following year, after which No 17 Joint Services Trials Unit arrived in connection with the Mk 2 version of the Bloodhound. After this unit departed in 1966 North Coates was again largely unused, but in October 1973 No 25 Squadron reformed here with Bloodhounds before moving to Bruggen. In December 1975 B Flight of No 85 Squadron was established here with Bloodhounds, staying in business until December 1990. This was the last RAF unit to leave North Coates, and the RAF vacated the entire site during the following year.

Although the station had supported missile operations for so many years, flying activity had effectively ended after the Second World War, apart from occasional arrivals and departures connected with the MU, and of course the helicopter detachment. Some communications aircraft were seen here over the years, but post-war North Coates was inevitably regarded as a missile base rather than an airfield. However, when the RAF left the airfield slowly returned to flying, and

A Bloodhound missile, part of C Flight 25 Squadron's battery at North Coates.

Bloodhound missiles look eastwards across the North Sea towards the enemy that never came.

private aircraft can now be seen here. The North Coates Flying Club operates from the old RAF airfield, although the old main runway has now been removed. Some of the original buildings survive, but most of the airfield is gone, and the Bloodhound launch pads have been obliterated.

Main features:
Runways: 248° 4,260 x 150 feet, concrete surface, NW/SE 4,350 x 150 feet, grass surface. *Hangars:* four Bellman, four F-Type, one B1. *Dispersals:* eight circular, seven loop. *Accommodation:* RAF: Officers 116, SNCOs 256, ORs 1,769; WAAF: 457.

NORTH KILLINGHOLME, Lincolnshire

53°38'09"N/0°17'31"W; TA130168; 26.5ft asl. 1 mile NW of Immingham on A1077 at South Killingholme

Unusually, two independent airfield sites are to be found at North Killingholme. The first can be found just to the north-east of North Killingholme village in an area of land that has long since been engulfed by industrial sprawl. Where endless acres of storage containers now stand, an airfield was created in 1914 intended for use by seaplanes, with a series of slipways running from the landing ground onto the adjacent beach. Some flight sheds and support buildings were also erected and the Royal Naval Air Service adopted the site for Home Defence duties from August 1914.

The United States Naval Air Service also came here in 1918 with Felixstowe flying boats, and the station was officially designated as a USNAS from 20 July. No 238 Squadron also established a detachment here, as did Nos 403 and 404 Flights, but these stayed only briefly and departed in August. In January 1919 the RAF took control of the site and a variety of units made brief stays here prior to disbandment. The last of these was No 229 Squadron, and in 1920 the base was abandoned.

It was not until 1941 that North Killingholme was again associated with military flying. An area of land just a mile from the seaplane base was deemed suitable for development into a much larger bomber airfield, as a satellite for nearby Goxhill. Surrounded by roads and railway lines, an area of farmland was duly acquired and cleared before three concrete runways were laid and hangars erected. Built to Class A standard, work started in August 1942 with John Laing & Son Ltd as the main contractor of this £810,000 project. The three runways were 04-22 at 2,000 yards, and 09-27 and 15-33 both at 1,400 yards. The thirty-six hardstandings were all loop-type.

A No 550 Squadron Lancaster departs from North Killingholme on its 100th raid.

The coastal site at Killingholme, pictured in 1947.

A T2 hangar stood by the technical site, south of runway head 22 and close to North Killingholme village. A second T2 was located on the south-west side of the airfield, between runway heads 04 and 33. A B1 hangar was erected later for the Ministry of Aircraft Production contractor engineer's use. Bomb stores were close to Skitter Beck between runway heads 15 and 09. The camp was to the east, between the village and Bass Garth, and consisted of single mess, communal and WAAF sites, six domestic sites and sick quarters, all dispersed in farmland. Accommodation availability was put at 1,939 males and 325 females.

Although still incomplete, the station opened in October 1943, and in January 1944 No 550 Squadron arrived with its Lancasters. Formed from C Flight of No 100 Squadron at Waltham, the unit became North Killingholme's only resident unit, eventually flying some 3,582 operational bomber missions. When the Squadron disbanded in July 1946 the airfield was assigned to No 35 Maintenance Unit some six months later. No 93 MU also used the site from 1948 but, with the exception of occasional communication flights, no aerial activity took place here after 1945.

Despite a relatively short and uneventful history, the station was an important part of Bomber Command, and its association with the nearby seaplane base certainly makes the site unusual. The runways, taxiways and hangars are all still here, although they are slowly falling into decay.

Main features:
Runways: 216° 6,000 x 150 feet, 272° 4,200 x 150 feet, 328° 4,200 x 150 feet, concrete surface. *Hangars:* two T2, one B1. *Dispersals:* thirty-six Heavy Bomber. *Accommodation:* RAF: Officers 112, SNCOs 322, ORs 1,250; WAAF: 335.

OUSTON, Northumberland

55°01'27"N/01°52'27"W; NZ081699; 462ft asl. 10 miles W of Newcastle off B6309, N of B6318

Designed as a replacement fighter airfield for Usworth (which had been reallocated), approval for land purchase here was given in 1939. Construction of the airfield proceeded without any difficulties and only hedge clearance and drainage were required before the runways were laid. Unusually, these were initially reduced in width towards their thresholds, but were subsequently widened to the standard 150 feet.

A glimpse of the main site at Ouston during the 1960s.

The station opened as a No 13 Group Sector Station on 10 March 1941 and a few weeks later the first aircraft arrived in the shape of Hurricanes from No 317 Squadron, which moved here from Acklington. This Polish unit was quickly engaged in operations, and in June it transferred to Colerne. Its place was taken by No 131 Squadron, which formed here on Spitfires, then moved to Catterick. No 234 Squadron formed on Spitfires here in June 1942, but after two months it shifted to Turnhouse. The rotation of fighter squadrons continued for some time, with units forming here prior to departure, or coming here to rest and train before resuming operations. This procedure lasted until 1943, when No 350 Squadron's Spitfires were the last to leave in July.

Acting as a base for air defence of the local region, and for maritime operations over the North Sea, night air defence was also an element of the station's activities, and No 410 Squadron operated Beaufighters and Defiants here in 1942. No 281 Squadron formed at Ouston in March 1942 to embark upon ASR operations with Defiants, Walruses and Ansons. Training and Army Cooperation was another role performed here, and No 613 Squadron spent some time at Ouston with Mustangs, as did No 657 Squadron with its Austers, the latter unit being associated with the large training area at nearby Otterburn. No 55 Operational Training Unit at Usworth dispatched aircraft to Ouston while using the airfield as its satellite, and in April 1943 No 62 Operational Training Unit moved here with a large fleet of Ansons, eventually absorbing No 1508 Flight with its Oxfords. These were all subsequently replaced by Wellingtons and Hurricanes, and the unit remained active until June 1945. One month later No 80 Operational Training Unit brought its Spitfires to Ouston, tasked with the training of French fighter pilots, remaining in business here until March 1946.

By this stage the first Royal Auxiliary Air Force unit had arrived, No 607 Squadron bringing its Spitfires here in May 1946. The unit received its first jet aircraft in the shape of Vampires in March 1951, and Ouston became a relatively active Vampire base for some years until the RAuxAF was wound down in 1957. More Army units continued to use the airfield, and No 1965 Flight was present from 1949 until 1954 with its fleet of Austers. Gliders were also based here from June 1948, No 27 Gliding School staying until November 1951. Durham Universities Air Squadron was present from September 1946 until May 1948 with Tiger Moths and Harvards. However, the airfield gradually became less active, and by the 1970s only the Chipmunks of No 11 Air Experience Flight and the Northumberland Universities Air Squadron were to be seen. These departed in September 1974 and the station was transferred to the Army, becoming Albermarle Barracks.

Thanks to the presence of the Army, the airfield and its runways survive in remarkably good condition, , including the main runway's post-war extension, which runs into an area of bushes and trees to the north. Most of the original buildings are also intact. Ouston is a remote site that was never associated with any particular role, but its history is by no means mundane.

Main features:
Runways: 090° 3,800 x 150 feet, 050° 3,600 x 150 feet, 140° 4,200 x 150 feet, concrete surface. *Hangars:* one J-Type, eight Blister. *Dispersals:* four fighter type. *Accommodation:* RAF: Officers 113, SNCOs 281, ORs 806; WAAF: 582.

OWTHORNE, Yorkshire

53°43'58"N/0°03'03"W; TA331281; 23.5ft asl. E outskirts of Withernsea on B1362 opposite Old Owthorne Vicarage

This coastal site was one of a number of similar landing fields brought into use during the First World War. This particular field was acquired in 1917, but it was not until June 1918 that No 506 (Special Duty) Flight formed here. This was one of No 251 Squadron's detachments that were maintained around the region at that time. Equipped with DH.6 aircraft, the unit was assigned to maritime patrol duties, and searching for German submarines was the unit's primary task. This activity continued into 1919, at which stage No 251 Squadron stood down its Flights and moved to Killingholme in preparation for disbandment. This resulted in the departure of No 506 Flight from Owthorne in March 1919 and the site appears to have been abandoned after that date. Being only temporary in nature, it is hardly surprising that no traces of the landing field have survived, although the basic proportions of the area can still be seen, as the field is still retained for agricultural use.

Main features:
Runways: none, grass surface. *Hangars:* none. *Dispersals:* none. *Accommodation:* not known.

PLAINVILLE, Yorkshire

54°2'11"N/01°07'17"W; SE576604; 56ft asl. 1 mile NW of York off B1363 on Plainville Lane, off Corban Lane

Although used as an airfield, this site was in effect little more than an allocated farm field that was cleared of obstructions and assigned to use by the Royal Air Force. Acquired in 1940, the landing field was used to accommodate aircraft that were dispersed from the key bomber airfields. No 58 Squadron arrived here in 1940 with its Whitley aircraft and remained until the end of 1941. Most activity here was of a relatively routine nature, although on 31 May 1941 Whitley Z6660 overshot the airfield when it returned from an operational mission. Only one member of the crew was injured, however, and the aircraft was recovered some days later. On 11 October 1941 another Whitley, serial Z9204, ditched in the North Sea off Skegness while returning from a mission. Again, there were no fatalities. In all other respects activities here resulted in no significant mishaps. The Squadron vacated Plainville at the end of 1941 and the site was then abandoned. The scattering of temporary support buildings was soon gone and today the site is now no more than a hedged farm field on the western side of Bull Lane.

Main features:
Runways: none, grass surface. *Hangars:* none. *Dispersals:* none. *Accommodation:* not known.

POCKLINGTON, Yorkshire

53°55'37"N/0°47'44"W; SE791485; 88.5ft asl. On A1079, 1 mile W of Pocklington, 10 miles NW of Market Weighton

Land at this site was first requisitioned in the summer of 1940, although it had been listed as a potential site for airfield development for some years. The area was certainly suitable for such a purpose, but the surrounding road network and a scattering of housing made the task of laying the airfield's runways difficult, and its resulting layout was unusual. Instead of just the standard three runways (all of which were shorter than usual), a fourth was also provided, positioned at an angle that was only a few degrees different from its counterparts. Why this was done remains unclear, although it is assumed that prevailing weather conditions dictated that even a short runway laid on the most suitable heading was better than none.

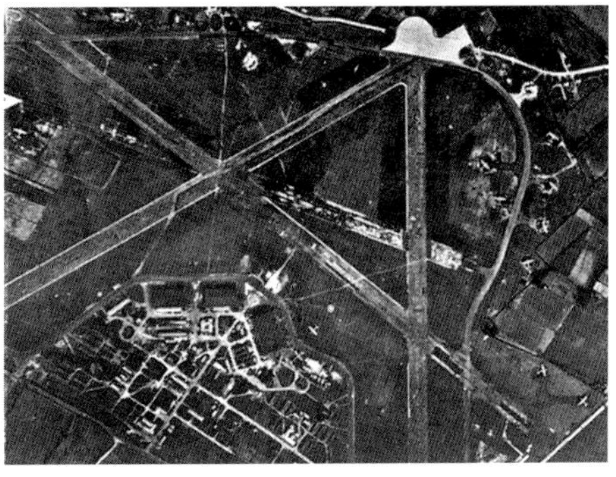

An aerial photograph of Pocklington, probably taken during 1944, illustrating the unusual runway layout and threshold extensions.

The airfield opened to flying in June 1941, and No 405 'Vancouver' Squadron's Wellingtons arrived from Driffield during that month. This unit (the first Canadian squadron in Bomber Command) began operations almost immediately, and remained busy until 1942, when the Wellingtons were gradually replaced with Halifaxes. When the unit had re-equipped it resumed its operational role, participating in the Thousand Bomber raid over Cologne on 30/31 May. However, in August the unit shifted to Topcliffe and was replaced by No 102 Squadron, which was in the process of converting onto the Halifax; once this process was complete the unit was quickly reassigned to operational missions.

At Pocklington, No 102 Squadron prepares to raid an armaments factory in France.

As a parent airfield for a group of airfields, Pocklington assumed control of Elvington and Melbourne, the latter being replaced by Full Sutton in April 1944. Later versions of the Halifax eventually arrived, and No 102 Squadron's activities continued, the night-bombing missions eventually widened to include day sorties and a number of specialised duties that included the supply of fuel to Belgium. Eventually the unit achieved 4,734 Halifax sorties for nearly 1,000 aircrew casualties.

A 1994 aerial view of the landscape around Pocklington. As can be seen, the airfield layout is still largely intact.

In May 1945 the station transferred to Transport Command and the Halifaxes began to be replaced by Liberators, but while this process was still taking place the unit transferred to Bassingbourn. Flying at Pocklington effectively ended in June, but the station survived as an Aircrew Holding Unit until the end of 1946, when the RAF finally left.

The site languished for many years, but today some flying has returned, albeit on a very modest scale. The Wolds Gliding Club uses parts of two old runways on the south-eastern side of the airfield. Most of the runway surfaces have survived, although there appears to be no trace of the early fourth runway. Almost all of the perimeter track and dispersals have gone, but some of the hangars can still be seen, the mighty J-Type still standing proud among the industrial and commercial buildings that have appeared in the old admin and technical site.

Main features:
Runways: 320° 5,400 x 150 feet, 352° 3,900 x 150 feet, 191° 3,900 x 150 feet, concrete and tarmac surface. *Hangars:* four T2, one B1, one J-Type. *Dispersals:* thirty-six Heavy Bomber. *Accommodation:* RAF: Officers 166, SNCOs 421, ORs 1,108; WAAF: 323.

PONTEFRACT, Yorkshire

53°42'10"N/01°20'01"W; SE440231; 88.5ft asl. At Pontefract on A639, south of M62

Like many such sites around the region, this small landing field was created in 1916, utilising an area of open land that was deemed suitable for aviation. The growing threat of German airship attacks required the maintenance of air defence cover over the region and the Royal Flying Corps adopted a wide range of small sites where its fighter aircraft could deploy. Horse-racing circuits were often identified as being ideal for this purpose, as they were inevitably flat and cleared of obstructions, and Pontefract's course was requisitioned for this purpose.

No 33 Squadron adopted the site in 1916 and its role was taken over by No 76 Squadron in October of that year. Activity continued on a sporadic basis until 1918, when the RAF vacated the site and horse-racing eventually returned, although it appears that sporting activities of this nature continued even throughout the years when the RFC was present. Bounded by the racecourse, the site is still here, and among the scrubland there are some patches of concrete and a few abandoned pits that may well be surviving evidence of the short-lived landing field.

Main features:
Runways: none, grass surface. *Hangars:* none. *Dispersals:* none. *Accommodation:* not known.

REDCAR, Yorkshire

54°35'46"N/01°05'21"W; NZ589227; 36ft asl. At Redcar on A1042 on Kirkleatham Lane

This site was, like many others, established on a horse-racing course but, instead of creating a landing strip on the actual course, an area of land adjacent to the course was cleared and prepared for use in 1918. Known locally as both Kirkleatham and Redcar, the Royal Naval Air Service operated here while mounting Home Defence duties, although much of the flying was concerned with maritime patrol and training. Various aircraft types were present, including HP.0/100s from No 7N Squadron, although the first flying machines to appear here were actually airships, as the site was allocated for use as an out-mooring station for Howden. Zero Class airships SSZ 54, 55 and 62 were the first to be seen here. No 510 Special Duty Flight formed here on 7 June 1918; it became part of No 68 (Operations) Wing and in November the unit transferred to West Ayton. The North Eastern Area Flying Instructors School was also established here in the summer of 1918, and in January 1919 it was joined by the North Western Area FIS, the two units combining during the following May. No 63 Training Squadron arrived in October 1918 and the airfield remained busy until the end of 1919, by which time the requirement for aircrew had rapidly diminished.

Above: *At Royal Naval Air Station Redcar in 1915, BE.2C 1109 lifts into the air at the start of a training sortie.* Phil Jarrett

Right: *A diagram showing the airfield site at Redcar in 1918. Unusually, the site was adjacent to the local racecourse rather than being situated on it.*

By October the airfield was largely dormant, the RAF vacated the site at the end of the year and the site was abandoned. The few airfield structures were eventually demolished and the landing field was eventually used for farming; the racecourse also disappeared in the process. Today the fields are still largely unchanged, and Cleveland Police maintain a facility on what was once the southern boundary of the landing field.

Main features:
Runways: none, grass surface. *Hangars:* four flight sheds. *Dispersals:* none. *Accommodation:* not known.

Below and right: *An Avro 504 on the grass landing field at Redcar, and soaring into the air.*

RICCALL, Yorkshire

53°49'34"N/01°01'34"W; SE642730; 29.5ft asl. 2 miles N of Selby off A19, N of A163, SE of Riccall village

One of many significant bomber airfields in this region, RAF Riccall opened in September 1942, although land was first acquired here in June 1940. The land was well suited for development, although laying the runways in a traditional triangular layout presented some problems as the surrounding roads, railway and land boundaries prevented a conventional design. The result was somewhat unusual, although all three runways were built to standard dimensions.

In October 1942 No 1658 Heavy Conversion Unit formed here, created from Nos 76 and 78 Conversion Flights which had arrived at Riccall in September when the station was still being completed. With a notional strength of some thirty-two Halifaxes, a variety of communications and support aircraft joined the unit, including Spitfires (used for gunnery tracking practice), Magisters and Tiger Moths. In November the HCU absorbed No 158 Conversion Flight (which had been based at Melbourne), and training operations expanded still further. The unit stayed very active at Riccall through the rest of the Second World War, and although many other aircraft visited the station it was the HCU that remained as sole resident. The work of the HCU did not enjoy the 'glamour' of the RAF's operational squadrons, but its training task was vital and hundreds of Halifax crews honed their skills in the skies above Riccall, before becoming a valued asset within Bomber Command.

The surviving remains of the Royal Air Force's presence buried in the woodland at Riccall.

When the war ended, the training role was no longer required and the HCU finally disbanded on 13 April 1945. However, Riccall survived and Transport Command took control of the station, No 1332 (Transport) Heavy Conversion Unit arriving from Nutts Corner just days after the bomber HCU stood down. With a fleet of Stirlings, Liberators, Yorks and Dakotas, the unit operated at Riccall until November, when it transferred to Dishforth. Flying effectively ended at the end of 1945 and Riccall was used as a sub-site for No 91 Maintenance Unit, becoming a storage facility for bombs. When this unit's role diminished, Riccall was abandoned by the RAF and lay unused for many years.

Almost all traces of the airfield were eventually obliterated and today the site is largely used for agriculture, although an industrial site occupies what was once the western side of the airfield. A T2 hangar survives, and some traces of a few dispersals, hangar bases and even one of the secondary runways can be seen, but perhaps the most obvious sign of the once busy airfield is the clearing in the trees and scrubland north of King Rudding Lane, where the vegetation's reluctance to flourish is caused by the remains of the old main runway, where the mighty Halifaxes once bounced onto terra firma.

Main features:

Runways: 223° 6,000 x 150 feet, 172° 4,200 x 150 feet, 292° 4,200 x 150 feet, concrete and asphalt surface. *Hangars:* four T2, one B1. *Dispersals:* thirty-six Heavy Bomber. *Accommodation:* RAF: Officers 165, SNCOs 520, ORs 761; WAAF: 341.

RINGWAY, Greater Manchester

53°21'11"N/02°16'27"W; SJ818840; 236ft asl. S of M56, 7 miles S of Manchester (Manchester International Airport)

One of the best-known centres of aviation in northern England, the first connections with flying began in the 1920s when attempts were made to develop civil aviation in the Manchester area. Two small flying fields were created, one at Barton and one at Wythenshawe, the latter site opening in 1929, followed by Barton in 1930. Both were less than ideal for large-scale commercial operations, and eventually another site was adopted at Ringway, construction work beginning in 1935. The new airport here opened on 25 June 1938 and within days commercial operations were under way with some 7,600 passengers passing through in the first year.

Parachute training under way at Ringway with Whitleys, Hotspurs and a Lysander on the airfield.

However, the dark days of the Second World War then arrived and on 1 September 1939 Ringway was requisitioned by the War Office. By this stage some military aviation had already arrived here, as Fairey Aviation had opened a factory on the site in June 1937, shortly before the airport opened. Created as a flight test facility for Fairey's Heaton Chapel works, Battles became common sights on the airfield, joined by Avro aircraft when that company opened its own centre here, and the prototype Manchester made its maiden flight at Ringway on 25 July 1939. Aircraft production was the main role at Ringway throughout the war, and in November 1940 No 3 FPP was formed here to supply the facility with ferry pilots. Becoming No 14 FPP three months later, the unit had its own fleet of Ansons, which enabled its crews to deliver aircraft around the UK.

Aircraft production continued for many more years after the war, and it was not until 1956 that the last aircraft was completed here, the final Firefly departing from Ringway in April of that year. Even before the site was requisitioned, Ringway was home to No 613 'City of Manchester' Squadron, whose Hinds were present from April 1939 until October. No 6 Anti-Aircraft Cooperation Unit formed here in March 1940 and stayed until March 1942, equipped with various types.

Ringway became well-known for its association with airborne forces, and the Central Landing School formed here on 31 August 1940. It later became the Central Landing Establishment with three component units devoted to parachute training, glider training and developmental work. Notionally equipped with Whitleys and Hectors, other types were also used by the unit, including Hinds and Overstrands, together with a variety of glider types. Much of the parachute training (and some of the glider activity) was conducted at nearby Tatton Park, but Ringway was inevitably busy with movements associated with the CLE, which became the Airborne Forces Establishment in January 1942. The glider element became No 296 Squadron before departing for Netheravon, after which another change of title resulted in the formation of the Airborne Forces Experimental Establishment, which reflected the increasing amount of trials work being undertaken by the unit. It finally left for Sherburn-in-Elmet in June 1942, although the Parachute Training School, formed at Ringway in February 1942, stayed until March 1946, by which time it was equipped with Dakotas.

Concrete runways were laid at Ringway late in 1941 in a typical triangular layout, although their position and length was compromised by the site's existing boundaries. Hangars also appeared in addition to the manufacturer's flight sheds.

After the Second World War the airport reopened, but military activity continued. No 613 Squadron arrived in May 1946 and acquired Spitfires towards the end of the year, before re-equipping with Vampires. No 663 Squadron also operated a Flight here for some time, with No 1951 Flight operating a fleet of Austers from Ringway until March 1957. That year marked the end of any significant military activity at Ringway, but the airfield remained very busy as an expanding airport. Its largely domestic function slowly grew to embrace a number of international destinations, and in 1951 the main runway was extended. The airport's purpose-built terminal was opened in 1961 and its familiar piers became a major attraction for spectators and enthusiasts who could stand on the pier rooftops and admire a huge variety of classic piston and jet airliners going about their business.

The runway was extended again in 1968, and the airport continued to expand until it reached a stage where another runway was required. This was laid parallel to, but south of, the existing main runway and, after opening in 2001, shared traffic with the old main runway. Its construction effectively destroyed the last traces of the hangars and flight sheds that had once stood here on the airfield's south-east boundary. They are now long gone under a massive runway from where the huge Boeing and Airbus airliners come and go. Military visitors are now rare, but a purpose-built viewing area contains a preserved example of the Nimrod (built nearby at Woodford). Oddly, this is perhaps the only surviving link with military aviation at this very busy site.

Main features:
Runways: 240° 4,300 x 150 feet, 200° 3,300 x 150 feet, 280° 3,000 x 150 feet, concrete and tarmac surface. *Hangars:* four Bellman, one Blister. *Dispersals:* twenty circular. *Accommodation:* RAF: Officers 40, SNCOs 36, ORs 1,845; WAAF: 0.

RIPON, Yorkshire

54°07'28"N/01°29'50"W; SE329699; 62ft asl. 1 mile SE of Ripon on B6265 (Ripon Racecourse)

Like many of the RFC's landing grounds in this region, this site was an existing horse-racing course adopted by the RFC in 1916 as a base for its fighter aircraft, which operated in the area on air defence duties. The landing field was established on flat land bounded by the race-track, and was occupied by a variety of aircraft, most being variants of the BE-series fighters operated by No 76 Squadron. That unit arrived in October 1916 and patrols were flown from the field on a sporadic basis. No 189 (Night) Training Squadron arrived in December 1917 but, like the other unit, its activity here is poorly documented and was probably only on a temporary basis until it departed again the following April. No 76 Squadron left the site somewhat later in March 1919.

After this the RFC soon abandoned the landing field and its temporary huts and tents quickly disappeared. The racecourse survived throughout this period and is still very much in business today. The area where the landing field was once located is still visible, although most of it is now part of a lake, and there is no physical evidence of any link with aviation here.

Main features:
Runways: none, grass surface. *Hangars:* none. *Dispersals:* none. *Accommodation:* not known.

RONALDSWAY, Isle of Man

54°04'58"N/04°37'37"W; SC282684; 32.5ft asl. 6 miles SW of Douglas on A5, N of Castletown

Aviation first came to this site in the late 1920s when Cobham's Air Circus operated its mixed fleet of aircraft here, and more civilian activity began to develop in the 1930s, with the establishment of commercial routes to Blackpool, Liverpool and other destinations. Activity abruptly ended when the Second World War began and the War Office requisitioned the airfield.

The main apron and multi-storey control tower at Ronaldsway.

No 1 Ground Defence Gunners School came to Ronaldsway in July 1940, having been formed some months previously at North Coates. The unit operated target facilities aircraft for local units, the main one being No 3 RAF Regiment School, which formed at Douglas in February 1942. The Wallaces, Harts and Gauntlets became a familiar sight at the former airport until they departed in February 1943. At this stage the airfield was being improved and expanded, with four concrete runways being laid and more hangars constructed in preparation for the arrival of the Royal Navy.

On 21 June 1944 the station became HMS *Urley* and RNAS Ronaldsway became the home to No 1N Operational Training Unit, hosting No 747 Squadron with Barracudas from July 1944. This torpedo bomber and reconnaissance type also equipped Nos 710 and 713 Squadrons, which arrived some months later. No 705 Squadron arrived in March 1945, acting as a replacement crew training unit, but also as an anti-submarine training unit, equipped with Swordfishes. It stayed for only three months, and in December 1945 the other resident units disbanded and the Fleet Air Arm vacated the site.

Civil flying then returned and civilian activity slowly expanded. The main runway was extended and new airport buildings emerged. A major airport terminal complex now occupies what was once merely a cluster of dispersals on the north-western airfield boundary, but the basic layout of the airfield is largely unchanged. The remains of the Fleet Air Arm aprons are still visible, and the large dispersal finger running to the north is still here, although it is now occupied by commercial and industrial buildings. Little military activity now takes place here, apart from occasional visitors among the daily flow of private aircraft and small airliners.

Main features:
Runways: 360° 3,000 x 90 feet, 041° 3,200 x 90 feet, 090° 4,200 x 90 feet, 129° 3,000 x 90 feet, tarmac surface. *Hangars:* six squadron type, two storage type, one B1. *Dispersals:* two aprons. *Accommodation:* RAF: Officers 267, ORs 1,377; WRNS: 480.

RUFFORTH, Yorkshire

53°56'58"N/01°11'04"W; SE536506, 62.5ft asl. 2 miles W of York on B1224 Wetherby Road

Situated just beyond the outskirts of York, land was acquired by the War Office in 1940, but the site was less than ideal for development into an airfield. Perched on a small ridge, the land became increasingly waterlogged as it fell away, and this required a significant amount of drainage work. Surrounding roads and buildings also confined the amount of available space, but despite this a standard triangular layout of three runways was laid, almost to a perfect 60-degree triangle. The perimeter track was littered with dispersals, most of which were to the west and south. Apart from the usual three hangars, only a few support buildings were constructed, with most of the admin and domestic buildings being off-site.

No 1663 HCU at Rufforth.

A No 1663 HCU Halifax lumbers home after a training sortie.

Opening in November 1942, the station had already hosted No 1652 Heavy Conversion Unit's Halifaxes for a short period in July (the unit's home base at Marston Moor was being repaired), and Nos 35 and 158 Conversion Flights (both with Halifaxes) for some weeks in September and October. Shortly after the station opened No 158 Squadron moved here from East Moor, and its Halifaxes were engaged in operations just a few days later, joining a long-range mission to Italy for an attack on Genoa. Three aircraft were lost on the mission and a fourth was abandoned on the return journey after running out of fuel.

The unit remained active here until February 1943, when it moved to Lissett. No 1663 Heavy Conversion Unit arrived a few weeks later and eventually acquired a large fleet of Halifaxes of various types, all geared towards the intensive training of new bomber crews. Late in 1944 the first Lancasters were delivered to the unit, and a few Spitfires were also obtained to assist with evasion and defensive training. Training activity continued until May 1945, by which time the need for new crews had rapidly dwindled, and the HCU disbanded.

Rufforth then fell silent, and it was not until 1948 that any flying returned when No 23 Gliding School arrived, becoming No 642 (Volunteer) Gliding School in 1955. It stayed here until July 1959, by which time more aircraft had arrived in the shape of Austers, operated by No 1964 Flight, part of No 664 Squadron. They came to Rufforth in February 1953 and stayed until March 1957, when the Royal Auxiliary Air Force was disbanded. No 60 Maintenance Unit also used Rufforth from November 1945 until July 1959, as did No 54 MU from 1952 until 1954, but little if any flying activity was ever associated with these units.

With the departure of the Austers and gliders, Rufforth's military flying days were over, and late in 1959 parts of the station were sold off, although some land was retained by the Ministry of Defence until the 1980s. The former bomber base is still largely intact, with the two secondary runways still complete, although a large centre section of the old main runway has been removed. The perimeter track, some of the dispersals and one hangar can still be seen, although the only flying activity here now is provided by civilian gliders and microlights.

Main features:
Runways: 240° 6,000 x 150 feet, 181° 4,200 x 150 feet, 294° 4,050 x 150 feet, tarmac surface. *Hangars:* two T2, one B1. *Dispersals:* thirty-six Heavy Bomber. *Accommodation:* RAF: Officers 124, SNCOs 568, ORs 839; WAAF: 251.

SAMLESBURY, Lancashire

53°46'27"N/02°34'10"W; SD625310; 246ft asl. 4 miles E of Preston on A677, S of A59

Adopted for development as a site for future civil aviation as far back as 1922, this airfield site was requisitioned by the Government in 1939 and allocated to the Ministry of Aircraft Production. Construction began in April 1939 and the airfield was laid between the confines of surrounding roads, rivers and rising ground, resulting in three moderately sized runways laid in a triangular pattern together with a series of five flight sheds. Once completed in 1940 the site was occupied by English Electric, a company already established in the area with its headquarters in nearby Preston and a factory airfield at Warton to the west.

Licence production of the Handley Page Hampden was the initial task to be undertaken here, and the first such aircraft was completed in February 1940, P2062 making its maiden flight on the 22nd. Production continued until March 1942, by which time some 770 aircraft had been manufactured. Halifax production began in 1941, the first aircraft (V9976) making its first flight on 25 August; a staggering 2,145 aircraft were subsequently built here.

No 9 Group's Communications Flight arrived in October 1940 (its headquarters being some 6 miles away at Barton Hall), and stayed here with a variety of aircraft types including Hurricanes and Oxfords until September 1944. Aircraft production continued after the end of the Second World War and many Vampire and Venoms rolled off Samlesbury's production

Canberra B6 interdictors under assembly at Samlesbury.

Canberra production at Samlesbury in 1952.

A Canberra PR9 low over Samlesbury, making a farewell flypast prior to the type's retirement from RAF service.

line. Canberra production followed, and Samlesbury remained associated with this magnificent bomber for many years, as many returned here for repair and modification. In the 1970s and 1980s Canberras also returned here for conversion into specialised versions or export models.

Samlesbury was also involved in production of the Lightning, Jaguar and Tornado, although most test flying associated with these aircraft took place at Warton. Parts for the Concorde supersonic airliner were also manufactured here. Samlesbury was also to become a major production site for the ill-fated TSR2 bomber, and some airframes were manufactured here before the project was cancelled in 1965. Their sorry carcasses were removed from the factory as scrap metal.

Flying from this site virtually ended in the 1980s although No 635 Volunteer Gliding Squadron continued to use the airfield until 2009. Production of the Typhoon fighter and parts for the T-45 Goshawk and F-35 Lightning has continued here over recent years, but the airfield is now closed and is likely to be built upon either by BAE Systems (as it expands its site) or by other commercial companies. At present the old main and secondary runways can still be seen, but for how long is open to speculation.

Main features:
Runways: 080° 6,000 x 150 feet, NW/SE 3,000 x 150 feet, N/S 3,000 x 150 feet, concrete and asphalt surface. *Hangars:* flight sheds. *Dispersals:* none. *Accommodation:* not known.

SANDTOFT, Yorkshire

53°33'50"N/0°51'50"W; SE753080; 10ft asl. 12 miles NE of Doncaster off A161on Sandtoft Road W of Belton village

Created as a satellite airfield for nearby Lindholme, RAF Sandtoft opened on 10 December 1943 as part of No 1 Group Bomber Command. Constructed on a very low-lying area of flat land, the site required a great deal of drainage work prior to the laying of three runways in standard bomber station pattern, together with the usual three hangars and scattering of dispersals. From December 1943 until November 1944 the airfield acted as a satellite for Lindholme, and most of the aircraft activity was provided by bomber and support aircraft from that station.

An aerial view of Sandtoft, illustrating how much of the original airfield still survives despite development.

In December 1944 it became part of No 7 Group, but remained in use by Lindholme's units until November 1945. In February 1944 No 1667 Heavy Conversion Unit moved here from Faldingworth with a large fleet of Halifaxes and a variety of fighter types, all allocated to aircrew training (the fighters being used for evasion and defensive training). Training flights resulted in a number of accidents, but the unit remained busy and received its first Lancasters towards the end of 1944. During August of the following year No 1656 Heavy Conversion Unit established a detachment here, and both units continued to operate until November, when they disbanded.

Flying came to an end, but No 35 Maintenance Unit arrived here from Heyworth in December 1945 and stayed until February 1946, followed immediately by No 61 Maintenance Unit, which stayed until 1947. In April 1953 the site was transferred to the United States Air Force for development into a fighter base, but the proposal never proceeded and after a couple of years the RAF resumed control and quickly vacated the site.

Much of the old airfield can still be seen today, although a huge portion of the north-east area has been allocated to car storage and industrial development. Belton Road now neatly crosses the old airfield, and to the south of this road a great deal of the old runways, perimeter track and dispersals can still be found. A section of the old southern perimeter track is now used for private flying and a variety of light aircraft can often be seen here, and even an occasional jet. To the west, the Trolleybus Museum occupies an area where one of the old T2 hangars once stood, so the only historic machines to manoeuvre here now are trolleybuses.

Main features:
Runways: 240° 6,000 x 150 feet, 120° 4,200 x 150 feet, 170° 4,200 x 150 feet, concrete surface with tarmac thresholds. *Hangars:* three T2. *Dispersals:* thirty-six spectacle. *Accommodation:* RAF: Officers 249, SNCOs 160, ORs 1,512; WAAF: 362.

SCALBY MILLS, Yorkshire

54°18'09"N/0°24'35"W; TA035908; 0ft asl. 1 mile N of Scarborough off Scalby Mills Road

A Blackburn Type L outside its hangar at Scalby Mills.

A small flight test facility was created here by the Leeds-based Blackburn company during 1914. The company was contracted to produce the Type L seaplane for participation in the Daily Mail Circuit of Britain race, but as the First World War approached the aircraft was acquired by the Admiralty. It was taken to the site at Scalby Mills and, after flight testing, was pressed into service with the Royal Naval Air Service here, flying patrols over the North Sea and keeping a lookout for German submarines. However, it was eventually involved in an accident and crashed into the cliffs at Speeton during 1915, being completely destroyed. This effectively ended flying activities at this site, as it was not used by the RNAS for any other purposes. Blackburn also abandoned its flight shed, although it remained here until the 1970s when it was finally demolished.

There is no evidence of the diminutive Type L's presence here now, although at low tide the remains of a long slipway can be seen at the very end of North Bay Promenade; it is thought that this was used by Blackburn during the company's brief stay here.

Main features:
Runways: none, seaplane base. *Hangars:* one flight shed. *Dispersals:* none. *Accommodation:* not known.

SCAMPTON, Lincolnshire

53°18'24"N/0°33'13"W; SK964798; 213ft asl. 5 miles N of Lincoln on A15, N of A1500

RAF Scampton as seen from the air in 1942.

A look inside the bomb bay of a No 57 Squadron Lancaster at Scampton. Visible are incendiary bombs and a 4,000lb 'Cookie'.

Undoubtedly one of the best-known and historic of the Royal Air Force's bases, flying first took place here in 1916. An area of land was acquired east of the village of Scampton, although the landing field was initially named as Brattleby, this being the larger village. No 49 Training (Ex Reserve) Squadron arrived here in October 1916, but stayed for only a month. No 37 Training Squadron then arrived from Catterick with a fleet of FK.3s, Avro 506s and RE.8s, remaining here until September 1917, when it was replaced by No 11 Training Squadron; the latter unit joined No 60 Training Squadron, which had moved here five months previously.

Training was the base's main role, although No 33 Squadron established a detachment here in December 1916 as part of its air defence duties. A month later No 81 Squadron arrived, ostensibly to work up prior to moving to France, but stayed at Brattleby and became part of No 34 Training Depot Station, which came into being in July 1918, combining all of the resident training units. A varied fleet of aircraft could be seen on the airfield, ranging from Camels and Dolphins to Avro 504s and Scouts, but during March 1919 the TDS merged with No 46 TDS at South Carlton and the landing field at Brattleby was closed down.

The site was an obvious choice for development as the Second World War approached, and a much larger area of land was requisitioned in 1935. This was completed as a grass airfield, dominated by the huge C-Type hangars that were erected next to a new admin, domestic and technical site which slowly emerged next to the Lincoln Road (Ermine Street, now the A15). The airfield opened as RAF Scampton on 27 August 1936, and two months later Nos 9 and 214 Squadrons arrived, the former flying Heyfords (coming from Aldergrove) and the latter Virginias (from Andover). The aged Virginias were replaced by Harrows in 1937, but in April of that year the unit moved to Feltwell. Its place was taken by No 148 Squadron, which arrived with its Wellesleys shortly before No 9 Squadron departed in March. In September No 49 Squadron transferred here from Worthy Down with Hampdens, and two months later No 83 Squadron arrived with more Hampdens.

These units were soon involved in operational missions, initially in the maritime environment, but land targets soon followed. In August 1940 one of No 49 Squadron's pilots was awarded a VC for his actions during an attack on Dortmund. A second VC was awarded to a pilot in No 83 Squadron for an attack on Antwerp just weeks later. No 83 Squadron received Manchesters in December 1941, as did No 49 four months later. The units had enjoyed considerable success with their Hampdens (2,636 sorties flown by No 49 Squadron and 1,987 by No 83), but the Manchester proved to be far less suitable. No 49 Squadron flew forty-seven sorties for the loss of six aircraft, while No 83 Squadron flew 152 sorties and lost nine Manchesters.

Peter Ward-Hunt was a pilot with No 49 Squadron at Scampton in 1939:

'I was aged 22, and was flying Hampdens with No 49 Squadron from Scampton. One night I returned to the mess about midnight. I was greeted by the CO, Wing Commander Wally Sheen, with the words, "Get to bed, briefing at 06.00." At the briefing it turned out that we were to try and bomb the *Scharnhorst*, which was thought to be sailing en route from Kiel to the Norwegian fjords. The raid was to be carried out by twelve of us from No 49 and twelve from No 44 Squadron at Waddington.

We took off about 7.30, around daybreak. The plan was to search the fjords, find the ships, and land back at Kinloss in Scotland. We set off in our formation, with No 44 vaguely following us. In those days, only the lead aircraft navigated – the rest just followed the leader. We came out of the cloud cover to see the northern coast of Denmark, so turned left a bit and headed up to Norway. We saw a fjord, and decided to go up the second one – only to meet No 44 coming back down it. We continued up the coast until we eventually turned for home. We had no idea where we were, so we circled a fishing boat that we saw. They were terrified, and all hid.

Well, then we saw Holy Island, and decided to land in the first field we came to. We were met by a farmer who offered to look after the plane while we were picked up by the Army. They found us somewhere to stay and a quantity of ale! The next day we went back to the craft and found we had enough fuel to reach the local airfield. Here we were told that the others had all returned apart from one plane, which hit the local church. We refuelled and returned to Scampton. The Waddington group had landed at the Firth of Forth, where the local Spitfires had been warned to expect them. That didn't stop two Spitfires opening fire on the returning planes, shooting one of them down.

This was typical of early war experience – flying in formations and getting lost. I flew fifty raids in all. After the introduction of the Pathfinders things got much easier! I did take part in one Thousand Bomber raid to Cologne. I was temporarily in charge of a training flight at Scampton at the time, training crews to transfer from Manchesters to Lancasters. We got the call that a special raid was on, and that all crews were to be involved. Well, I didn't get a Lancaster, I ended up with an old Manchester. Just as we were about to leave the Squadron Commander came and joined us as his plane was u/s.

Cologne was moonlit that night, and had been well marked. We did a lot of damage that night, but not to the cathedral. The plane got coned in the searchlights, so I had to drop the wing and dive in a spiral to about 1,000 feet. The CO wasn't too pleased with me as he didn't have anything to hold onto!

I was once hit, early on in the war while flying the Hampdens. I think we were on the way to Berlin, at least somewhere over Germany. The rear gunner, who in the Hampden was underneath the plane, announced that we were on fire. He then amended this to say that it was only petrol fumes. I checked the tanks, and three were empty. We turned back and landed near Norwich. I met a chap there who was an air gunner in his forties. He was ex-Army, but still insisted on flying. He found me a bed for the night and declared that he was off next morning to London for mass. Unfortunately he didn't survive the war. I found out later he had been shot down by our own gunners over Hull.'

This was a heavy loss rate, and one that would have been unsustainable, but the Manchester was soon replaced by the magnificent Lancaster within a matter of months, and by July both

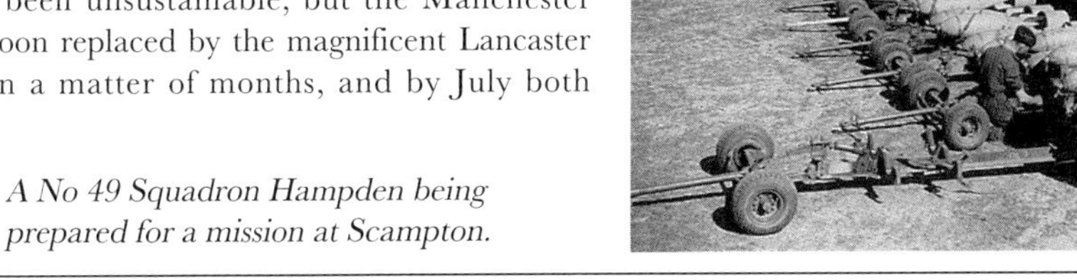

A No 49 Squadron Hampden being prepared for a mission at Scampton.

squadrons were fully equipped with the type. A fleet of Oxfords also came to Scampton in November 1941 when No 1518 Beam Approach Training Flight arrived, remaining here until June 1943. No 5 Group's Air Bomber Training Flight arrived in June 1942 but stayed for only two months. A more significant move was the departure of No 83 Squadron at the same time, its place being taken by No 57 Squadron from Feltwell. No 1661 Heavy Conversion Unit made a brief stay from November, as did No 467 Squadron (an Australian unit), but the next major change was the departure of No 49 Squadron in January 1943, its Lancasters leaving for Fiskerton.

Just a few weeks later, on 23 March 1943, Scampton's (and the RAF's) most famous squadron was formed here, No 617. Led by the legendary Guy Gibson, the unit slowly gathered personnel and aircraft for a mission that remained secret until only hours before it began. On the night of 16/17 May the unit's specially equipped Lancasters embarked on their historic mission. No 617 Squadron was formed by Wing Commander Guy Gibson from selected crews in No 5 Group, and the Squadron trained for six weeks for this special operation. Nineteen Lancasters were dispatched in three waves, each aircraft armed with the special 'bouncing bomb' developed by Barnes Wallis for attacking German dams. The entire operation was to be carried out at low level to escape attack from German night-fighters, the bombs being released just above the water in the dams.

One aircraft had to return early after it struck the sea a glancing blow, which tore off its bomb. Five further aircraft were shot down or crashed before reaching their targets, and one was so badly damaged by Flak that it had to turn back. This left twelve Lancasters available to bomb the dams. Wing Commander Gibson's aircraft and four other crews bombed the Mohne Dam and breached it, despite intense fire from light Flak defending it. Three aircraft went on to bomb the Eder Dam, which was also breached. Two aircraft bombed the Sorpe Dam and one the Schwelme Dam, but without causing breaches in their walls. The twelfth surviving aircraft could not find its target in misty conditions and returned to England without dropping its bomb. Three further Lancasters were shot down after they had bombed.

Total casualties were eight aircraft out of the nineteen dispatched. It is estimated that four were shot down by light Flak, one crashed after being damaged by the explosion of its own bomb, two crashed after hitting electricity cables and one after striking a tree when its pilot was dazzled by a searchlight. Of the fifty-six crew members in these planes, fifty-three were killed and three became prisoners of war, two of them badly injured. For his leadership of this amazing operation and for his courage in attacking Flak positions at the Mohne Dam after having carried out his own bombing run, Wing Commander Gibson was awarded the Victoria Cross. Thirty-four other men received decorations.

The breaching of the Mohne and Eder Dams was a major achievement. The Mohne reservoir contained nearly 140 million tons of water and was the major source of supply for the industrial Ruhr, 20 miles away. The water released caused widespread flooding and disruption of rail, road and canal communications and of the supply of electricity and water. The water supply network was particularly affected by the silting up of pumping stations by the flood water. It is not possible to state the effect of all this upon industrial production in precise terms, but there was certainly some disruption and water rationing was in force until the winter rains came and filled the reservoirs again.

The Eder was even larger than the Mohne, containing 210 million tons of water, but it was 60 miles from the Ruhr. The city of Kassel, 25 miles away, and the inland waterway system in the Kassel area were more affected by the attack on the Eder than was the Ruhr area. The German view is that if the aircraft that had been allocated to the Eder had been switched to the Sorpe Dam, the effect upon the Ruhr's industrial production would have been extremely serious, but the Sorpe's construction was of a nature that made it a difficult target for the Wallis bomb, hence its low priority in the raid. The Sorpe reservoir just managed to keep the Ruhr supplied with water until the Mohne Dam was repaired.

The number of people drowned has been calculated at 1,294, most of them near the Mohne Dam. The town of Neheim-Husten, 5 miles downstream of the Mohne Dam, took the full impact of the flood, and at least 859 people died there. It is believed that fifty-eight or more of the dead were around the Eder Dam.

Having completed their vital task, the Lancasters and the unit stayed for only a few more weeks

The once familiar sight and sound of Vulcan bombers thundering skywards from RAF Scampton.

efore leaving in August, together with No 57 quadron. This effectively left the airfield empty and vork began to lay three concrete runways, and it vas not until the following July that the site eopened. Designated as No 52 Base in 1943, it ecame No 15 Base from October 1944 with atellites at Dunholme Lodge, Fiskerton and Iemswell, and this arrangement continued until October 1945. No 1690 Bomber Defence Training light arrived and stayed for only a few weeks, being eplaced by No 1687 BDTF from December 1944 intil April of the following year. No 153 Squadron's ancasters arrived from Kirmington on 15 October 944 and stayed for a year, flying 1,041 operational nissions. No 625 Squadron moved to Scampton in April 1945, but flew only a few operational missions efore the end of the war. Training activity resumed luring the summer before No 153 disbanded in eptember followed by No 625 a month later.

Scampton survived beyond the Second World War and Nos 100 and 57 Squadrons arrived late in 945 for a stay of a few months. Various ground-based units also came to the base together with other nits with flying elements, including the Bomber Command Instructors School and the Bomber Command Instrument Rating & Examining Flight. The base also became a deployment site for the JSAF's Strategic Air Command, and the 28th Bomb Group brought its B-29s here from July to October 1948, replaced by the 301st Bomb Group until January 1949. No 230 Operational Conversion Jnit then arrived from Lindholme with a fleet of Lincolns, remaining here until October 1952, when its resence at Scampton was reduced to a Reserve Training Squadron. The Bomber Command Bombing School also came here in October 1952, but stayed for only a month before moving to Lindholme.

An unusual view of Scampton taken during a Red Arrows practice flight over the station.

Scampton then entered the jet age and four Canberra bomber squadrons – Nos 10, 18, 21 and 27 – came here from January until September 1953. However, shortly after the last of these units established itself here, they all departed again and within a year of the first arrivals the Canberras had gone again. The reason for this swift change was the decision to adopt Scampton as a base for the Vulcan, which required a major modification of the station's facilities. Apart from the new V-Bomber dispersals that were laid around the airfield, the most obvious change was the extension of the main runway, which, at 9,000 feet, reached out to the east across Ermine Street. Rather than simply close this major road, it was re-routed around the runway, and Scampton's station badge was eventually changed to represent this unique layout.

An unusual RAF photograph taken at Scampton (on the disused secondary runway) illustrating aircraft types operated by Support Command.

The station reopened early in 1958, and in May No 617 Squadron returned here, now equipped with Vulcans. No 83 Squadron arrived in October 1960 and Scampton's facilities were also modified to accommodate the Vulcan's Blue Steel stand-off nuclear missile – a temperamental weapon powered by an extremely volatile and lethal fuel that required sophisticated and expensive handling equipment. No 18 Joint Services Trials Unit was formed here in December 1961 for trials with the weapon.

When Finningley ended its Vulcan operations in December 1969 No 230 OCU moved to Scampton, and eventually a number of Hastings radar trainers were attached to the OCU as a Flight assuming the duties of the now disbanded Strike Command Bombing School. No 35 Squadron brought its Vulcans from Akrotiri in January 1975, and No 27 Squadron returned for a second period of operations with the Vulcan. Having first reformed on Vulcans here in April 1961, the unit disbanded in 1971 only to reform two years later with a specialised maritime radar reconnaissance version of the Vulcan. It was not until 1982 that the mighty Vulcan finally left Scampton and, with the OCU having disbanded in December 1990, the last of the operational aircraft were gone by 1983.

A year later the station re-emerged as a training base, and the home of the Central Flying School. The mixed fleet of Jet Provosts and Bulldogs also included the Meteor and Vampire of the CFS 'Vintage Pair' team, and these remained active until their eventual loss in two tragic accidents. Tucanos eventually replaced the faithful Jet Provosts with the CFS, and in 1984 the Red Arrows moved here from Kemble. Their world-famous red Hawks became a daily sight above the base, and Scampton remained busy until May 1995, when the CFS moved to Cranwell, followed by the Red Arrows a year later.

Nigger's grave, still lovingly cared for by the RAF.

This could have been the end of Scampton's existence, but the Red Arrows (which still practiced over Scampton) soon returned after the short-lived move to Cranwell was deemed to have been a logistical failure. However, no other flying units came to Scampton and the huge bomber airfield has been used almost exclusively by the Red Arrows ever since. The Red Arrows were expected to move south to Waddington but the team now looks set to stay at Scampton until their Hawk jets run out of airframe life in a few years' time. At that stage it seems likely that the remaining ground units as Scampton may well relocate, enabling the RAF to finally abandon this famous base for good.

Over on the northern perimeter of the airfield a small enclave is occupied by HHA, a civilian company that operates Hunters on contract work, using a flight shed that British Aerospace used for some years for flight testing Buccaneers and Phantoms modified at Brough, across the Humber. When the RAF finally does leave, it is probably inevitable that HHA will have to leave too, and flying at Scampton will finally come to an end. Much of the old station site is already in civilian hands (as a housing estate), but whether the rest of the airfield is ever developed remains to be seen. The historic Officers' Mess will hopefully be preserved, and of course the grave of Guy Gibson's faithful dog Nigger is still lovingly cherished, in front of No 617 Squadron's empty hangar.

Main features:
Runways: 230° 6,000 x 150 feet, 190° 4,200 x 150 feet, 290° 4,200 x 150 feet, concrete and tarmac surface. *Hangars:* one T2, five C-Type. *Dispersals:* thirty-six Heavy Bomber. *Accommodation:* RAF: Officers 198, SNCOs 1,044, ORs 1,084; WAAF: 268.

SCORTON, Yorkshire

54°23'57"N/01°37'49"W; NZ240004; 184ft asl. 3 miles E of Richmond, W of Scorton village on B6271

Although this region of Yorkshire was primarily occupied by bomber airfields during the Second World War, Scorton was designed for fighters. It opened to flying in 1939 as a satellite for Catterick, which was just 2 miles to the south. Laid on land between a railway and a river, the runways were somewhat shorter than those found at airfields in the area, but they were sufficient for fighter operations and, after a short period of operation as a grass field, hard runways were laid out in a standard triangular pattern, which fitted neatly within the available space.

No 219 Squadron sent the first aircraft here when a Flight of Blenheims arrived a few days after the station opened, the parent unit being at Catterick. Night-flying was the unit's primary task and it remained at Scorton until October 1940, when it moved to Redhill. Other units also used the airfield, particularly those based at Catterick, and in 1941 the three runways were laid, together with a small number of dispersed fighter pens around the airfield perimeter.

Flying resumed in October 1941 and No 122 Squadron arrived from Catterick with Spitfires, replaced by No 167 Squadron the following April, although the latter unit stayed for only a month before transferring to Castletown. Other units also made brief deployments here, but it was not until February 1942 that more intensive flying began, when No 406 'Lynx' Squadron brought the first of its Beaufighters here, the rest having arrived by June. This Canadian unit slowly converted onto a later version of the Beaufighter before moving to Predannack as an operational unit again, in September. By this stage No 410 'Cougar' Squadron had arrived and operations quickly began, the unit scoring the first operational success for Scorton on 6 September. No 410 did not stay long, however, and a month later it was gone.

No 122 Squadron's aircrew pictured during trap-shooting practice at Scorton.

In January 1944 the base became Headquarters for No 142 Airfield as part of No 85 Group controlling Nos 130 and 604 Squadrons. A month later No 56 Squadron with Typhoons and Spitfires arrived, then No 604 Squadron departed in April. In May 1944 the United States Army Air Force came to Scorton and the 422nd Night Fighter Squadron transferred here from Charmy Down with a fleet of P-61 Black Widow aircraft. The 425th Night Fighter Squadron followed in June to begin training for tactical operations in Europe. Their effectiveness (and a growing need for defences against the V-1 threat) led to both units leaving just a few weeks later, and by July the Black Widows were gone. This left Scorton without any resident units and the airfield reverted to being a satellite for Catterick. By the end of the year it had been transferred to Balloon Command for the storage of balloons and associated equipment. This effectively meant the end of flying from the site, and apart from occasional transport and communications movements the airfield fell silent.

No 224 Maintenance Unit came here in December 1944 and stayed for three years, after which No 91 MU used Scorton as a sub-site until December 1952. When that unit left, the RAF vacated the station and it was left abandoned for some years. Sadly, it later became a source of aggregates and huge gravel pits began to appear across the airfield site. This work eventually destroyed virtually every trace of the airfield, and now there is hardly anything to indicate that a fighter base was once here. The foundations of a secondary runway are still used for road transport on the site, but the only obvious reminder of the RAF (and USAAF) presence is a cluster of farm buildings, which includes one of the old Blister hangars.

Main features:
Runways: 258° 4,800 x 150 feet, 220° 3,600 x 150 feet, 351° 3,600 x 150 feet, tarmac surface. *Hangars:* eight Extra Over Blister, four Over Blister. *Dispersals:* six twin-engine type, four single-engine type. *Accommodation:* RAF: Officers 110, SNCOs 100, ORs 980; WAAF: 158.

SEACROFT, Yorkshire

53°49'28"N/01°27'42"W; SE355365; 358ft asl. 2 miles N of central Leeds west of A6120 on North Parkway

This small landing field was created early in 1916 as one of three sites around the city of Leeds assigned to the Royal Flying Corps. No 33 Squadron was responsible for the air defence of the region, mounting air patrols against the threat of German airship raids. Seacroft was an area of land simply cleared of obstructions, available to the squadron when required, particularly for night standby duties, when aircraft were housed on the airfield under armed guard. Few facilities were brought here other than tents, and by the end of 1916 the German threat had diminished and the landing site was no longer needed. By 1917 it was abandoned and eventually became part of the sprawling housing development that now covers this area. There is no evidence of the landing field today; indeed, it is almost impossible to be certain as to its precise location.

Main features:
Runways: none, grass field. *Hangars:* none. *Dispersals:* none. *Accommodation:* not known.

SEATON CAREW, County Durham

54°38'44"N/01°11'40"W; NZ520281; 16.5ft asl. 3 miles S of Hartlepool, N of junction of B127[illegible] and A178

Unusually, this airfield was in fact two separate sites, although they were separated by less than a mile. The grass landing field had first been surveyed in 1914, but it was some two years later that it was adopted by the Royal Flying Corps and became part of No 48 Wing. Unlike many of the RFC's landing grounds, this one was fairly substantial, with landing runs of up to 2,000 feet, and equipped with some permanent structures, not least three hangars adjacent to the eastern main road.

The monstrous Blackburn Kangaroo pictured at Seaton Carew.

No 36 Squadron's C Flight was established here with a gaggle of BE.2C aircraft shortly after the Squadron had formed at Cramlington in February 1916. Patrols over the local area were soon under way and, unlike many of the RFC's units, C Flight achieved some tangible success on the night of 27 November 1916 when one of its aircraft encountered German airship L34, which was heading towards Hartlepool in search of a suitable target. The airship was attacked and burst into flames, crashing into the sea off West Hartlepool. The airship threat had diminished by the end of 1916, and in 1918 the Squadron re-equipped with the Bristol F.2B, but the patrols from Seaton Carew saw little activity after 1916. However, the threat from German submarines steadily increased, and patrols over the North Sea became increasingly necessary.

It was this requirement that led to the creation of another site less than a mile south of the RFC airfield, on the nearby estuary. This became Seaton Carew II (the airfield being Seaton Carew I), and a slipway was constructed together with three small flight sheds and support buildings. It was opened in 1918 and a variety of Flights operated here until they were combined in August to become No 246 Squadron. A number of Shorts 184s and Sopwith Baby seaplanes used the base, which continued to develop through 1918 until the First World War ended and the seaplane base was suddenly redundant. Meanwhile, maritime patrols from the grass airfield had achieved yet more success, and No 495 Flight's Kangaroos were credited with eleven attacks, including partial credit for the sinking of UC-70 on 28 August 1918 (a task completed by HMS *Ouse*).

By 1919 the RFC had abandoned both the landing field and the seaplane base, although the scattering of buildings at both sites remained present for many years. Today the airfield site has been quarried and no traces of the old buildings can be seen along the A178 Tees Road. However, a small road running from the nearby roundabout leads south to the river estuary and, beyond a large industrial site where the road ends, a small track runs down to where the seaplane base's old slipway still stands, slowly crumbling, abandoned and forgotten.

Main features:

Runways: none, grass field. *Hangars:* two Bessonneau, one flight shed. *Dispersals:* none. *Accommodation:* not known.

SHERBURN-IN-ELMET, Yorkshire

53°47'13"N/01°13'0"W; SE517325; 23ft asl. 6 miles W of Selby, E of A162, E of Sherburn-in-Elme

This somewhat unusual airfield owes its origins to an Aircraft Acceptance Park, which wa established in 1917 when the Leeds-based Blackburn company created a site here. Variou aircraft types were built here, but the most numerous was the Cuckoo, some 132 machines bein assembled here. Although the aircraft was undoubtedly a significant design, its appearance late i the First World War inevitably meant that it was not to see any active service as a torpedo bomber Activity here abruptly ended when the First World War came to a close, but in January 1926 th Yorkshire Aeroplane Club began flying here and the site remained active as a centre for private an commercial flying until the outbreak of the Second World War.

An aerial view of Sherburn-in-Elmet in 1943.

A 1994 aerial view, showing the extended main runway and the grass strips that are now used for flying.

At that point the whole site was requisitioned by the War Office and designated as a fighter base. No 73 Squadron formed a Flight here with Hurricanes in June 1940, staying until September, and No 46 Squadron did likewise from March until May 1941. It is believed that some bomber aircraft also dispersed here at various stages from late 1939, when many of the RAF's bombers were scattered to

The magnificent First World War-vintage hangars at Sherburn-in-Elmet.

A fascinating look at one of Sherburn's hangars from the inside, illustrating the typically elegant construction.

sites across the country in anticipation of German attacks. Plans were also made for No 36 Elementary & Reserve Flying School to come to Sherburn, but the plan failed to materialise and in 1941 the site's association with operational flying ended.

However, the site was by no means redundant, and Blackburn Aircraft returned, setting up a repair and modification centre; the existing hangars and support buildings were expanded to embrace new factory and flight sheds and other buildings. The company was soon tasked with the production of Swordfish aircraft, and some 1,700 of these stalwart machines were eventually completed here, the first emerging in December 1940.

No 7 Ferry Pilots School was established in November 1940 in order to transport the new aircraft to their operational units, and it stayed active at Sherburn until October 1945. The concrete runway was eventually laid (together with a smaller grass strip), and when the Airborne Forces Experimental Establishment moved here from Ringway in June 1942 the runway was extended with an additional 700 feet of square-meshed track being added in order to accommodate the wide range of aircraft types that the unit used on testing and evaluation duties. The AFEE also created detachments at other airfields and remained extremely busy here until January 1945, when it moved to Beaulieu, leaving Sherburn almost empty, apart from the now resident Leeds University Air Squadron, which is believed to have operated only one Tiger Moth. No 30 Gliding School was also here from May 1944, but left late in 1945, together with the UAS, and Blackburn vacated its factory site to continue its activities at Brough, many miles to the south.

The airfield is still in use today, civil flying having slowly returned to the site following the war. This has resulted in a busy home for private and general aviation, which operates from a new tarmac runway and three grass strips; these are adjacent to the old factory runway, which is still intact but no longer used for flying. Today this once busy runway is only used for car brake testing, as part of the huge industrial complex that now borders the northern perimeter of the airfield.

Main features:
Runway: 050° 6,000 x 150 feet, concrete surface. *Hangars:* eight Blister. *Dispersals:* thirty-six circular. *Accommodation:* RAF: Officers 44, SNCOs 105, ORs 835; WAAF: 238.

SHIPTON, Yorkshire

54°01'33"N/01°08'36"W; SE551591; 59ft asl. 5 miles NW of York on A19, NW of Shipton village

As the A19 road runs north out of the village of Shipton, the scattering of houses makes way for the seemingly endless fields of crops, punctuated by only an occasional tree. To the west of the road, a seemingly unremarkable field reveals no hint to any causal observer that military flying once took place here. This same location was requisitioned by the Government in 1916 as a suitable landing field for the Royal Flying Corps' fighter aircraft, particularly those flown by No 76 Squadron, which was assigned to air defence duties in the region, protecting the skies against raids by German airships. The area was cleared of obstructions and supplied with personnel to guard the aircraft when they were present, but other than a scattering of tents there was no attempt to make the occasional visits by the RFC aircraft anything but temporary in nature. After little more than a year the site was no longer needed when the German aerial threat diminished, and the landing field was abandoned.

However, in 1939 the same site was once again requisitioned, this time for use by No 5 Salvage Centre, which was responsible for the recovery and possible repair of crashed and damaged aircraft in the Yorkshire region. This then became part of No 60 Maintenance Unit, and the RAF stayed in business here until November 1945, although a pseudo-military presence continued until 1991, thanks to a Sector Operations Centre that was built here in the early 1950s. This underground bunker is still here, but all traces of the RFC landing field are gone. However, adjacent to the field is the former No 60 MU site now used by industrial and commercial companies. Most of the RAF structure was destroyed by a fire in 1987, but the entrance is still here on Station Lane and some evidence of the RAF's presence is certainly still visible.

Main features:
Runways: none, grass surface. *Hangars:* none. *Dispersals:* none. *Accommodation:* not known.

SILLOTH, Cumbria

54°52'25"N/03°21'51"W; NY125540; 23ft asl. 15 miles W of Carlisle on B5302 at Barracks Bridge

Perched on the very edge of the Solway Firth, the RAF first came to Silloth in 1938, after the basic structure of a new airfield had been constructed here (begun the previous year). The site was designed for use by Maintenance Command and this plan affected the overall layout of the airfield, with only modestly sized runways; thanks to the confines of the surrounding roads and coastline, these intersected at the southern tip of the airfield boundary rather than in the centre of the airfield. There was also a large number of storage hangars (of various types) and an impressive scattering of dispersals, which surrounded the airfield and ran out on spurs into adjacent fields.

No 22 Maintenance Unit formed here on 5 June 1939 and was destined to remain here for more than 20 years – quite a lengthy stay for any RAF unit during this period. Eventually the unit also operated two satellite fields, at Hornby Hall (No 9 Satellite Landing Ground) and Hutton-in-the-Forest (No 8 SLG). Many of the unit's aircraft were shipped out to these sites for temporary storage until they were declared redundant in 1945.

Although the MU was expected to be Silloth's main resident unit, it quickly became only a lodger unit when the station was transferred to Coastal Command. From April 1940 No 1 (Coastal) Operational Training Unit was based here with a mixed fleet of aircraft engaged on training duties.

A wartime aerial photograph of Silloth, the countless dispersals around the field clearly visible.

The Coastal Command Landplane Pilots School was established here in November 1939, and by 1940 was operating Beauforts, Ansons, Bothas and Hudsons, the latter type becoming the most numerous and familiar aircraft at Silloth. The CCLPS became the Operational Training Unit from April 1940, and by 1942 had rationalised and increased its aircraft fleet until it boasted more than a hundred aircraft, some sixty-eight of which were Hudsons.

In March 1943 No 1 (C) Operational Training Unit departed for Thornaby, its place being taken by No 6 (C)OTU, which was largely equipped with Wellingtons, together with some Ansons and target-towing types such as the Martinet. Its Wellington training (which included Leigh Light operations) eventually expanded to include the Warwick, before the unit moved to Kinloss in July 1945. During the years of the Second World War some other units had also made brief stays at Silloth, including No 215 Squadron with its Harrow aircraft from September 1939 until April 1940, and No 320 Squadron, which maintained a detachment of Ansons here from October 1940 until January 1941.

A rare Luftwaffe target map from 1941, showing Silloth and the surrounding area.

A panoramic view of Silloth in 1941, showing the airfield's proximity to the coast.

However, Silloth's primary role was that of a Coastal Command training base, even though it was the Maintenance Unit that maintained a larger and longer presence here. Initially, the MU tended to handle Hurricanes and Sea Hurricanes, but a huge variety of aircraft types were eventually stored or processed here, and No 5 Ferry Pool was formed here in October 1945 in order to provide pilots for the collection and delivery of so many aircraft.

When the Second World War ended, Silloth remained in use as a Maintenance Unit site and even more post-war aircraft were sent here, many making their final flight before being stored and eventually scrapped. Perhaps the best-known (and numerous) type to have been processed at Silloth is the Gloster Meteor. Countless examples of this early post-war jet came to Silloth and spent many years languishing on dispersals around the site before being dismantled and scrapped.

By the end of 1945 the Coastal Command aircraft had gone from Silloth and the airfield remained mostly silent, with only delivery flights to the MU occasionally disturbing the area's tranquillity. The RAF finally vacated the station in 1960, and after this date only sporadic private flying took place. Even this limited amount of flying appears to have ended now, and the former

The Operational Training Unit at rest at Silloth during 1940.

The former station guardroom at Silloth, largely unchanged since it was first built.

RAF station is mostly occupied by holiday caravans. The runways have been partially dismantled but most of the concrete is still here, as are the perimeter tracks and many of the dispersals. The hangars are still very visible too, scattered around the airfield, largely empty.

Main features:
Runways: 053° 4,740 x 150 feet, 013° 3,390 x 150 feet, 103° 3,270 x 150 feet, concrete surface. *Hangars:* six E-Type, four D-Type, four L-Type, three C-Type. *Dispersals:* eighty, various. *Accommodation:* RAF: Officers 91, SNCOs 260, ORs 1,269; WAAF: 0.

SKIPTON-ON-SWALE, Yorkshire

54°13'32"N/01°25'50"W; SE372812; 78ft asl. 3 miles W of Thirsk, E of A167 at Sandhutton

One of many bomber airfields scattered through the vale of York, this site was first opened for flying late in 1942 as a satellite field for nearby Leeming, although Topcliffe was in fact the nearest airfield, just a couple of miles to the south-east. The surrounding road network and the adjacent River Swale placed restrictions on the overall size and shape of the new airfield, but despite this a fairly typical bomber airfield was created with an open triangular layout of three standard runways, together with the usual supply of thirty-six dispersals and three hangars.

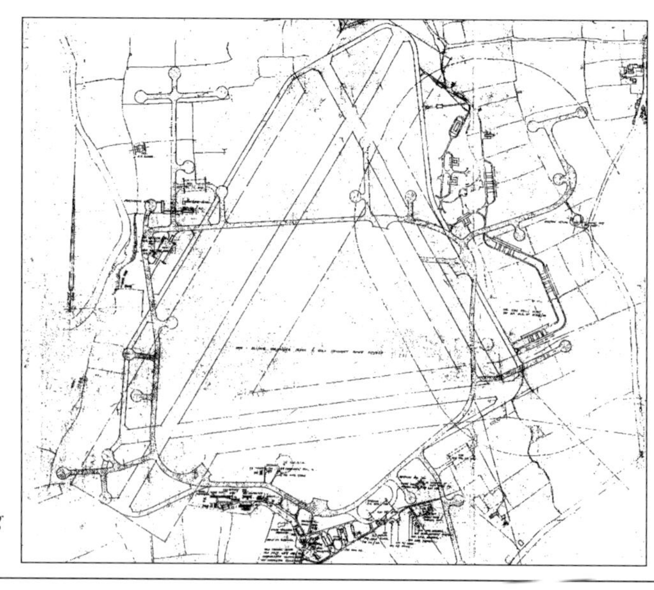

The airfield site plan drawing for RAF Skipton-on-Swale.

An RAF photograph of the newly completed airfield.

The first unit to arrive here was the Canadian No 420 'Snowy Owl' Squadron, which brought its Hampdens from Waddington on 7 August 1942, and immediately resumed the task of converting onto Wellingtons. Once the unit had completed conversion to the latter type, it departed for Middleton St George in October. It was not until the following May that the next unit arrived, this being No 432 'Leaside' Squadron, equipped with more Wellingtons. Within a matter of days the unit was back on operational duties, and by the time it departed to East Moor in September it had flown almost all of its 494 operational Wellington sorties from Skipton, prior to re-equipping with the Lancaster.

Just a few days after No 432 vacated Skipton, No 433 'Porcupine' Squadron arrived, but some weeks were to pass before the unit was equipped with the Halifax bomber. However, by January 1944 the Squadron was flying operational sorties, and had been joined at Skipton by No 424 'Tiger' Squadron, which had returned from the Middle East just before Christmas of the previous year. Also equipped with Halifaxes, the unit was also flying missions over Germany by January 1944.

John Beisley recalls his traumatic introduction to his time with No 433 Squadron:

'On 5 August 1944, having completed training at 1659 Heavy Conversion Unit, RCAF, Topcliffe, I was assigned a crew. We were posted to No 433 Squadron, Skipton-on-Swale.

Groundcrew prepare a No 420 Squadron Wellington for ops at Skipton-on-Swale.

A No 420 Squadron Wellington crew in front of their aircraft at Skipton-on-Swale.

Having reported to Squadron Offices, we proceeded towards our Nissen huts situated in a field about half a mile away next to the River Swale. Almost immediately as we approached the village, a Halifax returning from a bombing mission, damaged and on three engines, crashed at our feet about 50 yards away. The pilot had been given a red flare warning him to go around again, and while doing so a second engine failed, hence the crash. The pilot and flight engineer were killed, as well as a small boy in the village. The rest of the crew were seriously injured.

After a Squadron Reunion in Toronto in October 1982, a committee meticulously planned, with generous contributions, a dedication and plaque honouring the squadrons that served at Skipton-on-Swale, recording the crash incident. On 19 May 1984 the ceremony took place and many veterans attended. The plaque was placed on the spot where the Halifax had come to rest, having struck an elm tree. During the ceremony the last flying Lancaster flew over the site. The elm tree is no longer there but in its place stands a maple tree brought over from Canada. Probably one of many similar incidents – however, I felt it was worth placing on record.'

Suitably marked with its allocated unit, a refuelling bowser attends one of No 424 Squadron's Halifax bombers at Skipton-on-Swale. Ken Cothliff

After reorganisation in May, the station became a sub-station for No 63 Base (Leeming), but operations at Skipton were unaffected and the Canadians remained in business here with Halifaxes until January 1945, by which time No 424

Aircrew take a break at an Army canteen van at Skipton-on-Swale. Of particular interest is the personally embellished flying jacket. Ken Cothliff

Hector Patten Kopinski's magnificently adorned Halifax LV951 'The A Train' pictured at Skipton-on-Swale during August 1944, while it was with No 424 Squadron's H Flight. Ken Cothliff

Squadron had achieved 1,811 operational sorties for the loss of twenty-three aircraft, and No 433 Squadron had completed 1,926 missions, losing twenty-eight aircraft in the process. Their last mission took place on 25 April, by which time both units had re-equipped with Lancasters.

In August 1945 Skipton transferred to No 1 Group and the Canadian units shifted away from bombing sorties and began to concentrate on transport duties. However, in October 1945 they disbanded, leaving Skipton as an empty site, which the RAF vacated shortly afterwards. It remained unused for many years and all of the land was gradually sold off and slowly restored to agricultural use. The general outline of this once busy airfield is still very visible, but the runways are partially dismantled and overgrown, some parts occupied by farming buildings. The crumbling perimeter track is still here, but many of the dispersals are overgrown or removed. The hangar bases can still be found, but the airfield is slowly disappearing and is now probably more recognisable from the air, from where its ample proportions can still be appreciated.

Main features:
Runways: 222° 6,000 x 150 feet, 272° 4,200 x 150 feet, 338° 4,200 x 150 feet, tarmac surface. *Hangars:* two T2, one B1. *Dispersals:* thirty-six Heavy Bomber. *Accommodation:* RAF: Officers 225, SNCOs 457, ORs 1,242; WAAF: 166.

SNAITH, Yorkshire

53°40'56"N/01°05'08"W; SE604210; 39.5ft asl. 9 miles E of Pontefract, S of A645 on M62 at Pollington

Plans to build an airfield on a site adjacent to the village of Pollington were first established in the mid-1930s, but attention shifted to many other sites in the region and it was not until 1940 that construction of an airfield here began. Also unusual was the decision to take its name not from the nearest village at Pollington (which could easily be confused with Pocklington some miles away) but from Snaith, a couple of miles to the north-east. Three runways were laid and, despite the site being constrained by surrounding roads, two of them were completed to full bomber length standards. One runway straddled what is now the A645, and it is assumed that this road must have been closed for the duration of the RAF station's operational existence.

No 150 Squadron was the first unit to arrive here, bringing its Wellingtons from Newton in July 1941. Already operational, this unit was back in the air within a matter of days and remained active at Snaith until October 1942, when it moved to Kirmington. On 27 October 1942 No 51 Squadron arrived at Snaith with a fleet of Whitleys, ready to convert onto the far more capable Halifax. Once this task was complete, the Squadron rejoined the bomber offensive and continued in this role at Snaith until its last operational sortie on 25 April 1945, when it joined other RAF bombers on a mission to attack Wangarooge Island. By this stage the Squadron had flown 4,153 Halifax sorties, losing 108 aircraft in the process.

During January 1944 the unit's C Flight had been used to create what became No 578 Squadron. This unit built up at Snaith and flew its first operational missions from here, but moved to Burn during February. No 266 Squadron completed a one-week stay with its Typhoons in April, and No 266 Squadron stayed for a similarly brief period in April 1944 with its Spitfires. When No 51 Squadron left in April 1945 the airfield fell silent, although flying did return in September when Nos 1516 and 1508 Flights came to Snaith with their Oxfords, staying here until the following April. When they left, flying ended at Snaith and the RAF vacated the station just a few months later.

No 51 Squadron Halifax bombers at Snaith.

The airfield remained virtually intact for many years, but industrial, commercial and agricultural development has slowly encroached upon the site and now the old runways are largely overgrown and covered with buildings and junk, even though they are still mostly intact. The hangars have also survived, but most of the dispersals and the perimeter track have gone. Most noticeably, the M62 motorway now cuts through the former airfield, crossing the runways where two of the old runways once intersected. It is perhaps the surviving J-Type hangar that reminds some visitors of the days long before this area was littered with boxes and pallets, when the mighty Halifaxes stood here, preparing for war.

Main features:
Runways: 231° 4,200 x 150 feet, 273° 3,300 x 150 feet, 319° 6,000 x 150 feet, tarmac surface. *Hangars:* two T2, one J-Type. *Dispersals:* thirty-six Heavy Bomber. *Accommodation:* RAF: Officers 164, SNCOs 375, ORs 1,477; WAAF: 394.

SOUTH CARLTON, Lincolnshire

53°16'38"N/0°33'20"W; SK964765; 203ft asl. 1 mile N of Lincoln on B1398, E of South Carlton village, W of A15

A de Havilland DH.9A running its engine in front of one of South Carlton's hangars.

A Sopwith Camel pictured on a rainy day at South Carlton.

Heading north out of Lincoln along the A15, RAF Scampton's huge C-Type hangars soon come into view. However, in the fields just ahead of these buildings, another historic airfield once stood. Built for the Royal Flying Corps, South Carlton Training Depot Station opened in November 1916 with No 45 Reserve Squadron resident with a fleet of Farman FB.5 aircraft. The station was fairly simple but certainly more substantial than many of the temporary landing fields adopted by the RFC in this period. An area of farmland was cleared and levelled to enable take-off and landing runs of up to 2,000 feet to be achieved, and at least seven RFC hangars were constructed, together with a variety of brick-built support buildings. Some Bessonneau hangars were also erected and a number of wooden huts, which were used for accommodation.

Other units came and went after fairly short periods, including No 25 Squadron with DH.9As, No 69 Squadron with RE.8s, Nos 45, 61 and 39 Training Squadrons, and No 109 Squadron, but in July 1918 the resident Training Squadrons were combined to form No 46 Training Depot Station, with a mixed fleet of Camels, Dolphins and 504s. By July 1919 this much-reduced unit had become No 46 Training School RAF, and it remained active until April 1920. No 8 (Training) Wing's headquarters came to South Carlton in April 1920, but when it left again in August the station was wound down and abandoned by the end of that year.

South Carlton was commanded for some years by Lieutenant-Colonel Louis Strange, and his memoirs make mention of the fact that service life here was a serious business:

> 'Work in a Training Wing was no joke. The write-off of one machine for every 140 hours of flying meant losses of between thirty and forty aircraft per month in addition to seventy or eighty minor crashes. In May 1918 we had sixteen fatal crashes in the 23rd Wing ... but work had to go on at a feverish pace in order to cope with the overseas requirements.'

The old RFC airfield was swiftly returned to agriculture after 1920, but most of the permanent buildings survived for some time, being used by local farmers. Even today some traces of the old buildings are still here, lurking in a complex of farm buildings on a track that runs east from the B1398 at South Carlton village. This track turns south before resuming its original direction, marking the point at which this (the old station entrance road) once curved south of a line of RFC hangars.

Main features:
Runways: none, grass surface. *Hangars:* seven RFC pattern, six Bessonneau. *Dispersals:* none. *Accommodation:* not known.

SOUTH CAVE, Yorkshire

53°46'35"N/0°36'04"W; SE922320; 174ft asl. 10 miles W of Hull on A1034, N of South Cave village, S of Swinescalf Road

The short-lived landing site at South Cave was one of numerous such sites across the region that came into being during 1916. Adopted by the Royal Flying Corps, it was little more than an area of suitable farmland, cleared of all obstructions to enable fighter aircraft to land here as and when required. No 33 Squadron was assigned to air defence of the region, and with its headquarters at Tadcaster (Bramham Moor) it had a wide range of allocated landing fields scattered around the area. These could be used whenever necessary to mount patrols against the marauding German airships. When the threat of airship attack diminished at the end of 1916 the need for intensive air patrols also disappeared and the landing fields were no longer required.

A gathering of RE aircraft at South Cave.

It is not known how often the landing field at South Cave was used, but it is unlikely to have been frequented on more than a handful of occasions. No permanent structures were built here and the various tents and huts were removed when the RFC left. The fields immediately returned to agricultural use and today the landing field is a quiet and colourful sea of rapeseed.

Main features:
Runways: none, grass surface. *Hangars:* none. *Dispersals:* none. *Accommodation:* not known.

SOUTH OTTERINGTON, Yorkshire

54°16.54"N/01°24.27"W; SE386874; 118ft asl. 5 miles NW of Thirsk off A167, S of Station Road, E of South Otterington village

Like many other such sites around the region, a landing field was established here early in 1918 on farmland adjacent to South Otterington. It was essentially a reserved field simply cleared of obstructions, intended for occasional use by the Royal Flying Corps, most notably No 76 Squadron's Avro 504 and BE.2C aircraft. This unit was engaged in air defence operations over the region, and as part of its continual patrols to defend against German airships a series of landing fields was assigned to the unit, for use as required. The threat of German attack had diminished by the end of 1916 and this rendered the landing fields unnecessary, so they were swiftly abandoned.

The landing field at South Otterington comprised only a few tents, and these were soon gone, leaving the area to be returned to agricultural use. Today the site is a sprawling area of empty farmland, which contains no trace of the RFC's brief presence here so long ago.

Main features:
Runways: none, grass field. *Hangars:* none. *Dispersals:* none. *Accommodation:* not known.

SPEKE, Merseyside

53°20'44"N/02°52'54"W; SJ413835; 78.5ft asl. 5 miles E of Liverpool, S of A561 (Liverpool Airport)

Development of an airfield at Speke began in the 1920s, thanks to a far-sighted decision made by Liverpool Corporation to build an airport for the city. Constructed on 400 acres of the Speke Estate, it opened to flying in July 1930, and by 1933 was open to public use, something that was celebrated by an air show that year. In 1934 the Liverpool & District Aero Club arrived and the airport slowly expanded, becoming the second busiest airport in the country by 1939, when civil operations mostly stopped and the airfield was allocated to military use.

From 1940 a variety of fighter units came here for relatively short stays, most assigned to the air defence of the region and Liverpool in particular, as it was an obvious target for the Luftwaffe's bombers. Squadrons deployed to Speke included Nos 13, 229, 236, 303, 308, 312 and 315, operating types such as the Spitfire, Hurricane and Blenheim. The Royal Navy was also present at various stages, and the Merchant Ship Fighter Unit was formed here in May 1941, equipped with Hurricanes that were assigned to merchant ships and launched from their decks by catapult, providing shipping with some means of air defence when no other source was available. Not surprisingly, catapult launch gear was assembled

The spectacular launch of a Sea Hurricane from the catapult launcher at Speke.

at Speke, so the launch technique could be practised. Also here from 1941 was No 776 Squadron (a Fleet Requirements unit), which operated Blackburn Rocs together with a variety of other types and became one of the most active units to be based here.

Aside from the many units based here, Speke became home to a manufacturing base from 1938 when Rootes Securities constructed a factory here, embarking on the production of Blenheims, some 2,443 of which were built by the end of the Second World War. From 1942 Rootes began constructing Halifaxes, and more than 1,000 of these heavy bombers were built here during the war years. Rootes also became responsible for the assembly and overhaul of countless American aircraft, contracted to Lockheed initially for the assembly of Hudsons ordered for the RAF, but subsequently embracing a number of different types, and more than 2,500 fighter types operated by the USAF in Europe. No 1 Aircraft Assembly Unit was established at Speke to undertake the handling of the many American aircraft as they arrived from the States, or preparing them for their return some years later.

In 1944, as the end of the war approached, it was accepted that Liverpool was no longer under any direct threat from Germany and the airport site was duly handed back to civilian control. However, No 611 Squadron formed here on 10 May 1946, but stayed for only a few days before moving to Hooton Park; from June of that year the airfield was no longer host to any military units.

In the post-war years the Liverpool University Air Squadron came here for some time with Tiger Moths, as did No 186 Glider School (from June 1944 until March 1947), but by the 1950s Speke was developing into a large and busy airport and the only military aircraft to be seen here were occasional visitors engaged on transport or communications duties. The limitations of the airfield site were wisely recognised, and in 1966 a new runway was built to the south of the existing airfield, linked by a taxiway. This enabled the airport to grow still further, capable of handling aircraft of all types up to the weight and size of modern 'Jumbo' jets if necessary. Ultimately the gradual expansion of the airport led to the construction of a completely new terminal attached to the new runway, and the gradual wind-down of the original airport site.

Today Speke's original airport is gone and only the beautiful terminal building survives as a hotel. The runways have been bulldozed and new roads run across the old airfield. The new airport continues just to the south under the ownership of Peel Holdings, which (with typical flair for applying vulgar names to its airports) now refers to the facility as John Lennon International Airport.

Main features:
Runways: 261° 4,980 x 150 feet, 170° 3,090 x 150 feet, 221° 4,200 x 150 feet, concrete and tarmac surface. *Hangars:* two Bellman, six Blister. *Dispersals:* thirty-six spectacle. *Accommodation:* RAF: Officers 52, SNCOs 25, ORs 651; WAAF: 186.

SQUIRES GATE, Lancashire

53°46'23"N/03°02'0"W; SD319312; 39.5ft asl. 1 mile S of Blackpool on A5320 (Blackpool Airport)

Literally just a few minutes' walk from the end of Blackpool's famous promenade, Blackpool Airport is a thriving centre for private and commercial aviation that first appeared in the 1930s when a small grass field was developed here and used for short-range flights to destinations such as Liverpool and the Isle of Man. Indeed, there are claims that the very first Air Meeting (a forerunner of today's air shows) took place here in 1909 and was therefore the first such event in the UK, although this claim is generally accepted to belong to Doncaster. Oddly, the site used for these early meets was developed into a horse-racing course in 1910 – a reverse of a situation that applied to many sites across the UK where existing racecourses were adopted as airfields. However, horse-racing was not successful here and by 1915 the field was used only for flying, and when scheduled services shifted to Stanley Park in 1936 it became a centre for private and recreational aviation.

As the Second World War approached, Squires Gate became the home of No 42 Elementary & Reserve Flying Training School as part of No 50 Group. Operated on behalf of the RAF by Reid & Sigrist, the unit flew a fleet of Tiger Moths here from August 1939, but by the end of the year had disbanded, and operational RAF units began to appear here, largely tasked with air defence of the region. From 10 February 1940 the station became part of Fighter Command, and although a variety of training units came here (on the basis that the station was far away from any major threat) it was bombed by the Luftwaffe on a number of occasions in 1941, albeit without creating any significant damage.

No 256 Squadron Defiants scattered across the airfield at Squires Gate.

By this stage the airfield had three runways and a perimeter track, which held a small number of dispersals. The flat and uncluttered land enabled a conventional runway layout to be employed, the main runway running east-west with its western threshold just a few yards from the adjacent beach. The School of General Reconnaissance arrived in July 1940 from nearby Hooton Park with a fleet of Avro Ansons, but after only two months it was disbanded, its role having been assumed by No 2 School of General Reconnaissance, which had formed at Squires Gate the previous May, as part of No 17 Group. With a mixed fleet of Ansons and Bothas, it too was gone by the beginning of 1941, but was replaced by No 3 School of General Reconnaissance, and this unit became the most established unit at Squires Gate, flying Ansons, Bothas and (eventually) Spitfires. Its training activities resulted in a great deal of flying activity in the area, and many of the huge number of pupils were accommodated off-base in the

An aerial view of Squires Gate taken in 1946.

Hawker Hunter production under way at Squires Gate.

many hotels and guest houses in nearby Blackpool. The Spitfire element left late in 1942 to join No 8 Operational Training Unit, but the School remained in business at Squires Gate until the end of the Second World War, moving to Leuchars in August 1945. Also here from May 1943 were the Ansons operated by the School of Air Sea Rescue, and from August 1944 No 1510 Flight was here with its Oxfords for a one-year stay.

Squires Gate was perhaps best-known for its role as a factory site, and by 1940 a huge assembly shed had appeared adjacent to the airfield, under the ownership of Vickers-Armstrong. In September 1940 the first Wellington bomber was completed here, and by October 1945, when production ended, some 2,500 Wellingtons had emerged from the factory (with some work shared by a second factory at Stanley Park, although no flying was done from this small airfield through the war years).

From 1946 Squires Gate was back under civilian control, although military aircraft were still to be seen here, thanks to the presence of the aircraft factory. Many Hawker Hunters were built and test-flown here, the performance of this agile jet requiring the old main runway to be replaced by a new 6,000-foot one, which was laid across the existing airfield site, extending out to the east. The older runways remained available for use, however, and two of the strips are still occasionally used even today.

After the factory closed in 1957 the airfield became a more traditional airport site, and has continued to survive over successive decades, accommodating many private aircraft and a number of commercial operators. Links with military aviation still continue, and RAF aircraft can often be seen here while engaged on training flights, or as a staging point for air show appearances. Sadly, the magnificent Vulcan bomber that greeted airport visitors at the site entrance for many years has now gone, its owner having abandoned it to the ravages of the sea air.

Main features:
Runways: 080° 3,900 x 150 feet, 140° 3,450 x 150 feet, 020° 3,300 x 150 feet, bitumen surface. *Hangars:* four Bellman, two Boulton Paul. *Dispersals:* sixteen circular. *Accommodation:* RAF: Officers 58, SNCOs 23, ORs 357; WAAF: 0.

STRETTON, Lancashire

53°20'46"N/02°31'24"W; SJ652833; 230ft asl. 3 miles SE of Warrington on M56, W of M6 junction

Development of an airfield at Stretton began in 1940 when plans were first made to build a Royal Air Force station that could host fighter aircraft. The risk of German attacks on major industrial targets in the region (particularly Liverpool and Manchester) necessitated a significant local air defence capability, and Stretton was identified as a suitable location. However, as the airfield was being developed it became increasingly obvious that the German threat was rapidly diminishing and that there would probably be no need for a fighter base here by the time it was completed. Consequently, the site was transferred to the Royal Navy (in what was essentially an exchange for the RAF's use of facilities at Machrihanish), and when it finally opened on 1 June 1942 it was as a Fleet Air Arm station – HMS *Blackcap*.

Over the following 16 years a staggering total of some forty-one naval units were based here at various stages. Most stayed only briefly (having been deployed here for training), although Nos 1831 and 1842 Squadrons (Royal Naval Volunteer Reserve) became longer-term residents and stayed until they were disbanded in 1957. No 1831 Squadron, formed here on 1 June 1947, was assigned to fighter operations, while No 1841 Squadron, established for anti-submarine operations on 18 August 1952, formed the Navy's Northern Air Division. A major Aircraft Maintenance Yard was also established here, and eventually became a major unit, handling up to a third of the Fleet Air Arm's maintenance requirements, and accommodating aircraft that came from carriers engaged on convoy duties in the Atlantic, berthed at nearby Liverpool.

When the Second World War ended a large number of redundant types (particularly those supplied by the United States) were delivered here for disposal, and most were eventually cut up on site. Fairey Aviation also set up a facility here, accepting aircraft (predominantly Barrucudas) from the company's factory at Heaton Chapel for flight testing prior to delivery to operational squadrons.

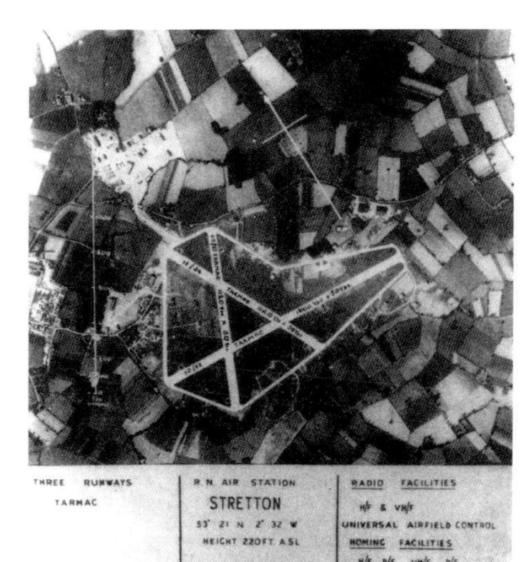

An official wartime photograph of RNAS Stretton.

A Boulton Paul Sea Balliol taxiing to the runway at Stretton.

Situated close to the USAF's major base at Burtonwood, the airports at Manchester and Liverpool, RNAS Burscough, and the factory airfield at Warton, this was a very busy area for air traffic, and to compound problems still further much of the RNVR flying was conducted at weekends when civilian traffic was generally even higher. To combat the congestion a Joint Radar Control Unit was established a few miles south of Stretton, staffed by civilian and naval controllers.

No 767 Naval Air Squadron brought its aircraft to the base in September 1952. Initially equipped with Fireflies, the unit also flew Sea Furies and subsequently acquired Avengers, Sea Hawks and Attackers. As a Landing Signal Officer Training Squadron, the unit had its own 'dummy deck' where aircraft could be operated as on a real aircraft carrier. The unit achieved more than 2,000 'deck landings' per year. The last unit to arrive at Stretton was No 728B Naval Air Squadron, which formed here in January 1958 before moving to Malta a few weeks later.

HRH The Duke of Edinburgh arrives at Stretton for an official visit in May 1956.

Attackers on parade during the Duke's visit.

RNAS Stretton closed on 4 November 1958 and the site was quickly abandoned. The construction of the M56 motorway eventually resulted in the disused airfield being split into two portions, the motorway running almost parallel to the old main runway. Almost all of the airfield site to the north of the runway was eventually consumed by industrial development, but to the south of the motorway very little development has taken place. The main runway is still here, as are significant portions of the secondary runways, together with the old perimeter track. Shell used the runways for high-performance car testing for some years, and the only flying that has taken place here since the 1950s has been the occasional visit of crop-dusting aircraft. The site's future seems unclear but, with some police driver training taking place, it seems likely that cars will be the final users of the runways, rather than aircraft. Crowley Lane now runs along what was once part of this taxiway, leading to the eastern threshold of the main runway.

Main features:
Runways: 027° 3,360 x 150 feet, 097° 4,800 x 150 feet, 157° 3,360 x 150 feet, tarmac surface. *Hangars:* four squadron type, four storage type. *Dispersals:* six fighter pens, ten hardstandings. *Accommodation:* FAA: Officers 329, ORs 1,016; WRNS: 117.

STRUBBY, Lincolnshire

53°18'24"N/0°10'32E; TF450810; 36ft asl. 6 miles S of York, W of B1122

This area of flat farmland was first acquired for military use in 1941. Chosen as a suitable site for the creation of a satellite airfield for East Kirkby, construction proceeded without any significant problems. Although a road had to be closed and a variety of field hedges and farm buildings had to be demolished, the land was ideal for the task at hand and three concrete runways were laid, the two secondary runways even being afforded some additional length thanks to the generous amount of land available. The standard hangars were erected and the usual supply of bomber dispersals, but the airfield's relatively late development meant

Invasion-striped Beaufighters on dispersal at Strubby.

A fascinating wartime image of Strubby's Beaufighters in action against German shipping in the North Sea.

that there was no immediate use for the site when it was completed, so it was placed under Care and Maintenance from January 1944 until it opened for flying in the middle of April.

On 19 April 1944 No 280 Squadron arrived and its Warwicks, and was based here as part of its contribution to the D-Day operation. When its work was done, the unit moved to Langham in September. Strubby was responsible for the control of operational activity in the area of the North Sea adjacent to the base for the duration of the Allied invasion of Europe. As part of this commitment Nos 144 and 404 Squadrons moved here from Davidstow Moor on 1 July 1944 with their fleet of Beaufighter strike aircraft, forming what became the Strubby Strike Wing. For two months the Wing's Beaufighters ranged far and wide across the North Sea, attacking German naval targets and ensuring that the Allies had total superiority in the region. The Beaufighters moved to Banff in September and Strubby transferred to Bomber Command. No 619 Squadron arrived from Dunholme Lodge with its Lancasters soon after the Beaufighters had left. As part of No 55 Base (HQ at East Kirkby) the unit stayed at Strubby until the Second World War ended. No 227 Squadron arrived here in April 1945, but by this time both units had no operational tasks and concentrated on training and transportation duties before vacating Strubby during the summer. However, the station remained active for a while as since the middle of 1944 it had also been allocated as a Relief Landing Ground for nearby RAF Manby's Empire Air Armament School, which operated a large and varied fleet of aircraft.

College of Air warfare Jet Provosts pictured during a formation display rehearsal at Strubby during the late 1960s.

Strubby's main runway in 2011, where Varsities once chugged their way into the air.

T2 hangars and control tower, all surviving in good condition at Strubby.

Despite this, routine flying activity ended on 7 September 1945 and the only movements here were the occasional arrivals and departures of Lancasters, which were stored on the airfield's dispersals and secondary runways prior to being disposed of. Strubby was allocated as a sub-site for No 381 Maintenance Unit (storing aircraft for the Tiger Force) and No 384 MU, followed by Nos 35 and 93 MUs, which maintained a presence until 1950.

In June 1949 the station became a satellite field for RAF Manby, following the formation of the RAF Flying College there, and eventually the College's No 2 Squadron based most of its Vampires, Meteors and Canberras at Strubby, where there was plenty of available air space and a much longer runway compared to the perilously small runways available at their home base. This association with nearby Manby continued for some years, the RAF Flying College becoming the College of Air Warfare, and the Canberras, Vampires and Meteors eventually making way for Varsities, Jet Provosts and Dominies. With only a very modest runway at Manby, the CAW made daily use of Strubby's facilities and the brightly painted Varsities could often be seen (and heard!) lumbering around the circuit or chugging and squealing along the taxiway. When the CAW formed its own aerobatic team, much of its formation training was conducted over Strubby, and the famous Macaws Jet Provost team owed much of its success to many hours of activity over Strubby.

Flying activity continued until Manby's activities wound down, and in August 1972 a final mixed formation of CAW aircraft roared over the base to bid farewell. The RAF abandoned the site and it lay unused for some years, although a few civil helicopters did operate here for some time in support of offshore oil rigs. However, the Strubby Gliding Club (now the Lincolnshire Gliding Club) established a presence here in 1978 and remains active at Strubby to this day. Oddly, the club's gliders and aircraft operate from part of the old perimeter track on the northern edge of the airfield, the old main runway now being completely disused, with half of its length having been dug up and returned to agriculture. But the hangars remain (now used by commercial and retail companies) and the once derelict control tower is now beautifully restored as a private residence. Strubby was a late addition to the RAF's inventory, but it played a vital part in its offensive operations and provided a useful contribution to the RAF's post-war training capabilities, supporting a nearby station that was arguably incapable of operating efficiently in isolation.

After many years of abandonment, the control tower at Strubby has been restored to top-class condition and is now a private dwelling.

Main features:
Runways: 086° 6,000 x 150 feet, 146° 4,730 x 150 feet, 026° 4,410 x 150 feet, concrete and tarmac surface. *Hangars:* two T2, one B1. *Dispersals:* thirty-six Heavy Bomber. *Accommodation:* RAF: Officers 112, SNCOs 370, ORs 1,261; WAAF: 369.

STURGATE, Lincolnshire

53°22'47"N/0°40'56"W; SK877878; 59ft asl. 3 miles SE of Gainsborough, S of A631 on Hill Road

The F-84 Thunderjet, which was scheduled to be based at Sturgate as part of the USAF's expansion plans.

One of the later bomber airfields to be developed, RAF Sturgate opened on 1 March 1944, although it did not fully open for flying operations until 12 April. The area comprised flat farmland and preparing it for flying was a straightforward task, although it did require the closure of two roads, one being a fairly significant link between the villages of Upton and Heapham. Three full-length bomber runways were laid, together with the standard arrangement of thirty-six dispersals, perimeter track and hangars. Once completed it was not considered to be of immediate use, and it was not until September 1944 that the first aircraft arrived when No 1520 Beam Approach Training Flight brought its fleet of Oxfords here from Leconfield. The Flight was also associated with the Aircrew Training School, whose crews also used the same Oxfords for circuit flying practice while they awaited postings to Heavy Conversion Units.

Other units also used the airfield briefly, but none was relocated here, and the most frequent visitors were aircraft from Hemswell's No 1 Lancaster Finishing School. It was not until June 1945 that Nos 50 and 61 Squadrons relocated to Sturgate from Skellingthorpe, but their stay was brief and in January 1946 they left for Waddington. This was probably due to plans for reassigning the airfield to the United States Air Force, and after a period of inactivity Sturgate was duly transferred to the USAF on 30 July 1952 as part of Strategic Air Command. The 2928th Air Base Group assumed control and the station was used to host temporary deployments of USAF fighter units, mostly equipped with F-84 Thunderjets. This activity continued until 1964, when the Americans vacated the base and the site was sold off.

Civilian flying gradually developed here and now the airfield is operated by Eastern Air Executive Ltd, mostly in support of business flying but also private and recreational activity. A limited amount of flying takes place from half of the old main runway, which is maintained in good condition – the eastern portion has sadly been removed. Almost all of the other parts of the airfield have also gone, with only a small portion of a secondary runway still to be seen, and a small stretch of perimeter track. The old road to Upton (Common Lane) is now restored and runs across the threshold of the remaining section of active runway, intersecting part of the old perimeter track. From here it is possible to obtain a good view of the largely empty airfield, and the runway from where the RAF's bombers briefly operated, and the USAF's jet fighters screeched into the skies.

Main features:
Runways: 078° 6,000 x 150 feet, 205° 4,200 x 150 feet, 141° 4,200 x 150 feet, concrete and asphalt surface. *Hangars:* two T2, one B1. *Dispersals:* thirty-six Heavy Bomber. *Accommodation:* RAF: Officers 75, SNCOs 190, ORs 420; WAAF: 286.

TADCASTER, Yorkshire

53°51'57"N/01°19'31"W; SE444412; 164ft asl. 3 miles SW of Tadcaster, N of A64 on Spen Common Lane

The historic RFC hangar still stands proudly at Tadcaster.

This historic site first appeared in 1916. Initially known as Bramham Moor (and subsequently as Tadcaster), it was created for use by the Royal Flying Corps in response to a growing need for air defence cover around the country. The emergence of a very significant threat from German airships (which made repeated attempts to attack industrial areas and cities across parts of the UK) led to the expansion of the RFC's Home Defence activities, and numerous landing fields were created, either as headquarters sites or from where fighter aircraft could be maintained on alert, or simply refuelled and rearmed.

No 33 Squadron was tasked with the air defence of a major portion of Yorkshire, particularly the cities of Sheffield and Leeds, and the unit came to Tadcaster in March 1916 to establish its headquarters here, with a series of Flights being set up at other airfields, together with a scattering of simple landing fields that could be used as required. The unit stayed here for some six months with a fleet of BE.2C, BE.12 and Scout aircraft. No 57 Squadron arrived in June 1916 and stayed for a similar six-month period, and in October No 33 Squadron established what became No 75 Squadron, although this unit moved out once it was established, taking its parent unit with it. No 41 Training (Ex Reserve) Squadron came here in July 1916 with a fleet of Maurice Farman trainers, but stayed for only a few weeks before moving to Doncaster. However, No 46 Training Squadron shifted in the opposite direction during December, staying at Tadcaster until July 1917, when it was replaced by No 14 Training Squadron from Catterick, joined by No 68 Training Squadron a few weeks previously.

By this stage the airfield at Tadcaster had become well-established with a variety of support, admin and technical buildings, together with some substantial wooden-framed hangars. In July 1918 the two training units were amalgamated to form No 38 Training Depot Station as part of the RFC's 8th Wing, largely now equipped with Avro 504s together with a smaller number of Sopwith Pups and SE.5As. By August 1919 the unit had been renamed as No 38 Training Squadron. In November 1918 No 94 Squadron arrived from France, followed by No 76 Squadron from Ripon during March 1919, but with the First World War ending it was inevitable that all the units would soon be disbanded. In June 1919 they duly wound down and the newly formed Royal Air Force vacated the airfield.

Over the following years the site was largely abandoned, although all of the landing area quickly returned to agriculture. All of the station's buildings were eventually demolished, but one hangar survived and this is now a listed building, remaining here as the only significant survivor of a short period in Britain's history when this site was a very busy centre for military aviation.

Main features:
Runways: none, grass field. *Hangars:* four RFC General Service type. *Dispersals:* none. *Accommodation:* not known.

TATTON PARK, Cheshire

53°20'28"N/02°22'03"W; SJ756827; 144ft asl. 2 miles S of Altrincham, S of M56, E of Ashley Road

It was during July 1940 that the RAF's Squadron Leader Louis Strange asked his former First World War colleague and fellow aviator Maurice Egerton (who became Lord Egerton and owned the huge estate at Tatton Park) if part of this land could be used for military purposes, particularly the training of parachutists destined for the airborne forces. Edgerton agreed and the first trial jumps were made here during that same month, using a tethered barrage balloon as a suitable drop platform. The trials were successful and No 1 Parachute Training School (based at nearby Ringway) quickly adopted the site for intensive training. Numerous barrage balloons soon appeared, each carrying a metal cage from which the troops could practise their jump, deployment and landing techniques. More than 60,000 personnel were eventually trained here over a period that lasted into 1946, many drops being made from aircraft (mostly Whitleys), which transported the troops directly from Ringway so that drops could be made in groups of ten or twenty personnel, normally from heights of 800-1,000 feet. Most drops were made onto a cleared area, although some (particularly those performed by Special Forces) were made directly into the woodlands surrounding the drop zone.

Although many aircraft were regularly seen over the site, only a few small communications and liaison aircraft landed here, until the site was also adopted for use as a dispersed storage site from 1940. Tatton Park became No 13 Satellite Landing Ground, operated by No 48 Maintenance Unit based at Hawarden, and a variety of aircraft types were brought here for short-term storage, held either on the open landing field or hidden in the surrounding trees. Military activity continued until 1957, at which stage the MU vacated the site and parachute training ended, leaving the landing field and drop zone to return to a more peaceful status. Today there is no evidence of the area's links with military aviation, other than a memorial adjacent to the former drop zone.

Main features:
Runways: none, grass field. *Hangars:* none. *Dispersals:* none. *Accommodation:* not known.

THEDDLETHORPE, Lincolnshire

53°23'35"N/0°13'14E; TF477907; 2ft asl. 1 mile N of Mablethorpe off A1031, N of Crabtree Lane on coast

Theddlethorpe's first connections with aviation began in 1927 when the beach was used for occasional gunnery practice by units based at nearby North Coates. Summer armament practice camps were operated at the latter base and, although some activity was performed in the skies over North Coates, the nearby beaches were a more suitable alternative. Eventually, the RAF's presence at Theddlethorpe became more substantial as the requirement for bombing ranges increased, and a dedicated site was established here during the Second World War. It had a purpose-built control tower, some accommodation for personnel just inland, and a cleared landing field adjacent to the Saltfleet Road; the latter was used occasionally for liaison flights, although it was also adopted as an emergency landing ground for RAF Manby.

Perhaps not surprisingly, the use of the bombing range did not always proceed without the occasional 'interesting moment', as evidenced by David Day, who recalls his father's use of the range:

'After service with No 223 Squadron in North Africa, my father was sent to RAF Manby to be commissioned and undertake further training as a bomb aimer. I believe he was converting to Lancasters at the time. One of the ranges used was offshore near Mablethorpe and my father was, on this particular day in 1943, selected as the bomb aimer on this live training flight. I don't understand much about the mechanics of bomb-sights, drift, offsetting and other such terms, but I gather that there are many variables that come into play when dropping a bomb. You don't just wait until the target is below you as they seem to in American movies!

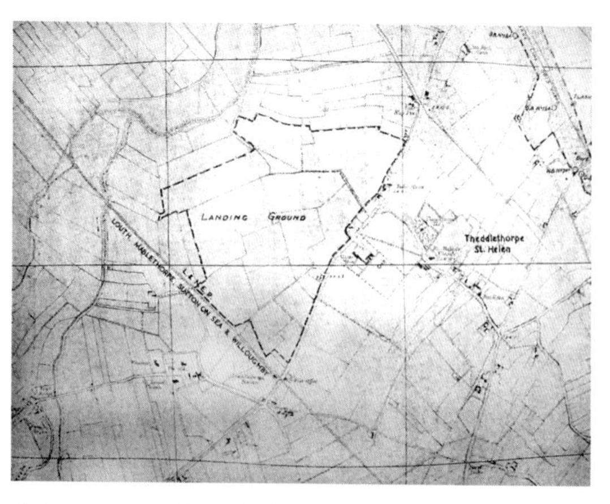

A 1948 map of Theddlethorpe landing ground.

The once-familiar visual target on the beach at Theddlethorpe.

Working in tandem with the navigator, my father pressed the tit and had the satisfaction of knowing he'd done his job well. However, on arrival back at the airfield the crew were met, escorted to a hut and kept incommunicado while each was questioned at length about the events leading up to the release of one particular 500-pounder on this training flight. All were mystified about the reason for such treatment.

To find the cause for this investigation it is necessary to change the focus to a small row of cottages lying parallel to the coast quite close to the bombing range. Earlier in the morning one particular chap had just emerged from the little wooden latrine at the bottom of his garden. It was a clear morning and he was aware of the sound of heavy bombers high overhead. No doubt he was used to it! He had just reached the door of his cottage when the world – as he knew it – ceased to exist. His thoughts, when eventually he was able to make sense of the scene around him, can only be imagined – although various comments have been ascribed to him in the telling! His outside toilet – which he had vacated only a minute before – had ceased to exist and the spot was marked by a very large crater! All the cottages were damaged by blast but amazingly no person or animal had been injured.

This, it was subsequently decided at the tribunal that followed, was the result of my father's aim on that particular day. I understand that as the junior member of the crew he "carried the can" for the mistakes of others, but perhaps such matters should remain in the "cloud cover of time". This can be told, suitably embellished, as a very humorous tale. In reality it could have been very serious and obviously the RAF treated it as such.'

A radar van in the sand dunes at Theddlethorpe during 1956.

After the war the range became a sub-site controlled by RAF Binbrook, and Canberras from there became regular users of the range, often performing their bombing runs from altitudes of up to 40,000 feet. As tactics changed, more low-level weapons delivery became common, and by the late 1960s Theddlethorpe was busy with a steady stream of 'customers' from the RAF, USAF and occasionally other NATO forces. However, the growing popularity of nearby Mablethorpe as a holiday resort began to cause difficulties, and holidaymakers were unimpressed by the regular appearances of types such as Buccaneers, Phantoms and Hunters, commencing their bombing dives directly over the busy beaches.

However, it was the development of a gas terminal just half a mile from the range that finally sealed Theddlethorpe's fate, and in 1973 it was closed. By this stage Donna Nook (just a few miles further up the coast) was an established weapons range, and activities were concentrated there. Today, there are still a few surviving buildings at Theddlethorpe although the beach is now open again to casual visitors – some of whom might be surprised to see the carcass of a tank (a former target) still rusting here. Sadly, the once familiar 'bandstand' target marker has long since been demolished.

The tower and mobile radar van at Theddlethorpe, also in 1956.

Main features:
Runways: none, weapons range. *Hangars:* none. *Dispersals:* none. *Accommodation:* not known.

THIRSK, Yorkshire

54°13'55"N/01°21'29"W; SE419819; 114ft asl. W of central Thirsk on A61 (Thirsk Racecourse)

Like many other racecourses around the country, that at Thirsk was adopted by the Royal Flying Corps early in 1916 as a suitable site for aviation. The large, flat areas of land in the centre of racecourses were obvious choices for easy conversion into landing fields, and the grass field at Thirsk was a particularly suitable site that required no significant improvements. No 76 Squadron brought its Avro 504s here and operated from the site for a couple of years, largely assigned to air defence duties around the region. No permanent structures were built, as the racecourse buildings were ideal for this purpose. However, by 1918 the site was largely redundant and the RFC abandoned it, leaving the area to return to the more peaceful sporting pursuits that continue to this day.

No traces of the RFC's presence are to be found today. It should be noted that another minor airfield is now active to the south-east of Thirsk, but this has no connections with military aviation.

Main features:
Runways: none, grass field. *Hangars:* none. *Dispersals:* none. *Accommodation:* not known.

THOLTHORPE, Yorkshire

54°06'21"N/01°15'54"W; SE481680; 62ft asl. 10 miles NW of York, W of A19 on Hag Lane, NE of Tholthorpe village

First identified as suitable for development during the mid-1930s, the airfield here was constructed in 1940, and was only partially complete by August when the RAF station first opened. As part of No 4 Group it was a satellite for Linton-on-Ouse, and it was from this station that the first aircraft came when No 77 Squadron brought its Whitleys for a brief stay in September. However, at this stage the airfield was little more than a grass field site with few permanent buildings, and a major construction programme then got under way, transforming the site into a fully equipped bomber station. It is not known whether Tholthorpe was to have been completed to this standard from the outset, but it seems that the decision may have been made retrospectively, as it was not until June 1943 that the airfield was reopened for use. However, this long development period may also have been due to difficulties in draining the site, and laying runways within the available space. Three concrete strips were duly built, although one was slightly shorter than standard. However, in all other respects a standard bomber airfield emerged as a sub-station for No 62 Base (Linton).

An RAF reconnaissance image of Tholthorpe taken shortly after the airfield's completion.

No 434 'Bluenose' Squadron formed here on 13 June, a Canadian unit equipped with Halifaxes, followed by No 431 'Iroguois' Squadron in July, this unit being in the process of converting from Wellingtons onto the Halifax. Operations began in August 1943 when No 434 Squadron participated in a long-range raid over Milan on the 12th/13th. Both units continued on operations here until December when they were replaced by two more Canadian units, No 425 'Alouette' Squadron from Dishforth and No 420 'Snowy Owl' Squadron from Dalton, both with Halifaxes. By February these units were operational at Tholthorpe and No 420 was in action over Berlin by the 15th. They continued in this role until 20 April 1945, when No 420 Squadron launched an abortive attack on Berlin, followed by No 425 Squadron's last mission on the 25th, to Wangarooge. Ultimately, No 420 Squadron flew 2,477 sorties and No 425 2,455, for a surprisingly low number of casualties. Of particular note is that Tholthorpe was one of few RAF stations where the George Cross and George Medal were awarded, the citation being as follows:

A No 420 Squadron Halifax pictured between sorties at Tholthorpe.

A Halifax bomber taxies out from its dispersal at Tholthorpe, en route to Germany.

A Halifax receives attention to one of its engines at Tholthorpe.

'One night in June 1944 an aircraft, while attempting to land, crashed into another which was parked in the dispersal area and fully loaded with bombs. The former aircraft had broken into three parts and was burning furiously. Air Commodore Ross was at the airfield to attend the return of aircraft from operations and the interrogation of aircrews. Flight Sergeant St Germain, a bomb aimer, had just returned from an operational sortie and Corporal Marquet was in charge of the night groundcrew, while Leading Aircraftmen McKenzie and Wolfe were members of the crew of the crash tender. Air Commodore Ross, with the assistance of Corporal Marquet, extricated the pilot who had sustained severe injuries.

Halifaxes at rest on their dispersals at Tholthorpe.

The remains of a No 425 Squadron Halifax that crashed during June 1944.

At that moment ten 5,000lb bombs exploded and this officer and airman were hurled to the ground. When the hail of debris had subsided, cries were heard from the rear turret of the crashed aircraft. Despite further explosions from bombs and petrol tanks which might have occurred, Ross and Marquet returned to the blazing wreckage and endeavoured in vain to swing the turret to release the rear gunner. Although the port tail plane was burning furiously, Ross hacked at the Perspex with an axe, then handed the axe through the turret to the rear gunner who enlarged the aperture. Taking the axe again, the Air Commodore, assisted now by Flight Sergeant St Germain as well as Corporal Marquet, finally broke the Perspex and steel frame supports and extricated the rear gunner. Another 500lb bomb exploded, which threw the three rescuers to the ground.

FS St Germain quickly rose and threw himself upon a victim in order to shield him from flying debris. Ross's arm was practically severed between the wrist and elbow by the second explosion. He calmly walked to the ambulance and an emergency amputation was performed on arrival at Station Sick Quarters. Meanwhile Marquet had inspected the surroundings and, seeing petrol running down towards two nearby aircraft, directed their removal from the vicinity by tractor. McKenzie and Wolfe rendered valuable assistance in trying to bring the fire under control and also helped to extricate the trapped rear gunner, both being seriously injured by flying debris.

Air Commodore Ross showed fine leadership and great heroism in an action which resulted in the saving of the lives of the pilot and rear gunner. He was ably assisted by St Germain and Marquet, who both displayed courage of a high order. Valuable service was also rendered by McKenzie and Wolfe in circumstances of great danger.'

Crews head off towards the scattered dispersals aboard a rather rudimentary 'crew bus' at Tholthorpe during 1944. Ken Cothliff

Ross (the Station Commander) received the George Cross, and Marquet and St Germain the George Medal, while McKenzie and Wolfe received the British Empire Medal.

In June 1945 the two resident squadrons returned to Canada in preparation for service in the Far East, and this marked the end of Tholthorpe as an operational station. The RAF vacated the site shortly afterwards and it lay abandoned for many years. The site is now restored to agricultural use, although parts of the old runways can still be seen, together with a couple of the hangars, the perimeter track, and some of the original airfield buildings, including the control tower, which has been restored to good condition.

Halifax bombers at Tholthorpe during February 1945.

Main features:
Runways: 281° 6,000 x 150 feet, 056° 4,200 x 150 feet, 350° 4,200 x 150 feet, concrete and tarmac surface. *Hangars:* two T2, one B1. *Dispersals:* thirty-six Heavy Bomber. *Accommodation:* RAF: Officers 131, SNCOs 468, ORs 902; WAAF: 233.

THORNABY, Yorkshire

54°31'56"N/01°18'05"W; NZ453154; 70ft asl. 2 miles W of Middlesbrough, N of A174, W of A19 at Thornaby

Aviation first came to this site in 1916, when an area of land was acquired for use by the Royal Flying Corps. The landing field created here was relatively basic, although it was more substantial than the many fields that the RFC adopted for occasional use by the unit responsible for air defence of the region. Adjacent to Stainsby Beck, the 34-acre site was used by No 36 Squadron, although there are few records to indicate how frequently it was used, or whether any other RFC units were ever based here.

This aerial view of Thornaby in 1940 provides an excellent illustration of how RAF airfields were camouflaged during the wartime period. The huge hangars are all adorned with camouflage paint and even the runway surfaces have been painted in order to render them less visible from the air. Phil Jarrett

The field was abandoned by 1919 and remained unused as open countryside until 1930, when the RAF returned here and re-established the original landing field. No 608 (County of York North Riding) Squadron formed here on 17 March and its Westland Wapiti bombers became a familiar sight around the local area, particularly at weekends when most of the reservist flying took place. In 1937 the unit exchanged its Wapitis for Hawker Demon fighters and Thornaby became part of No 1 (Air Defence) Group, but by the end of the decade it was part of No 18 Group Coastal Command. No 608 Squadron operated Demons for a couple of years after which it returned to offensive operations with Bothas (and some Ansons) and eventually Blenheim during 1941. The Squadron then acquired Lockheed Hudsons and remained in business at Thornaby until January 1942, when it moved to Wick. Other units had also come to the station by this stage including No 9 Flying Training School with Battles and Ansons, and No 42 Squadron with the rathe less familiar Vildebeeste torpedo bomber.

Vampire T11 WZ419: the ubiquitous Vampire trainer was a common sight at Thornaby.

When the Second World War began the airfield was hosting No 608 Squadron, wit Blenheims, and No 220 Squadron, flying Ansons in the maritime reconnaissance role and slowly converting onto Hudsons. By this time the airfield had been improved quite considerably, with thre concrete runways having been laid in 1941, together with a perimeter track and some dispersals The standard triangular runway layout was achieved but the constraints of the adjacent River Tees a smaller beck, woodland and roads meant that some careful design work was required. Even so, th longest runway was still only a modest 3,600 feet in length. It was not until some years later that th runways were extended, requiring more road closures, drainage work and levelling.

No 6 (Coastal) Operational Training Unit formed at Thornaby in July 1941 with a fleet o Hudsons, Ansons and Oxfords together with a handful of target facilities types. The unit's intensiv training role eventually led to nearby West Hartlepool being adopted as a satellite field, followed by Longtown. In October the unit's Hudsons were replaced by Wellingtons, but in 1943 a switch wa made with No 1 (Coastal) Operational Training Unit, which transferred from Silloth. Still operatin Hudsons, the type returned to Thornaby, although the unit also operated Halifaxes, Fortresses an Liberators, converting pilots onto these aircraft for the maritime role. After just seven months th OTU disbanded in October, and the Air Sea Rescue Training Unit moved in, having formed a week previously at Bircham Newton, equipped with Warwicks and other types.

No 280 Squadron also arrived (also operating Warwicks), and No 281 Squadron formed in November with yet more Warwicks. These units operated detachments at various sites, so thei presence at Thornaby was often quite limited, and it was aircraft from other units that became mor regular sights here, including detachments from Nos 306, 332, 401, 455 and 608 Squadrons, wit various types such as the Spitfire and Beaufighter. By the summer of 1944 only No 280 Squadro was present, and its detachment was replaced by aircraft from No 279 Squadron in October continuing Air Sea Rescue cover over the region. The last significant wartime presence here was N 455 Squadron, and when it departed in May 1945 the station was almost silent.

However, unlike many RAF stations Thornaby survived and after a short period of inactivity N 608 Squadron returned, initially with Spitfires but eventually acquiring Mosquitoes and finally a fleet o Vampire jets. No 275 Squadron, equipped with Sycamore helicopters, moved to Thornaby fron Linton-on-Ouse in November 1954, and stayed until October 1957, when it moved to Leconfield. Th end of the Royal Auxiliary Air Force in 1957 marked the end of 608 Squadron's presence, and by Apri its Vampires had gone. Five months later the Hunters of No 92 Squadron arrived, but after a stay of jus a year they departed and Thornaby was closed as a Royal Air Force station in October 1958.

The airfield laid unused for many years but the gradual expansion of Middlesbrough's industrial, commercial and domestic development eventually saw the airfield disappear under houses, factories and shops. It is quite remarkable to see how this large and well-equipped station has now been completely obliterated – not a hint remains of the RAF's presence here. It is only the local street names and a stone monument that remind visitors of the area's long association with military aviation.

Main features:
Runways: 227° 5,940 x 150 feet, 178° 4,200 x 150 feet, 287° 4,260 x 150 feet, concrete surface. *Hangars:* three Blister, two C-Type, two aeroplane sheds, one Bellman. *Dispersals:* thirty-six finger type. *Accommodation:* RAF: Officers 130, SNCOs 350, ORs 1,153; WAAF: 359.

THORNE, Yorkshire

53°37'28"N/0°55'55"W; SE707147; 3ft asl. 7 miles SW of Goole, E of M18, SE of Moorends

This relatively unknown site was acquired for the Royal Flying Corps early in 1916. One of many such sites around the region, it was little more than an allocated area of farmland, cleared of obstructions and secured for occasional use by fighter aircraft tasked with the air defence of the local region. No 33 Squadron used the site on a sporadic basis while maintaining patrols. However, when the threat of German airship attacks diminished these sites were no longer needed and, like many others, it was abandoned in 1917. Although it was examined for reuse as the basis for a much larger bomber airfield in the 1930s it was not selected, and never returned to military service. As no permanent structures had ever been built here, there are no surviving traces of the RFC's brief presence here so long ago.

Main features:
Runways: none, grass field. *Hangars:* none. *Dispersals:* none. *Accommodation:* not known.

TOPCLIFFE, Yorkshire

54°12'19"N/01°22'51"W; SE404790; 78.5ft asl. 4 miles SW of Thirsk on A167, W of A168, N of Topcliffe village

One of many sites in the Vale of York developed for military use, the airfield at Topcliffe emerged in 1938 and opened in September 1939 as part of No 4 Group. Located in an area that was remarkably close to two other major airfields (Dalton and Linton-on-Ouse), the site was geographically sound, but, like the other airfields here, it suffered from a risk of waterlogging, which required significant drainage work, and an ever-present risk of mist and fog. Although there were no major restrictions on the airfield's overall size and layout, the completed grass landing field did feature some unusual arrangements, particularly the five C-Type hangars; instead of a traditional semi-circle, these were laid out in two pairs at an intersecting angle to a central single hangar. Why they were built in this fashion remains unclear.

No 77 Squadron brought its Whitley bombers here on 5 October, joined a few days later by No 102 Squadron's aircraft, both units soon being declared operational; they remained active here until November 1941, when they moved out so that the airfield could be improved. Three concrete runways

An RAF reconnaissance photograph of Topcliffe taken shortly after the airfield's completion.

This oblique view of Topcliffe in 2006 was taken from one of the resident Tucano trainers. Two aircraft can be seen in flight towards the bottom of the picture and four aircraft are also parked on the airfield in front of the surviving wartime hangars.

were duly laid in a standard triangular pattern (the main runway being subsequently extended stil further) and a perimeter track appeared, along which a series of circular dispersals were constructed.

By the summer of 1942 the airfield was ready to accept aircraft and No 102 Squadron returned or 7 June, now equipped with Halifaxes. The unit stayed for only a few weeks, then moved to Pocklington after which the Canadian No 405 'Vancouver' Squadron came to Topcliffe from Beaulieu with its flee of newly acquired Wellingtons. Also in August No 419 'Moose' Squadron reformed here with Wellingtons, but it stayed for only a few weeks before shifting to Croft, to convert onto Halifaxes. As a replacement, No 424 'Tiger' Squadron arrived on 15 October with its fleet of Wellingtons. By January this unit was operational and the station was transferred to No 6 (RCAF) Group.

In 1943 Topcliffe became the parent station for No 61 (RCAF) Base, with Dalton, Dishforth and Wombleton under its control, but in November 1944 it became No 76 (RCAF) Base, losing control of Dalton in the process. Other units made short stays here, including Nos 1516 and 151? Flights (both flying Oxfords), but by the summer of 1943 the squadrons had gone, and Topcliffe was assigned to training duties, No 1659 Heavy Conversion Unit having moved here from Leeming in March. Tasked with the conversion of Canadian pilots onto the Halifax, the unit eventually transferred onto Lancasters and remained in business here until September 1945, accounting for the training of hundreds of aircrew.

When the Canadians left, Topcliffe returned to the control of No 4 Group and, as part o Transport Command, the first post-war unit to arrive here was in fact a training unit, in the shape o No 5 Air Navigation School, which transferred its Wellingtons, Oxfords and Ansons from Jurby in September 1946. It became No 1 Air Navigation School the following April and trained a significant number of navigators here until it left for Hullavington in July 1949.

Lockheed Neptune maritime patrol aircraft were a familiar sight at Topcliffe until they were ultimately replaced by Shackletons, based in Scotland and Cornwall.

Topcliffe then became home to four squadrons operating the Hastings transport – a direct derivative of the wartime Halifax. Nos 24, 47, 53 and 297 Squadrons moved here from February 1951 and the base became busy with the sight and sound of the big, heavy transports going about their business transporting troops and equipment around the UK, Europe and beyond. The Hastings transports were joined by a more unusual aircraft from 1952 when Coastal Command assigned a fleet of maritime patrol aircraft to the base. Pending the arrival of the purpose-built Shackleton, the Lockheed Neptune was leased from the US and three squadrons (Nos 36, 203 and 210) were formed at Leeming to operate the type. They stayed until 1957, by which stage No 1453 Flight had been formed to undertake airborne early warning trials on the type.

After their departure (and the loss of the Hastings squadrons), Topcliffe returned to training and in March 1957 No 1 Air Navigation School returned, now equipped with Marathons and Varsities. The unit stayed until December 1961, after which the Air Electronics School arrived from Hullavington, eventually becoming the Air Electronics & Air Engineers School, equipped with a fleet of Varsities and some Argosies. The Northern Communications Squadron came to Topcliffe in 1964 and stayed for some five years, initially equipped with Devons and Pembrokes but acquiring Bassets in 1965. The resident units slowly disappeared around 1970, and by 1972 the station was largely redundant. Topcliffe was then handed to the Army, becoming Allanbrooke Barracks in the process.

However, the airfield was retained in good condition and, although used only for Army communications flights and some helicopter training, the RAF restored a presence here, the Joint Elementary Flying Training School appearing in 1993 with its fleet of Fireflies. The unit moved to Barkston Heath in 1995 and the Central Flying School's Tucano Squadron transferred here from Scampton, joined by a Tucano Low Level Training element from Cranwell. These units then became part of No 1 FTS (based at nearby Linton), and Topcliffe acted as both a base for these units and as a Relief Landing Ground for No 1 FTS as a whole. The airfield is now still part of the Army's Allanbrooke Barracks, but RAF Topcliffe is still here, and Tucano trainers buzz around the airfield circuit on a daily basis. Two of the three runways are still in use although (rather oddly) much of the perimeter track is separated from the runways by a fence.

A freshly painted Lockheed Neptune pictured shortly after delivery from the United States.

A Neptune chugs along Topcliffe's perimeter track, ready for a maritime patrol mission over the North Sea.

Topcliffe in 2011, still active and largely unchanged since the Second World War.

Main features:
Runways: 314° 6,000 x 150 feet, 266° 4,200 x 150 feet, 318° 4,200 x 150 feet, concrete surface. *Hangars:* five C-Type. *Dispersals:* thirty-six Heavy Bomber. *Accommodation:* RAF: Officers 242, SNCOs 537, ORs 1,260; WAAF: 420.

USWORTH, Tyne & Wear

54°54'50"N/01°28'17"W; NZ339578; 118ft asl. 1 mile E of Washington on A1231 at junction with A19

This site first became associated with military aviation in 1916 when the Royal Flying Corps established a simple landing field here for use by No 36 Squadron, the unit responsible for air defence of the local area. At this stage the airfield was referred to as Hylton, and less commonly as West Town Moor, this being the name of the land on which the landing field was built. No 36 Squadron eventually moved its Headquarters here, operating detachments at Ashington and Seaton Carew. The unit's BE series aircraft were supplemented by Sopwith Pups and eventually Bristol Fighters, and these aircraft were seen around the region on a daily basis, patrolling the air space for any signs of marauding German airships. However, by 1919 the newly formed RAF was rapidly contracting and, like many other former RFC units, No 36 Squadron disbanded and the airfield was closed down.

It was a decade later that the RAF returned when No 607 Squadron Royal Auxiliary Air Force came here on 17 March 1930, although it was 1932 before the unit's aircraft arrived in the shape of Westland Wapitis, followed by Hawker Demons and Gloster Gladiators. They stayed until October 1939, when the unit moved to Acklington, although brief return visits were made for two periods in 1940, by which time the Squadron had re-equipped with Hurricanes. In February 1937 No 103 Squadron moved to Usworth with a fleet of Hawker Hinds, re-equipping with Battles in 1938 before moving to Abingdon.

Following the outbreak of the Second World War, RAF Usworth was rebuilt, two concrete runways being laid here, although the constraints of roads and railway lines restricted their lengths, and much of the station's domestic, admin and technical site straddled what is now the A1290. Fighter dispersals were also built, and both Nos 43 and 64 Squadrons were stationed here in 1940 with Hurricanes and Spitfires, flying air defence sorties around the region. The airfield's significance was not overlooked by the Luftwaffe, and on 15 August 1940 an attempt was made to bomb the site, but the RAF's fighters (flown from Usworth, Catterick and Acklington) repelled the attack.

Usworth airfield is seen shortly after completion, with evidence of construction work still visible.

The 1934 Empire Air Day at Usworth, with Wapitis and Harts clearly visible.

Bristol Fighter J6612 pictured at Usworth.

Usworth shifted to training activities in 1941 and No 55 Operational Training Unit arrived from Aston Down in February, equipped with more than a hundred aircraft, most of which were Hurricanes. Such was the unit's activity that both Ouston and Woolsington were used as satellite fields, but after only a year the unit moved to Annan, and in June 1942 No 62 Operational Training Unit arrived with a fleet of Ansons equipped for air radar training. However, the presence of large barrage balloon defences around Sunderland quickly encouraged the unit to shift most of its training to Ouston, and in July 1943 the whole unit moved there. By this stage the short-lived presence of No 416 NFS USAAF had also ended, its Beaufighter detachment lasting for just a few weeks from May 1943. With these units gone, the airfield remained largely silent, save for some gliding activities performed by No 31 Gliding School.

After the war No 23 Reserve Flying School came to Usworth in February 1949 and stayed for some four years. No 2 Basic Air Navigation School flew Ansons here for two years from April 1951, and No 664 Squadron was present from February 1954 until March 1957 with its Austers. The last regular users of RAF Usworth were the Chipmunks of the Durham University Air Squadron; arriving in May 1949, they stayed here until October 1957. However, by 1960 the station was effectively redundant and the RAF vacated the site.

It was purchased by Sunderland Corporation in 1962 and reopened as a regional airport, a wide variety of civil aircraft using the site, ranging from small private aircraft to much larger Dakotas on scheduled services. Sadly, like many such airfields, it was not sufficiently popular to cover its operating costs and by 1984 it had been written off as a commercial operation. The airfield closed on 31 May of that year, and although profitability was cited as the reason for the airfield's abandonment it was almost certainly the promise of a large car manufacturing facility that finally secured the airfield's fate. Nissan duly built its huge factory here and destroyed all of the former airfield site in the process.

Today only a small trace of one runway survives adjacent to Nissan Way, just east of a small roundabout. Thankfully, the North East Aviation Museum survived the airport's closure and its collection of aircraft were moved to a new site just to the north of the car factory. Among the exhibits is a Vulcan bomber, which must have been one of the biggest and heaviest aircraft to have ever landed here.

Main features:
Runways: 236° 2,400 x 150 feet, 192° 2,400 x 150 feet, concrete and tarmac surface. *Hangars:* one Lamella, one Calendar, three Blister. *Dispersals:* sixteen square pen, fourteen circular. *Accommodation:* RAF: Officers 39, SNCOs 44, ORs 940; WAAF: 115.

WALTHAM/GRIMSBY, Lincolnshire

53°30'06"N/0°04'25"W; TA278022; 62.5ft asl. S of Grimsby on A19, S of Holton-le-Clay

During the early 1930s a private flying club was established on meadowland 3 miles from the town in the parish of Waltham. The Lincolnshire Aero Club became a popular attraction here, with a clubhouse and two wooden hangars, joined by a larger, more substantial hangar built in 1937. In 1938 Waltham aerodrome was selected for the establishment of one of the Elementary & Reserve Flying Training Schools set up by the Air Ministry and run under civilian contracts to provide instruction for would-be RAF pilots. A variety of training aircraft types were used by the school during 14 months of activity, but the principal type was the Tiger Moth.

A No 100 Squadron Lancaster at Waltham in 1945.

Apart from an occasional visitor, the airfield was largely devoid of flying tenants from September 1939, and late the following year it was temporarily closed, the site having been surveyed and found suitable for development. However, work began in the winter of 1940/41 to extend the flying field into the parish of Holton-le-Clay, taking in part of the A16 Louth to Grimsby road on the north-east side. The three runways were 18-36 at 1,200 yards, 06-24 at 1,400 yards, and 12-30 at 1,100 yards. Pan hardstandings, thirty-six in number, were built off the encircling perimeter track. Two T2 hangars and one B1 were eventually provided.

Initially opened as a satellite airfield for Binbrook in November 1941, the new station was officially named Grimsby, although the local name Waltham persisted among locals and servicemen on the station. This may have led to some confusion elsewhere, as there was already a White Waltham airfield near Maidenhead. No 142 Squadron's Wellingtons arrived from Binbrook in November 1941 and carried out bombing operations from Grimsby until December 1942. With an urgent need for more night-bombers to support the 'Torch' invasion, No 142 was split, half going to North Africa and the remainder to Kirmington, where it was used as the basis to form another squadron in the New Year. With the break-up of No 142 Squadron, No 1 Group used Grimsby to add a new Lancaster unit to its strength, and No 100 Squadron (reformed in mid-December) commenced operations on the night of 8/9 March 1943.

From early 1942, Gee, Walker & Slater Ltd had been involved in extending runways 18-36 to 2,000 yards and 12-30 to 1,400 yards across the A16. During this work some nineteen hardstandings were lost and replaced with loops. Additional domestic sites gave accommodation for a maximum of 2,203 males and 254 females. In November 1943 No 100 Squadron's C Flight became No 550 Squadron, and by mid-January its growth brought a move to North Killingholme.

However, No 100 made Grimsby its home until 2 April 1945 when, owing to deterioration of the runways, a move was made to Elsham Wolds. This marked the end of Bomber Command flying units at the station. Operations from Grimsby cost 164 bombers missing in action or crashing in the UK, forty-eight being Wellingtons and 116 Lancasters.

In the immediate post-war years the hangars were used by No 35 Maintenance Unit for storage, and the flying field reverted to agricultural use. In later years improvements to the A16 – with a bypass for Holton-le-Clay – reclaimed part of the eastern side of the field, where a memorial to the men of No 100 Squadron can be seen. Today, a great deal of the airfield is gone, although significant portions of the runways have survived and the old B1 hangar still stands next to the crumbling paving of the old main runway extension.

Main features:
Runways: 181° 6,000 x 150 feet, 241° 4,200 x 150 feet, 301° 3,300 x 150 feet, concrete and tarmac surface. *Hangars:* two T2, one B1. *Dispersals:* thirty-six Heavy Bomber. *Accommodation:* RAF: Officers 110, SNCOs 319, ORs 967; WAAF: 286.

WARTON, Lancashire

53°44'34"N/02°53'27"W; SD413277; 32.5ft asl. 7 miles W of Preston on A584, S of Freckleton

The long-established airfield at Warton has enjoyed two markedly different periods of operation. Its first began in the late 1930s when an area of land was acquired here for development into a satellite airfield for nearby Squires Gate. The flat land of the Ribble Estuary was certainly suited to aviation, but it was also prone to flooding, and seabirds were also an ever-present danger. Despite this, work continued slowly, then during construction it was decided to hand the airfield to the United States Army Air Force as a suitable site for a new Base Air Depot. This plan duly went ahead and although the basic airfield layout was completed as per the original RAF plans (with three runways in a standard triangular layout, together with dispersals and perimeter track), the USAAF required large workshop sheds and these were constructed on the northern side of the airfield.

Warton airfield seen from the east, illustrating the site's proximity to the Ribble estuary.

The USAAF accepted the site as Station 582 on 17 July 1943, although it had been present since the previous January. Tasked with the repair and overhaul of aircraft and equipment, Warton gradually built up to a frenzy of activity, with more than 10,000 aircraft being handled here, almost half of which were P-51 Mustangs. Thankfully the threat from Luftwaffe bombers had largely gone by mid-1943, as Warton would undoubtedly have been one of the most lucrative targets with huge numbers of bomber, fighter and specialised aircraft parked all around the airfield on dispersals and on inactive runways. However, having being such an incredibly busy and important site, by 1945 it was quickly becoming redundant and on 19 November it was handed back to the RAF, becoming a sub-site for No 90 Maintenance Unit. The USAF opted to concentrate its support facilities at Burtonwood, and with the Americans gone the airfield was suddenly silent.

It was perhaps fortuitous for Warton's future that the English Electric company had its headquarters in nearby Preston, and was actively searching for a suitable airfield to supplement activities already based at Samlesbury, from where they could begin flight-test operations on various aircraft, particularly its emerging A1 design. In 1947 English Electric acquired the airfield from the RAF and immediately set up the beginnings of what became a large and famous facility.

Warton's control tower captured on film on 26 October 1943.

Above: *A rare image of P-47 Thunderbolts at Warton on 10 March 1944.*

Left: *The prototype A1 bomber pictured at Warton shortly before its maiden flight in May 1949.*

Lightning interceptors nearing completion at Warton.

Above and right: *The legendary TSR2 XR219 arrives at Warton during 1965, greeted by countless members of the English Electric team who were responsible for much of the aircraft's design.*

A trio of English Electric's iconic designs – the TSR2, Lightning and Canberra – outside the famous red-painted flight sheds at Warton. Author/Phil Jarrett

The bomber design duly made its first flight here on 13 May 1949 and became the world-famous Canberra. It was followed by the equally legendary P1, which matured into the Lightning interceptor. Not surprisingly the main runway was extended in order to accommodate the Lightning's performance, and other developments also took place, not least the creation of new manufacturing facilities. But despite this the airfield remained largely unchanged and today is much the same as when it first opened to flying. The magnificent TSR2 was also to have been based at Warton for test flying, but sadly only one aircraft (XR219) flew before the project was cancelled. By this stage the aircraft had flown (supersonically) to Warton from Boscombe Down, but after a few test flights the programme was cancelled, and the aircraft left Warton by road.

Far happier times followed, however, and the Jaguar, Jet Provost/Strikemaster, Tornado and Typhoon all emerged from Warton's factory, joined more recently by the Hawk and the short-lived Nimrod MRA4, which was to have been test flown here. For the future, the Typhoon will undoubtedly keep Warton's facilities busy, and with growing interest in Unmanned Aerial Vehicles it may well be that some particularly unusual flying machines take to the air here in the future.

Eurofighter Typhoon production at Warton.

Main features:
Runways: 080° 5,630 x 150 feet, 200° 4,180 x 150 feet, 330° 3,960 x 150 feet, concrete and tarmac surface. *Hangars:* one storage type, one repair shop type. *Dispersals:* thirty-six loop type. *Accommodation:* Officers 660, ORs 15,242.

WATH HEAD, Cumbria

54°49'18"N/03°05'18"W; NY297479; 233ft asl. 7 miles SW of Carlisle on A595 at Spain Wood

The RAF adopted this site in 1941 as a dispersal sub-site for No 12 Maintenance Unit based at Kirkbride. Known locally as Jenkins Cross, it was hardly an airfield at all, being simply an allocated stretch of open land cleared of obstructions and secured by RAF personnel. A grass strip (approximately heading 06/24) was maintained in good condition, enabling aircraft to be flown in and out of the site as required, most being accommodated in cleared areas within the surrounding woodlands, ensuring that the site was almost impossible to spot from the air without a suitable map location. Few permanent structures appeared here, and only a few huts were erected, the common appearance of Robin hangars not taking place as there was sufficient tree cover to house all of the unit's aircraft within the adjacent woods.

Various types were seen here, including Hampdens, Bostons, Halifaxes, Wellingtons and Oxfords. Designated as No 10 Satellite Landing Ground, it remained in use by No 12 MU until June 1944, when No 18 Maintenance Unit took over (headquartered at Dumfries) and, although activity remained much the same, more time was allocated to the acceptance of Wellingtons, which were brought here on their final flights for disposal. This activity continued until December 1945, when the RAF finally abandoned the site; its huts and equipment had soon disappeared, leaving the area to return to its former tranquil state. There seems to be no remaining trace of the site's association with aviation, although a careful search of the woodland may well reveal one or two hints of the RAF's presence here.

Main features:
Runways: none, grass surface. *Hangars:* none. *Dispersals:* none. *Accommodation:* not known.

WEST AYTON, Yorkshire

54°13'34"N/0°29'34"W; SE989822; 78.5ft asl. 6 miles S of York, W of B1122

This small grass landing field is believed to have been built on what was Scarborough's horse-racing course, although maps of the landing field do not indicate any presence of a racecourse, and its location is quite some way from Racecourse Road. The Royal Naval Air Service came here early in 1916 equipped with BE.2C fighters, and established a base for air defence operations around the region. Initially the airfield had been set up to counter the very real risk of attack from German airships, but when this risk diminished operations shifted to maritime patrols, and the endless search for German submarines out in the North Sea. No 251 Squadron established a Flight here for this purpose, designated as No 510 (Special Duty) Flight, flying DH.6s, but when the unit moved here from Redcar late in 1918 the First World War was almost over, and by March 1919 the Flight had disbanded and the airfield was abandoned.

Today there are no traces of any aviation activity here; indeed, much of the land has been flooded, presumably as part of the agricultural activity that dominates the area.

Main features:
Runways: none, grass field. *Hangars:* none. *Dispersals:* none. *Accommodation:* not known.

WEST COMMON, Yorkshire

53°14'18"N/0°33'39"W; SK961722; 32.5ft asl. W of Lincoln, S of A46 on A57, N of Hewson Road

Handley Page 0/400 serial D4562 at West Common.

Another 0/400, serial 8344, pictured at West Common.

Visitors to the city of Lincoln will be familiar with the old racecourse to the west of the city, and the grandstand that sits across the A57 road. Like many such courses, it was adopted for aviation use during the First World War, and the centre of the field was cleared of obstructions for use by both Royal Flying Corps aircraft and other machines being flight tested by a variety of aircraft companies situated in the local area. In 1915 it was designated as No 4 Aircraft Acceptance Park and a number of racecourse buildings were requisitioned for use, while two large flight sheds were erected on the airfield. Various aircraft types were seen here, including many fighter types and larger aircraft such as the HP.0/400. Further development took place over the years, but by 1918 a significant amount of activity had shifted to Bracebridge Heath just a few miles to the south-east. However, the airfield remained in use until 1919, at which time it was abandoned and horse-racing returned to the site.

Today the area shows no signs of its links with aviation, and golfing now takes place on this once flat and cleared area. Aviation is still very much in the vicinity, however, and with Scampton just to the north the world-famous Red Arrows are often seen nearby while performing their daily practice routines.

Main features:
Runways: none, grass field. *Hangars:* two flight sheds. *Dispersals:* none. *Accommodation:* not known.

WEST HARTLEPOOL, County Durham

54°39'01"N/01°12'40"W; NZ509286; 52.5ft asl. S of Hartlepool, S of A689, W of B1277

Although referred to as West Hartlepool, this landing field was created at Greatham on the southern outskirts of Hartlepool. Established prior to the Second World War, it was designated as a satellite airfield for Thornaby in 1940, although there is little evidence to suggest that it was ever used by that unit. The first recorded aircraft here were Spitfires from Nos 243 and 403 Squadrons, which were based here in the summer of 1942, tasked with air defence of the local region, particularly the protection of the Hartlepool, Middlesbrough and Newcastle area. Other units were also based here on a temporary basis, but there was no significant presence here, as there were no permanent facilities other than a watch office. No 1613 (Anti-Aircraft Cooperation) Flight

A 1950s aerial view of West Hartlepool.

formed here on 1 November 1942, operating Henleys in support of No 15 LAAPC at nearby Whitby, but the unit moved to Driffield just four months later.

After the unit's departure the airfield was virtually abandoned, although it remained under the control of Thornaby and was not finally abandoned until 1945. With no permanent structures to remove other than the rudimentary watch office, the site was quickly restored to agricultural use, but as the expansion of Hartlepool continued the site gradually disappeared under industrial development. The field is still recognisable even today, but a large warehouse complex sits on the site and various access roads and a parking area are scattered across the field where Spitfires once raced skywards.

Main features:
Runways: none, grass field. *Hangars:* none. *Dispersals:* none. *Accommodation:* RAF: Officers 36, SNCOs 182, ORs 446; WAAF: 0.

WICKENBY, Lincolnshire

53°18'51"N/0°20'57"W; TF100809; 85.5ft asl. 8 miles NE of Lincoln, S of A46, W of B1202

An aerial photograph of Wickenby in September 1942.

Crews prepare a Lancaster at Wickenby for a night bombing mission.

Land was first acquired at this site in 1941 and construction of an airfield here began later that year. Designed as a satellite for Market Stainton, the airfield was completed to a mid-war standard with a typical layout comprising three runways in a triangular pattern, flanked by a perimeter track, thirty-six dispersals, and three hangars. Other airfield buildings were gradually added, although few were of any substance and most were little more than wooden huts; despite this, a surprising number have survived to this day.

This Class A bomber airfield was begun by McAlpine in late 1941. The site took in land in the parishes of Snelland and Holton, necessitating the closure of the road between the two villages. The concrete runways were 09-27 at 2,000 yards, and 04-22 and 16-34, both at 1,400 yards long. Thirty-six pan-type hardstandings were provided, one neutralised by a T2 hangar adjacent to the technical site on the north-east side, and a lone B1 at the opposite extremity of the flying field near the end of the south-western runway. Domestic sites, mostly Nissen huts, were dispersed in fields on the eastern side of the B1399. Available accommodation was put at 1,788 male and 287 female. Additional construction work on the airfield was carried out by Laing in early 1943.

It was September 1942 when the first aircraft arrived here, these being the Wellingtons of No 12 Squadron, and although the unit swiftly resumed operations after transferring from Binbrook the Wellingtons were replaced by Lancasters after only a few weeks. The Squadron stayed here through the rest of the Second World War, and completed some 3,882 Lancaster sorties for the loss of 111 aircraft – the second highest loss rate in Bomber Command.

During November 1943 C Flight of No 12 Squadron became No 626 Squadron, also equipped with Lancasters; this unit also stayed at Wickenby, eventually completing 2,728 sorties for the loss of forty-nine aircraft. Bob Bennet was a pilot with the unit while at Wickenby and his records include accounts of the Squadron's operational sorties, including this excerpt from May 1944:

> '27 May 1944: Op 2: Lancaster Y2: F/O Bennet/Crew
>
> Pilot's comments: Aachen. Combat with Ju 88 – Awarded DFC (Distinguished Flying Cross) as a result of this trip. (Ju 88 shot down having attacked it while it was attacking another Lancaster that was damaged.) Several close calls but all returned in our crew unwounded.
>
> Navigator's comments: Aachen, Germany: Railway Marshalling Yards; Bomb Load 13,000lb HE. Time airborne 4 hours 55 mins. I recollect that our return flight, we saw fires burning on the ground, south of Brussels, at Bourge Leopold, a large German Infantry Barracks, which had been attacked by other aircraft of Bomber Command. Near Dunkirk, as we crossed the enemy coastline, we were attacked by a night-fighter. I was standing in my astro-dome at the time and I clearly observed heavy machine gun and cannon fire storming towards us. In my mind's eye I still see the trace as being red, blue, mauve and green. Bob started to "corkscrew" violently, and the fighter swerved away to port. Our Mid-Upper and Rear Gunners had opened fire and I saw fire being returned by another Lanc above us. I was hanging on in the astro-dome when he came in for a second attack and I recollect seeing the fire from our turrets clearly hitting the fighter, which I thought was either an Me 210 or Me 410. Our corkscrew gyrations were so violent at this time that I was thrown to the floor and crawled back into my seat. I heard the gunners shout over the intercom that they had hit the fighter, which was not seen again, and which they claimed as a "probable". We had no damage to our aircraft (which was UM-Y2 as we had not taken delivery of UM-R2).
>
> Mid-Upper Gunner's comments: Aachen, Germany. This was our first operational sortie. Bombed at 14,000 feet. Carried 14,000lb of HE. Objective was a marshalling yard and what I saw of the target it was well and truly plastered. The flak in this area was really intense; half the time I couldn't see because of the numerous flashes. About sixty miles from the target we were attacked twice by a Ju 88. Although he didn't get us the experience sort of shook the crew. Both Gerry and I fired about 150 rounds apiece at him. We were credited with a probable in this engagement. Crew in our hut was shot down. We lost two on this raid.
>
> Bomb Aimer's recollections: We were attacked by an enemy fighter from behind, and as I was up in the nose I was helpless and could only watch, as great streams of green (German, ours were red) tracer bullets passed under our wings. It seemed to last an eternity but could only have been seconds really before Bob threw the plane into a violent dive and by means of weaving sharply from side to side managed to shake the fighter off. (Big sighs of relief all round!) A huge photograph of the raid was stuck up on the wall of the mess but disappeared soon. I think our gunners snaffled it.
>
> Incidental: The railway lines at the yards, which were not seriously hit in the raid of two nights earlier, were now seriously damaged and all through traffic was halted. A large proportion of delayed-action bombs were dropped. The bombing also hit the nearby Aachen suburb of Forst, which, in the words of our local expert Hubert Beckers, "was razed to the ground". The local hospital, an army barracks, an army stores office, 2 police posts

and 21 industrial buildings were hit, as well as 603 houses. 167 people were killed and 164 injured. The local people were impressed that the whole raid only lasted 12 minutes.'

Further recollections come from pilot Les Landells, who recalls the raid on Essen, flown on 23 October 1944:

'A trip to Essen was one to remember, because of a problem on take-off, and one which we were lucky to survive. We taxied out of our dispersal point and down the perimeter track to use the full length of the 2,000-yard main runway, then onto the grass to gain extra yards. The other aircraft started their run from where the two runways met, which could save 200-300 yards on take-off. Les revved up the engine with the brakes on, so that the plane was more or less jumping up and down on the spot. We received the green Aldis Lamp signal from the caravan parked at the side of the runway. There was radio silence before a raid, as the RAF did not want to give the Germans an early warning. We were now charging down the runway to build up our speed to 115 knots for take-off with our gross weight of 68,400lb. As we got past the point of no return, Les called out that he had a speed problem for take-off. At the end of the runway all Les could do was to pull back on the stick; our speed was about 95 knots, and we had no alternative. The plane lifted off the ground, but was flying like a brick. We were now over the farm fields, in the dark of night, when we hit the ground again, then skipped back into the air. We had six touch-downs before the plane stayed in the air. By the time we had reached Lincoln, which was 8 miles away, we had then reached the magnificent height of 500 feet. We then set course for Essen, which we reached and bombed successfully. The return trip was uneventful and we had a safe return.

There was a sequel to this trip, which was not exactly a morale-booster for the crew. Next day George, Fred and myself were at our dispersal point discussing the future of UM-M2, the Lancaster, which had nearly cost us our lives. Our plan was to get Les (Pilot) to set course on automatic pilot; we would all bail out over England and let the plane fly on its own and crash in Germany. We did not let our pilot know our plan, and our next trip on UM-M2 solved that problem. I have in more recent days found the cause of the take-off problem. Before UM-M2 went to Essen it was bombed and fuelled up for a trip to East Germany. The long trip meant full fuel tanks and fewer bombs to get maximum weight. When the trip was changed to West Germany it meant more bombs and less fuel (500 gallons). The bomb load was topped up and there were 500 gallons too much left in the tanks. We then had 4,500lb above the maximum weight and this caused the lift problem. Our gross weight was 4,500lb above the maximum and about 20,000lb above the normal maximum landing weight.

Also there could have been a problem with the undercarriage collapsing. If the undercarriage had collapsed on one of our take-off bounces, the plane would probably have blown up. I do not know it there was an official follow-up on the failure to take the 500 gallons out of the tanks. I have spoken to Les about the full use of the 2,000-yard runway. He said he knew that the plane had a take-off problem with the normal 68,400lb load, but was most certainly not aware of the extra 500 gallons of fuel, which had not been removed from the fuel tanks when the additional bombs were loaded.'

Even more Lancasters frequented the airfield after Wickenby became a satellite for No 14 Base at Ludford Magna. Both Lancaster squadrons disbanded in 1945, No 12 Squadron in September and No 626 in November. The only other unit to operate from Wickenby was No 109 Squadron, which came here with Mosquitoes in October 1945, staying only for a few weeks.

After the war the airfield was much less active and, with the operational squadrons gone, the station was adopted by No 61 Maintenance Unit based at Handforth. No 233 MU assumed control in September 1948 and a month later Wickenby became the Headquarters site for No 93 Maintenance Unit. However, few if any aircraft used the airfield, and No 93 MU was primarily associated with ammunition storage. With sub-stations at Bottesford and Fulbeck, the Headquarters stayed here until October 1951, when it moved to Newton, although the MU retained Wickenby until April 1956 when the RAF vacated the site.

After a long period of abandonment the airfield slowly crumbled and returned to agriculture, but private flying eventually appeared here and continues to this day. Much of the airfield is deserted and overgrown, with most of the perimeter track and dispersals having long gone. The Snelland road off the B1399 now cuts directly across the airfield, but to the north of the road the northern portions of two runways are still in use. Two hangars have survived together with a few other buildings, including the control tower. Sadly, only Cessnas and similar light aircraft now bumble around Wickenby's circuit, the roar of Merlin engines being heard only occasionally when BBMF aircraft can be heard in the distance, flying from Coningsby.

Main features:
Runways: 273° 6,000 x 150 feet, 217° 4,200 x 150 feet, 162° 4,200 x 150 feet, concrete surface with wood chippings. *Hangars:* two T2, one B1. *Dispersals:* thirty-six Heavy Bomber. *Accommodation:* RAF: Officers 131, SNCOs 412, ORs 1,245; WAAF: 287.

WINDERMERE, Cumbria

54°23'45"N/02°56'07"W; NY393004; 164ft asl. 1 mile N of Windermere, W of A591 at White Cross Bay

During the Second World War the risk of attack from German bombers led to many military units and production facilities being shifted as far away from the south-east of the country as possible. The well-known aircraft manufacturer Shorts was mindful of the risks associated with its facility at Rochester in Kent and a new site for seaplane production was eagerly sought. Windermere was studied and tested as a potential seaplane base in 1940, but it was not until 1941 that construction of a factory site began here. Two large factory sheds emerged, together with a huge hangar that was at the time the largest single-span construction in the UK, capable of accommodating three complete Sunderland flying boats. The first Sunderland to be completed here rolled out of the factory and down the purpose-built slipway in September 1942; thirty-five Sunderlands were eventually manufactured here.

Left: *Short Sunderland flying boat production at Windermere.*

Below: *Sunderland wing and engine cowling structures on the production line at Windermere.*

In 1944 the factory became a Civilian Repair Organisation site, and in addition to the refurbishment of a further twenty Sunderlands other aircraft were repaired and modified here. When the war ended the factory was soon redundant, and Shorts abandoned the site in 1949. Most of the buildings had gone by 1951, although the huge hangar reappeared in Liverpool, housing a chemical factory.

A Sunderland flying boat emerges from the factory at Windermere

Today the White Cross Holiday Park occupies the old factory site and caravans and holiday chalets stand on the concrete where the Sunderland hangar was once erected. The slipway is still here, but only boats sit at anchor where the mighty Sunderlands were seen so many years ago.

Main features:
Runways: none, factory site. *Hangars:* one factory flight shed. *Dispersals:* none. *Accommodation:* not known.

WOMBLETON, Yorkshire

54°13'59"N/0°58'13"W; SE672824; 118ft asl. 8 miles W of Pickering, S of A170, S of Wombleton village

Aviation first came to this site in the 1930s, when a small private airfield emerged on an area of land known locally as Welburn Hall. When the Second World War began, the site was identified as being suitable for development into a bomber airfield and construction of what became RAF Wombleton commenced in 1942. Creating an airfield on an existing site was relatively easy, although a local road out of Wombleton had to be closed and some farm buildings were demolished, but eventually a perfectly proportioned bomber airfield was completed. It had three full-length runways in a symmetrical triangular layout, surrounded by a perimeter track leading to the usual thirty-six dispersals and three hangars.

An RAF reconnaissance photo of Wombleton taken shortly after the airfield's completion.

The airfield opened for flying in October 1943 as part of No 6 Group RCAF as a sub-station for No 61 Base at Topcliffe. By November the first aircraft had arrived and No 1666 Heavy Conversion Unit was established here after having moved from Dalton. This unit's Halifaxes were joined a month later by the Lancasters of No 1679 Heavy Conversion Unit, and in January 1944 the two units were combined to become No 1666 HCU, the Lancasters being sent to other units so that No 6 Group could concentrate on Halifax operations. In June 1944 the name 'Mohawk' was bestowed on the unit and intensive Halifax training activities continued here until January 1945, when the first Lancasters began to replace the Halifaxes. However, the unit's training activities were already starting to wind down; by the summer the HCU was effectively redundant and in August it disbanded.

A Lancaster bomber at rest between missions at Wombleton. Of interest are the unusually painted spinners.

A 1990 aerial view of Wombleton, illustrating how much of the airfield has survived, but with a public road now cutting across the site.

The RAF retained the station for a variety of ground units well into the 1950s, but the airfield was no longer used after the HCU left. Hungerhill Lane – the road that was closed when the airfield was built – is now back in its original location, running north to south across the site. To the east almost all traces of the airfield have been obliterated, with trees occupying what was once the main runway. But to the west of the lane the airfield is virtually intact, although it is slowly crumbling away. Looking along the grass-lined paved slabs along the old main runway it is still possible to picture the countless Halifaxes and Lancasters that lumbered into the air here so many decades ago.

Main features:
Runways: 104° 6,000 x 150 feet, 046° 4,200 x 150 feet, 167° 4,200 x 150 feet, concrete surface. *Hangars:* two T2, one B1. *Dispersals:* thirty-six Heavy Bomber. *Accommodation:* RAF: Officers 285, SNCOs 466, ORs 910; WAAF: 204.

WOMBWELL, Yorkshire

53°31'23"N/01°22'58"W; SE410031; 95ft asl. 1 mile SE of Barnsley, N of A633, N of Everill Gate Lane

This austere site was one of many such landing fields created for the Royal Flying Corps in 1916. Although used as a landing field, it was in effect simply an allocated area of farmland cleared of obstructions and secured for use as required. No 33 Squadron's A Flight used the field for deployment as part of its responsibility for the air defence of the region. The site came into use in March 1916, but by November it was no longer required and the Royal Flying Corps abandoned it. Within weeks there were no traces of the RFC's presence, and today the area is part of Waterside Park, an area of open scrubland where walkers can enjoy the countryside, unaware that RFC fighter aircraft once drifted onto the same grassland.

Main features:
Runways: none, grass field. *Hangars:* none. *Dispersals:* none. *Accommodation:* not known.

WOODFORD, Cheshire

53°20'06"N/02°09'22"W; SJ896820; 295.5ft asl. 2 miles E of Wilmslow on A5102 at Woodford

Although this airfield cannot be regarded as a military site as such, its connections with military aviation are so strong that it cannot be overlooked. It was in 1924 that an area of countryside at New Hall Farm was purchased by A. V. Roe & Co, the famous aircraft manufacturer. Keen to develop a landing field from where new designs could be flight-tested, the area of acquired land was levelled and cleared, and two small hangars were constructed. The airfield opened in 1925, an event celebrated by a large air show. Various aircraft types were tested here, but it was the Second World War that led to the site's expansion.

The landing field was drastically enlarged and two concrete runways were laid, together with aprons, perimeter track and a handful of parking dispersals. A huge manufacturing facility emerged on the northern side of the airfield and a new flight shed was erected to the south, close to the two original hangars, which were retained. Production of the Avro Anson was the first large-scale programme to be conducted here, but as the war progressed it was the Manchester and ultimately the immortal Lancaster that occupied Woodford's workforce, which peaked at around 3,000 by 1944.

Avro Ansons pictured on Woodford's production line as they approach completion. In the distance the Tudor production line is also visible. Phil Jarrett

An Avro Tudor on the production line at Woodford, still awaiting its landing gear and about to receive its piston engines. Phil Jarrett

The Avro Athena prototype pictured during ground trials at Woodford, its Armstrong-Siddeley power plant pictured during a test run. Phil Jarrett

The Avro Athena pre-production trials fleet is seen on the flight line at Woodford. In the distance are the aged First World War-era hangars that were assembled when the airfield first opened. Phil Jarrett

A well-known Avro publicity photograph of Lancasters undergoing test flights at Woodford.

A busy scene at Woodford during the 1960s with a mix of Vulcan, Valiant and Victor aircraft assigned to the Blue Steel missile trails programme.

The mighty Vulcan pictured during Skybolt missile trials at Woodford.

Although inevitably associated with Avro designs, Woodford was also responsible for the manufacture of the Nimrod maritime patrol aircraft.

After the war the factory site continued to remain very busy and production moved on from the Lancaster to its successor, the Lincoln, followed by its maritime derivative, the Shackleton. By this stage the airfield's main runway was being extended in anticipation of even larger and heavier aircraft, and the shorter secondary runway was eventually redesignated as a taxiway, linking the north and south sites. The magnificent Vulcan bomber was test flown and assembled here, after which the factory concentrated on the manufacture of Nimrods. Military aircraft production then began to slow down, but the decision to convert the RAF's fleet of Victor bombers into refuelling tankers required a company capable of handling the task. With Handley Page gone, it was with some irony that the Victors were converted at Woodford by Handley Page's competitor company, Avro.

Other well-known aircraft types were built and flown here, such as the Andover and the BAe 146 regional jet, after this programme was moved to Woodford from Hatfield. But when even civil aircraft contracts began to dwindle, Woodford's future looked uncertain. The RAF's new Nimrod MRA4 was to have kept the site in business, but when the Ministry of Defence abruptly cancelled the programme the Nimrod airframes awaiting conversion were simply scrapped on site at Woodford and no further manufacturing activity took place here after that tragic event. With all of British Aerospace's work having shifted to Warton, the Woodford site was closed, with all flying activity ending in August 2011. The site was sold off towards the end of the year and now awaits an uncertain future.

Housing and commercial development looks certain to engulf the airfield site, but tentative plans have been made to create a Heritage Centre here, and this will hopefully embrace the two hangars erected here when the site first opened back in 1925. It is also hoped that the surviving Vulcan XM603 will be retained and placed on public display, reminding visitors of the site's proud history. There can be only a few sites that can claim to have contributed so much to Britain's military capability.

Main features:
Runways: 067° 6,700 x 150 feet, NE/SW 6,000 x 150 feet, concrete and asphalt surface. *Hangars:* four flight sheds, one factory shed. *Dispersals:* six various. *Accommodation:* not known.

WOODVALE, Merseyside

53°34'49"N/03°03'21"W; SD301098; 32.5ft asl. 5 miles SW of Southport on A565, S of Woodvale

Designed as a fighter station for units operating over the North West region, this airfield was one of various sites created in response to the Luftwaffe's demonstrable capability to attack targets with a military or industrial significance far from the more obvious regions in the South East. Cities such as Liverpool were clearly at significant risk, and Woodvale was an ideal location with flat land close to the beach, and sufficient space to lay modestly sized runways within an area constrained by both a railway line and roads. The land was acquired in 1940, but by the time construction had been completed the threat to the region had diminished quite significantly.

A THUM Flight Spitfire PR.19. Woodvale's Spitfires were the very last of the type to operate in RAF service, and this aircraft is now part of the RAF's Battle of Britain Memorial Flight.

However, the RAF moved No 308 Squadron here from Northolt in December 1941, just days after the airfield opened. The Polish unit quickly resumed patrols with its Spitfires after settling in, but with very little action in which to participate it was no surprise that the unit departed for Exeter in April 1942. Further units followed, including Nos 63, 195, 198, 219, 256, 285, 308, 313, 315, 316, 317 and 322 Squadrons, but most stayed for only a matter of weeks, operating Spitfires, Mustangs, Typhoons, Beaufighters and Defiants. Some of the short-stay units were not engaged on fighter operations, and were primarily associated with support tasks such as target-towing. No 12 (Pilot) Advanced Flying Unit was dedicated to training, flying Blenheims here from January until August 1944. Despite the relative brevity of each unit's stay, the station was always busy throughout the war, but by 1945 it soon became far less active.

Early in 1945 the airfield was transferred to the Fleet Air Arm and, as HMS *Ringtail II*, it was designated as a satellite airfield for Burscough. Fireflies and Hellcats became common sights here, but in January 1946 the station was returned to the RAF and the Navy's presence was soon gone, apart from the dummy flight deck markings painted on one of the runways. In July 1946 No 611

Woodvale as seen from an AEF Tutor in 2010.

West Lancashire Squadron brought its Spitfires here, and from May 1951 these were replaced by Meteor jets, necessitating an extension to the main runway. Rather comically, by the time the extension was completed the unit was ready for departure, and in July 1951 the Squadron moved to Hooton Park just a few miles away.

The airfield then became a far quieter place, with only gliders from Nos 192, 190 and 186 Gliding Schools, together with the Liverpool University Air Squadron, which brought its Tiger Moths here, followed by the Manchester University Air Squadron in 1953. The Tiger Moths were eventually replaced by Chipmunks. However, Woodvale earned a place in the RAF's history when the Temperature & Humidity Flight moved here from Hooton Park in July 1951, bringing with its mixed fleet of aircraft. Among these were the last Spitfires in RAF service, and it was from Woodvale that the last RAF Spitfire sortie took place on 10 June 1957, after which the unit operated Mosquitoes and Meteors. After moving on to the Battle of Britain Flight, three of these Mk 19 Spitfires are still active more than 50 years later. The THUM Flight was absorbed into No 5 Civilian Anti-Aircraft Cooperation Unit in January 1958, and when this unit disbanded in September 1971 it was flying some of the very last Meteors in RAF service.

Woodvale's history from this date onwards was less varied, and most flying here was conducted by the Manchester & Salford UAS, Liverpool UAS and No 10 Air Experience Flight, which eventually exchanged Chipmunks for Bulldogs, then Tutors. The Tutors still fly regularly from the airfield, but few other aircraft are seen here other than civilian flying club aircraft, police helicopters and occasional visitors. The airfield remains virtually unchanged from the days when it first opened, other than the runway extension created for the Meteors.

Main features:
Runways: 223° 4,800 x 150 feet, 351° 3,300 x 150 feet, 273° 3,450 x 150 feet, tarmac surface. *Hangars:* nine Extra Over Blister, three Bellman. *Dispersals:* six twin-engine, six single-engine. *Accommodation:* RAF: Officers 122, SNCOs 100, ORs 776; WAAF: 638.

WOOLSINGTON, Northumberland

55°02'15"N/01°41'28"W; NZ198714; 235ft asl. 6 miles NE of Newcastle off A696 (Newcastle Airport)

As with many major cities across the UK, Newcastle Corporation expressed an interest in aviation as early as the 1930s, and an area of land was acquired at Woolsington during 1933, resulting in a municipal airport opening here on 26 July 1935. The site was very basic when it first opened and comprised little more than a cleared grass landing field together with a few huts.

Woolsington in the 1950s, starkly different from the huge international airport that now occupies this site.

The first military aircraft to arrive here were civilian operated, these being the Hinds and Tiger Moths of No 42 Elementary & Reserve Flying Training School. They operated from the site from June 1939, training fledgling pilots in anticipation of the coming war. The unit disbanded on 3 September and only a modest amount of military activity followed, most for only short periods by training and support units, although Acklington's No 72 Squadron did occasionally base some of its Spitfires here. By contrast the Durham University Air Squadron came here in February 1941 equipped with just one Tiger Moth.

The next significant arrival was No 278 Squadron, which came here early in 1942 with Ansons and Walruses, engaged on Air Sea Rescue duties, Woolsington being relatively close to the North Sea. No 83 Maintenance Unit moved to Woolsington in July 1940 and, as a Repair & Salvage Unit, a variety of aircraft were brought to the station, mostly by road, after which parts would be removed and the remaining carcasses disposed of. This activity continued until April 1946 when the unit vacated the site.

No 62 Operational Training Unit (base at Ouston) adopted the airfield as a satellite in November 1943, and the unit's Ansons regularly used the grass field, which was, rather oddly, divided into two separate landing areas within the airfield boundaries. However, by the end of 1945 there was no longer any military need for the site, and civilian flying resumed in the early 1950s. A large concrete runway was laid and airport terminal facilities were constructed. Today the airfield is a busy and popular international airport. Military aircraft can still be seen here quite regularly, with various types appearing here on training sorties, or on transport and communications flights, not least in connection with the Otterburn and Spadeadam ranges that lay to the west.

Main features:
Runways: N/S 3,450 x 150 feet, NE/SW 4,050 x 150 feet, E/W 3,450 x 150 feet, NW/SE 3,150 x 150 feet, grass surface. *Hangars:* four Blister, two civil flight sheds, one Bellman. *Dispersals:* one apron hardstanding. *Accommodation:* RAF: Officers 24, SNCOs 60, ORs 517; WAAF: 92.

WORKSOP, Nottinghamshire

53°19'34"N/01°03'53"W; SK623814; 134.5ft asl. 1 mile NE of Worksop, E of B6045, adjacent to Scofton village

Situated at the western extremities of what could be referred to as 'Bomber Country', this RAF airfield was one of the last to be built during the Second World War. Land was acquired in 1942, but it was not until November 1943 that the airfield opened for flying. Construction was laborious thanks to the tress, roads and buildings that had to make way for the three concrete runways, but eventually a complete bomber airfield emerged and was immediately assigned to nearby Finningley as a satellite field.

No 18 Operational Training Unit based many of its Wellingtons here, and in December 1944 the unit's Polish element was switched to No 10 OTU, while the rest of the unit's activities were relocated to Worksop to complete training, after which the unit disbanded a few weeks later. This was the only connection that the station had with wartime operations, and from March 1945 the Engine Control Demonstration Unit came here for a six-month stay, operating a mixed fleet of aircraft for test and research duties. No 1 Group's Communications Flight also came here in 1946, but stayed for only a few months. The Central Night Vision Training School also stayed here for a couple of years from the summer of 1946.

It was not until August 1952 that more regular activity began when No 211 Advanced Flying School formed here, largely in response to a perceived need for more aircrew, following developments in Korea. Meteor trainers (both twin-seat T7s and single-seat F8s) became a familiar sight and sound at Worksop, and in June 1954 the unit was renamed as No 211 Flying Training School. More Meteors arrived in May 1955 when No 616 Squadron brought its aircraft here from Finningley so that the latter base could be redeveloped for the new V-Force. However, like all Royal Auxiliary Air Force units, the Squadron disbanded in March 1957. Meanwhile, in June 1956 No 211 FTS was merged with No 4 FTS, additional Meteors and some Vampires joining the unit, and flying training continued here until June 1956 when the FTS departed and the RAF vacated the site.

After a couple of years the airfield and the few station facilities were sold off and no further flying activity took place here other than occasional flights by small private aircraft. The airfield slowly decayed, although the main runway remained in good condition into the 1980s and was sometimes used for vehicle testing. Today, however, almost every trace of the airfield is gone, the runways, perimeter tracks and dispersals having been dug up and grassed over. Only a tiny portion of the old main runway remains, adjacent to Thievesdale Lane, together with a small stretch of perimeter track that leads to the remnants of the aprons where the many Meteors and Vampires once stood.

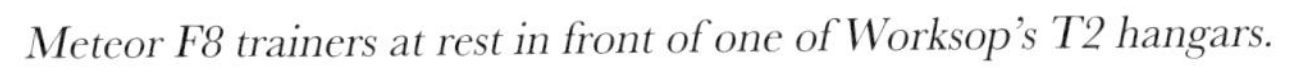

Meteor F8 trainers at rest in front of one of Worksop's T2 hangars.

A Meteor T7 pictured between training sorties at Worksop.

Main features:
Runways: 280° 6,000 x 150 feet, 220° 4,200 x 150 feet, 340° 4,200 x 150 feet, concrete surface. *Hangars:* two T2. *Dispersals:* thirty-six spectacle. *Accommodation:* RAF: Officers 102, SNCOs 296, ORs 1,340; WAAF: 330.

YEADON, Yorkshire

53°51'57"N/01°39'27"W; SE226411; 659.5ft asl. 6 miles NW of Leeds on A658 (Leeds Bradford Airport)

Aviation first came to Yeadon in 1931 when a small airfield opened here for private and recreational flying. The 60-acre grass field had few facilities, but by 1935 the site had grown and was accommodating some commercial flights. A year later the first military aircraft appeared when No 609 (West Riding) Squadron brought its Hawker Harts here, flying mostly at weekends. These were replaced by Hinds in January 1938, and only a year later these gave way to Spitfires, but in August 1939 the Squadron moved to Catterick.

No 13 Group Fighter Command took over the base at the beginning of the Second World War, then shifted to No 12 Group. A variety of units made short stays here, including some bomber squadrons, as the airfield was designated as a 'scatter field' for Bomber Command. In March 1941 No 51 Group Flying Training Command assumed control, and No 20 Elementary & Flying Training School was formed here with Tiger Moths. Tasked with the evaluation of pilots destined for aircrew training, the unit remained active here until January 1942. Also at Yeadon was No 51 Group's Communications Flight (the unit's Headquarters being at Moorfield House in Leeds), but by the end of 1944 all of the military units had gone apart from the gliders of No 23 Gliding School.

The airfield was then transferred to the Ministry of Aircraft Production and a shadow aircraft factory was constructed here together with two modestly sized paved runways (and an unusual underground production hall). Avro adopted the site and production of Ansons and Lancasters got under way, with almost 11,000 personnel assigned to the factory. As each new aircraft was completed it was towed from the factory along a purpose-built link road, enabling the aircraft to be test flown from the airfield.

PRODUCTION AT YEADON

WELL HIDDEN.—The dotted line shows the location of the giant Avro shadow factory at Yeadon, near Leeds in Yorkshire. Externally the structure blends into the surrounding countryside. Some details of production during the War are given in the article below.

A newspaper clipping illustrating the disguised Avro factory at Yeadon.

The carefully disguised Avro factory at Yeadon. Seventy years later it remains in use as a distribution depot.

Eventually some 3,881 Ansons were built here, together with 688 Lancasters and, after the war, Yorks and Lincolns. No 608 Squadron returned in November 1946, now equipped with night-fighter Mosquitoes, although these were replaced by Spitfires a few months later. The unit left again in October 1950 (to Church Fenton), after which No 1964 Flight's Austers were the only military aircraft present until they disappeared as part of the Royal Auxiliary Air Force's untimely end in 1957.

No 9 Air Experience Flight arrived in 1958, and the Leeds University Air Squadron moved here in March 1954, both eventually standardising on the Chipmunk, but by the end of 1959 they had departed, and Yeadon's association with military aviation was at an end. Civil flying immediately resumed, and by 1965 a completely new concrete runway had been laid, running across the old RAF runways that were now no more than linking tracks to a new airport terminal, completed in 1968. The airport enjoyed steady success and, with typical foresight, the runway was extended in 1984 to enable airliners of all sizes and weights to use it (the adjacent A658 passing under the runway in a tunnel).

Lancasters pictured outside the Avro facility at Yeadon, awaiting delivery to the RAF.

A 1955 aerial photograph of the main site at Yeadon.

Leeds Bradford is now a busy international facility that has handled aircraft such as the Boeing 747 and Concorde, although most of the activity here is in the form of regional aircraft, charter flights and private flying. Some military aircraft are still seen here, however, RAF aircraft often visiting or performing practice approaches, but the days when literally hundreds of military aircraft could be seen here are long gone.

Main features:
Runways: none, grass field. *Hangars:* two flight sheds. *Dispersals:* none. *Accommodation:* not known.

YORK/CLIFTON, Yorkshire

53°59'32"N/01°05'34"W; SE595555; 49.5ft asl. NW of York at junction of A1237 with B1363

A small municipal airport was opened at Clifton in July 1936. Like many similar facilities around the country it was little more than a cleared grass field with only a few huts and tents to support the civilian flying training that took place, but a large hangar and clubhouse were constructed here, making the site rather more substantial. Not surprisingly, the airfield (and its aircraft) were acquired by the military when the Second World War began, and although plans were made to form No 52 Elementary Flying Training School here, the airfield was initially used as a scatter field for Bomber Command, when expectations of large-scale German attacks on RAF airfields were at their highest.

The surviving T2 hangar at York.

No 4 Squadron's Lysanders arrived in August 1940 and began a long association with Army Cooperation, with many Army units based both in the city of York and the surrounding areas. No 4 (Bomber) Group's Communications Flight had also arrived by this stage, and its mixed fleet of aircraft stayed in residence until the summer of 1942. Despite some damage caused by a Luftwaffe attack on York, the station avoided the direct attention of the enemy and the site was developed to include a standard layout of three concrete runways together with a perimeter track, dispersals and hangars, destined for use by No 48 Maintenance Unit, which came here in 1941, tasked with the repair of Halifax bombers.

It was the appearance of the runways (which extended out over Clifton Moor) that eventually led to the site often being officially referred to as Clifton, rather than York. Other units based here included No 231 Squadron, operating Tomahawks and Lysanders from March until July 1943, at which time the unit was re-equipping with Mustangs. Nos 430 and 613 Squadrons also operated Mustangs from the airfield for short periods in 1942 onwards. Auster units also came to York, with No 657 Squadron staying here from June until August 1943, together with Nos 658 and 659 Squadrons, present from August 1943 until early 1944. Even the Fleet Air Arm had a brief presence here, the Seafires of No 809 Squadron being here from December 1942 for some five months.

The last significant presence here was No 4 Air Delivery Flight, which came to York in March 1944 tasked with the delivery of aircraft to and from various locations on behalf of No 12 Group. When this unit departed in June 1945 the station wound down and the airfield was used briefly for civilian flying, but even this did not last for too long. The ever-expanding outskirts of York slowly encroached on the airfield until it was engulfed by housing, commercial and industrial development. Today the site is almost completely obliterated and lost under countless buildings and roads – even though many of the roads have suitably aviation-related names. But on the B1363 Wigginton Road there is a surviving area of undeveloped land where the remains of the main runway's threshold can still be seen, seemingly lost among the urban sprawl.

Main features:

Runways: 245° 4,800 x 150 feet, 353° 4,200 x 150 feet, 308° 4,200 x 150 feet, concrete and tarmac surface. *Hangars:* twelve Blister, one T1, one civil flight shed. *Dispersals:* fifteen, various. *Accommodation:* RAF Officers 39, SNCOs 59, ORs 286; WAAF: 119.

Index

Airports, civilian

Airships

Squadrons, FAA and RNVR continued…

Squadrons, RAF and allies